Student Study Guide/Solutions Manual for use with

BIOCHEMISTRY
THE MOLECULAR BASIS OF LIFE
Fourth Edition

Patricia DePra

Trudy McKee

James R. McKee

New York   Oxford
OXFORD UNIVERSITY PRESS
2009

Oxford University Press, Inc., publishes works that further Oxford University's
objective of excellence in research, scholarship, and education.

Oxford   New York
Auckland   Cape Town   Dar es Salaam   Hong Kong   Karachi
Kuala Lumpur   Madrid   Melbourne   Mexico City   Nairobi
New Delhi   Shanghai   Taipei   Toronto

With offices in
Argentina   Austria   Brazil   Chile   Czech Republic   France   Greece
Guatemala   Hungary   Italy   Japan   Poland   Portugal   Singapore
South Korea   Switzerland   Thailand   Turkey   Ukraine   Vietnam

Copyright © 2009 by Oxford University Press, Inc.

Published by Oxford University Press, Inc.
198 Madison Avenue, New York, New York 10016
http://www.oup.com

Oxford is a registered trademark of Oxford University Press

ISBN: 978-0-19-534292-5

Printing number: 9 8 7 6 5 4 3 2 1

Printed in the United States of America
on acid-free paper

Contents

1 Biochemistry: An Introduction

Brief Outline of Key Terms and Concepts

OVERVIEW
CHEMOSYNTHESIS; PHOTOSYNTHESIS

1.1 WHAT IS LIFE?
All living organisms obey the chemical and physical laws. Life is complex, dynamic, organized, and self-sustaining.
Life is cellular and information-based. Life adapts and evolves.

BIOMOLECULE; MACROMOLECULE; ENZYME; METABOLISM; HOMEOSTASIS GENE; MUTATION

1.2 THE LIVING WORLD
PROKARYOTIC CELL; EUKARYOTIC CELL
Classification of living organisms into three domains: BACTERIA, ARCHAEA, and EUKARYA

BIOGEOCHEMICAL CYCLE; EXTREMOPHILE; EXTREMOZYME; BIOREMEDIATION; ORGANELLE

1.3 BIOMOLECULES
In living organisms, most molecules are organic. Organic FUNCTIONAL GROUPS determine the chemical properties of organic molecules. The four classes of small biomolecules are amino acids, sugars, nucleotides, and fatty acids.

HYDROPHOBIC, HYDROPHILIC

BIOPOLYMERS	SMALL BIOMOLECULES
PROTEINS	AMINO ACIDS
POLYSACCHARIDES	SUGARS
NUCLEIC ACIDS	NUCLEOTIDES

Several types of LIPIDS have FATTY ACID components.

AMINO ACIDS AND PROTEINS: POLYPEPTIDE (PEPTIDE, OLIGOPEPTIDE); PEPTIDE BOND; NEUROTRANSMITTER

CARBOHYDRATES: SUGARS, MONOSACCHARIDES

FATTY ACIDS: SATURATED, UNSATURATED

DNA, RNA; PURINE, PYRIMIDINE; GENOME
DNA is an ANTIPARALLEL double helix.
RNA; TRANSCRIPTION; NONCODING RNA

BIOCHEMISTRY IN THE LAB
GENOMICS; GENE EXPRESSION; FUNCTIONAL GENOMICS; PROTEOMICS; BIOINFORMATICS

1.4 IS THE LIVING CELL A CHEMICAL FACTORY?
SYSTEM; AUTOPOIESIS

BIOCHEMICAL REACTIONS
NUCLEOPHILIC SUBSTITUTION: NUCLEOPHILE, ELECTROPHILE, LEAVING GROUP; HYDROLYSIS
ELIMINATION REACTION: DEHYDRATION
ADDITION REACTION: HYDRATION
ISOMERIZATION REACTION
OXIDATION-REDUCTION (REDOX) REACTIONS
OXIDIZE, REDUCE
OXIDIZING AGENT, REDUCING AGENT

ENERGY: THE CAPACITY TO MOVE MATTER
In living organisms, energy is usually generated by REDOX REACTIONS.
AUTOTROPH; HETEROTROPH
PHOTOAUTOTROPH; CHEMOAUTOTROPH
CHEMOHETEROTROPH; PHOTOHETEROTROPH

METABOLISM is the sum of all enzyme-catalyzed reactions in a living organism.
Classes of biochemical pathway:
METABOLIC (ANABOLIC and CATABOLIC)
ENERGY TRANSFER
SIGNAL TRANSDUCTION

BIOLOGICAL ORDER
In living organisms, a constant input of energy is needed to sustain processes of highly ordered complexity. Examples include: synthesis of biomolecules, transport across membranes, active transport, cell movement, and waste removal.

1.5 SYSTEMS BIOLOGY is an attempt to reveal the functional properties of living organisms by developing mathematical models of interactions from available data sets. The systems approach has provided insights into the EMERGENCE, ROBUSTNESS, and MODULARITY of living organisms.
Compare the systems approach to REDUCTIONISM.
EMERGENT PROPERTY; DEGENERACY; FEEDBACK CONTROL; NEGATIVE FEEDBACK; POSITIVE FEEDBACK; MODULES

1.1 WHAT IS LIFE? LIFE IS:

1. Complex and dynamic: biomolecules, biochemical reactions

2. Organized and self-sustaining, characterized by
   – hierarchical order, from atoms to multicellular organisms, that requires a constant influx of energy and matter
   – enzyme-catalyzed reactions; metabolic pathways that can be regulated; (*homeostasis*)

3. Cellular: cell membranes control transport into and out of the cell

4. Information-based

5. Able to adapt and evolve (*mutations*)

1.2 THE LIVING WORLD: LIVING CELLS ARE EITHER PROKARYOTIC OR EUKARYOTIC

DOMAINS:	BACTERIA, ARCHAEA *AND*	EUKARYA
	single-celled organisms	many Eukarya are multicellular
	have	*have*
	PROKARYOTIC CELLS	EUKARYOTIC CELLS
	that are	*that are*
	MUCH SMALLER	MUCH LARGER
	with	*with*
	NO ORGANELLES, NO NUCLEUS	MEMBRANE-BOUND ORGANELLES, NUCLEUS WITH GENETIC MATERIAL

ADVANTAGES OF BEING MULTICELLULAR (AS OPPOSED TO BEING SINGLE-CELLED):
- able to provide a relatively stable environment for most of the organism's cells
- able to be much more complex in form and function
- able to exploit environmental resources more effectively

1.3 BIOMOLECULES

ORGANIC REVIEW: FAMILIES (FUNCTIONAL GROUPS)

alkene (C=C)	aldehyde (carbonyl)	amine (amino)
alcohol (hydroxyl)	ketone (carbonyl)	amide (amido)
thiol (thiol, –SH)	ester (ester)	carboxylic acid (carboxyl)

NOTE: The α–carbon in carbonyl, carboxyl, and amido groups is the carbon *adjacent to* a carbonyl carbon.

SMALL BIOMOLECULES: MAJOR CLASSES

SMALL BIOMOLECULES ARE BUILDING BLOCKS FOR LARGE BIOMOLECULES:

amino acids	$\longrightarrow$	peptides, polypeptides, proteins
monosaccharides	$\longrightarrow$	carbohydrates; polyglycans
fatty acids	$\longrightarrow$	(components of lipids)
nucleotides	$\longrightarrow$	nucleic acids (RNA, DNA)

Small biomolecules also carry out special biological functions (e.g., as neurotransmitters or hormones), serve as energy sources, and/or take part in complex reaction pathways.

AMINO ACIDS AND PROTEINS

- Amino acids contain an amino group, a carboxylic acid group, and a side chain (R group). In α-amino acids, the amino group is attached to the α-carbon.

- Amino acids are linked together by peptide (amide) bonds, which have double-bond character that impacts the overall structure with its rigidity.

- Amino acid *residues* in proteins

- Polypeptides: peptides (up to 50 amino acids), proteins (longer)

SUGARS AND CARBOHYDRATES: MONOSACCHARIDES, POLYSACCHARIDES

- Sugars contain alcohol groups and either aldehydes (in aldoses) or ketones (in ketoses).

- Names may also indicate number of carbons, such as "aldohexose."

- Polysaccharides (polyglycans) include starch and cellulose (plants), glycogen (animals).

- Nucleotides contain either ribose or deoxyribose.

- Glycoproteins and glycolipids are proteins and lipids that contain carbohydrates.

- Functions: important source of energy, structural support, and participation in intracellular and intercellular communication.

FATTY ACIDS

- Fatty acids contain one carboxylic acid with a long hydrocarbon chain. Fatty acids are usually unbranched with an even number of carbon atoms.

- Unsaturated fatty acids have at least one double bond in the hydrocarbon chain. Saturated fatty acids have only C–C single bonds (they're "saturated" with hydrogen atoms).[1]

- Fatty acids are components of lipid molecules. Lipids are not water-soluble.

- Triacylglycerols (three fatty acids + glycerol) store energy.

- Phosphoglycerides (two fatty acids + glycerol + phosphate + a polar compound) are major structural components of cell membranes.

NUCLEOTIDES AND NUCLEIC ACIDS (DNA AND RNA)

- Nucleotides contain a five-carbon sugar (ribose or deoxyribose), a purine or pyrimidine base, and one or more phosphate groups.

  – Purine bases: adenine and guanine

  – Pyrimidine bases: thymine, cytosine (DNA only), and uracil (RNA only)

- Nucleotides are essential in many biosynthetic and energy-generating reactions. ATP is a nucleotide.

- Nucleotides join together via phosphodiester linkages to form nucleic acids – DNA or RNA. The nucleic acid sugar-phosphate backbones alternate ...sugar-

---

[1] Note the difference between the biochemistry definition of a saturated fatty acid and the organic chemistry definition of a saturated molecule: a saturated fatty acid contains no alkenes, but still has a C=O in its carboxyl group.

phosphate-sugar-phosphate… A purine or pyrimidine base is connected to each sugar.

- The specific and unique order of the bases – the base sequence – holds genetic information. Genetic information flows from DNA to RNA to proteins.
- Hydrogen bonding occurs between specific base pairs: DNA: AT, GC

    RNA: AU, GC

- This hydrogen bonding stabilizes the DNA double helix and allows **TRANSCRIPTION** – the synthesis of RNA from DNA.
- Types of RNA: **mRNA** (messenger RNA), **rRNA**(ribosomal RNA), and **tRNA** (transfer RNA). All three types work together to synthesize polypeptides. Other types of RNA molecules: ncRNA, siRNA, miRNA, snRNA, snoRNA (noncoding, short interfering, micro, small nuclear, and small nucleolar RNA, respectively).

EXAMPLES OF EACH CLASS OF SMALL BIOMOLECULE

AMINO ACID	SUGAR	NUCLEOTIDE

FATTY ACID	

ORGANIC FUNCTIONAL GROUPS THAT MAY BE NEW TO YOU:

PHOSPHOESTER	PHOSPHODIESTER	PHOSPHOANHYDRIDE	Compare with an ANHYDRIDE:

1.4   IS THE LIVING CELL A CHEMICAL FACTORY?

BIOCHEMICAL REACTIONS, ENERGY, METABOLISM, AND BIOLOGICAL ORDER

AUTOPOIESIS and AUTOPOIETIC SYSTEMS:
autonomous, self-organizing, and self-maintaining

BIOCHEMICAL REACTIONS ARE CATALYZED BY ENZYMES *(BUT THE REACTIONS THEMSELVES ARE THE SAME ONES LEARNED IN ORGANIC CHEMISTRY)*

MOST COMMON REACTION TYPES (AND SOME EXAMPLES)
- Nucleophilic Substitution (hydrolysis of peptides to form amino acids); **NUCLEOPHILE, ELECTROPHILE, LEAVING GROUP**

- Elimination (removal of water to form an alkene)

- Addition (hydration adds H_2O to an alkene; hydrogenation adds H_2)

- Isomerization (converting glucose-6-phosphate to fructose-6-phosphate; both have the same molecular formula but a different arrangement of atoms)

- Oxidation-Reduction (redox) reactions (oxidizing an alcohol to an aldehyde)

BIOCHEMICAL OXIDATION AND REDUCTION: REMEMBER THE ORGANIC DEFINITIONS

In general chemistry, you learned that oxidation is the removal of electrons and that reduction adds electrons (resulting in a lower, or *reduced*, charge or oxidation number). In organic, you learned that oxidation is the addition of oxygen (and/or the removal of hydrogen) and that reduction is the removal of oxygen (and/or the addition of hydrogen). In biochemistry, we'll tap into both of these definitions.

Redox reactions always involve a *transfer* of electrons, so when anything is oxidized, something else must be reduced. An **OXIDIZING AGENT** does the oxidizing (and becomes reduced in the process); a **REDUCING AGENT** does the reducing (and becomes oxidized).

Biochemical redox reactions involve transferring one or two electrons at a time. Usually an H^+ rides along, so what appears to be transferred is an H atom (H•) or a hydride ion (H:⁻). (Think of the H atom as an H^+ with 1 electron, and the hydride ion as an H^+ with 2 electrons.)

Let's look at an essential example: the reduction of the coenzyme NAD^+ to produce NADH. With its additional H, we can identify NADH as the reduced form. With its extra + charge, NAD^+ is the oxidized form. When NAD^+ is reduced, something else must be oxidized. NAD^+ is the oxidizing agent, because NAD^+ oxidizes another molecule. Here's a specific example: NAD^+ oxidizes ethanol to form acetaldehyde:

$$NAD^+ \ + \ CH_3CH_2OH \longrightarrow NADH \ + \ H_3C-\overset{\displaystyle O}{\underset{\displaystyle H}{C}} \ + \ H^+$$

| oxidized
form | reduced
form | reduced
form | oxidized
form |

We could also say that ethanol reduces NAD^+ to form NADH. No matter how we look at it, two electrons (and an H^+) were transferred from ethanol to NAD^+.

ENERGY

Cells obtain energy by oxidizing biomolecules (or certain minerals, or by photosynthesis). Energy is in the form of electrons. ***The more reduced the molecule, the more energy it contains.*** Take another look at the classes of biomolecules. Can you see why fatty acids, with all of those $-CH_2-$ groups, produce more energy (ATP) when they're metabolized than the oxygen-rich sugars do?

Organisms can be classified by how they obtain energy: autotrophs (photoautotrophs or chemoautotrophs), heterotrophs (chemoheterotrophs or photoheterotrophs).

METABOLISM = SUM OF ALL ENZYME-CATALYZED REACTIONS IN A LIVING ORGANISM

METABOLIC PATHWAYS synthesize and degrade biomolecules.

ANABOLIC PATHWAYS	CATABOLIC PATHWAYS
BIOSYNTHESIS	DEGRADATION
(*Anna* builds biomolecules...	...and the *cat* claws them apart.)
ANABOLISM REQUIRES ENERGY $ATP \rightarrow ADP$	CATABOLISM STORES ENERGY (as ATP) $ADP \rightarrow ATP$
Uses electrons *(typically)* from NADH, FADH$_2$, or NADPH to form NAD$^+$, FAD, or NADP$^+$ and larger biomolecules	Electrons are captured by NAD$^+$, FAD, or NADP$^+$ to form NADH, FADH$_2$, or NADPH and smaller, more oxidized molecules

ENERGY TRANSFER PATHWAYS: CAPTURE AND TRANSFORM ENERGY
Example: Photosynthesis captures light energy and converts it into chemical bond energy (in sugar molecules).

SIGNAL TRANSDUCTION PATHWAYS: RECEIVE AND RESPOND TO SIGNALS

Steps: 1. Reception
2. Transduction
3. Response

BIOLOGICAL ORDER IS ACCOMPLISHED BY:

1. Synthesis and degradation of biomolecules

2. Transport of ions and molecules across cell membranes (**ACTIVE TRANSPORT** requires energy provided by ATP hydrolysis)

3. Production of motion

4. Removal of metabolic waste products and other toxic substances

LIVING CELLS NEED A CONSTANT FLOW OF ENERGY TO MAINTAIN THIS HIGH DEGREE OF ORDER.

1.5 SYSTEMS BIOLOGY: LIVING SYSTEMS AS INTEGRATED SYSTEMS

EMERGENCE
Certain characteristics result, or *emerge*, from interactions between the components of a system. These **EMERGENT PROPERTIES** cannot be explained by studying the individual parts separately.[2]

ROBUSTNESS: WHAT MAKES A SYSTEM ROBUST?
DEGENERACY: the same or similar functions can be performed by different components
COMPLEX CONTROL MECHANISMS such as the regulation of enzymes via **POSITIVE FEEDBACK** or **NEGATIVE FEEDBACK**

MODULARITY: Subsystems, or modules, perform specific functions.

---

[2] The opposite of emergence is reductionism (to understand the whole, understand the parts), an incomplete approach to understanding living systems.

IN THE LAB: 1.1 AN INTRODUCTION

THE HUMAN GENOME PROJECT; GENOMICS; FUNCTIONAL GENOMICS; PROTEOMICS; BIOINFORMATICS

HINT: *NOW* is the best time to be *SURE* that you have a SOLID understanding of these concepts from organic:

Polar vs. Nonpolar	Hydrogen Bonding
Hydrophilic vs. Hydrophobic	Dipole-Dipole Interactions
Acidic vs. Basic	van der Waals Forces

All of these are *essential* in developing a deep sense of why biomolecules "behave" the way they do, and will help you to move from memorization to *learning* much more readily.

Whether a biomolecule is polar or nonpolar has a HUGE impact on its properties, including how it interacts and reacts with other molecules. For example, how a protein's amino acid R groups interact determines its overall shape, and its specific shape determines whether that protein will be fibrous and give structural support, or a globular enzyme that catalyzes a very specific reaction.

If you feel the least bit rusty, invest the time *now* to help your understanding throughout the semester.

AFTER STUDYING THIS CHAPTER, YOU SHOULD BE ABLE TO:

- Identify functional groups in a given molecule.

- Recognize the structures of the four classes of small biomolecules. Classify a given compound as an amino acid, sugar, fatty acid, or nucleotide, and describe its function in living organisms.

- Recognize the four types of reactions: nucleophilic substitution, elimination, oxidation-reduction, and addition. In oxidation-reduction reactions, recognize which reactant was oxidized and which was reduced.

- Describe the differences between Archaea, Bacteria, and Eukarya; prokaryotic cells and eukaryotic cells; prokaryotes and eukaryotes.

- Demonstrate an understanding of the general characteristics of life, living organisms, and the study of biochemistry, including recent discoveries and the scientific insights that have resulted.

Use this space to note any additional objectives provided by your instructor.

CHAPTER 1: SOLUTIONS TO REVIEW QUESTIONS

1.1   The discovery of living organisms in the sunless hydrothermal vents challenged the traditional concept that life depends upon sunlight (either directly or indirectly) to exist, and expanded our understanding of living processes, views of biodiversity, and the capacities of living organisms. The fact that new species discovered in these hydrothermal vents shared much in common with previously known species supports the concept of the unity of all living organisms.

1.2   The three domains of living organisms are the Archaea, the Bacteria and the Eukarya. The Archaea are best known for the capacity of many of their species to thrive in extremely hostile environments. Bacterial species are characterized by their vast biochemical diversity. The Eukarya are extraordinarily complex organisms, many of which are multicellular.

1.4   The functional group(s) in each molecule are:

a. aldehyde group

e. alkene

b. carboxylic acid and amino groups

f. amide group

c. sulfhydryl group

g. ketone group

d. ester group

h. alcohol group

1.5   Amino acids occur in peptides and proteins. Sugars occur in oligosaccharides and polysaccharides. Nucleotides are the components of nucleic acids. Fatty acids are components of several types of lipid molecules, e.g., triacylglycerols and phospholipids.

1.7   a.   The functions of fatty acids include energy storage and membrane components. Fatty acids are components of larger lipid molecules within cell membranes. Some fatty acids are also precursors to hormones.

b.   Sugars function as energy sources and as components of polysaccharides (such as starch, cellulose, glycogen, and chitin) which function as energy sources and/or as structural elements. Nucleotides contain the sugars ribose or deoxyribose. Glycoproteins and glycolipids located on cell membranes also contain sugars, and play critical roles in many cellular interactions.

c.   Nucleotides are involved in energy transformations. They are also components of DNA and RNA.

d.   Most amino acids are the building blocks of proteins. Some have special functions as neurotransmitters or as precursors to other molecules (such as vitamins). Peptides and proteins have a variety of functions, including structural support or catalytic activity.

1.8   DNA is the repository of each organism's genetic information. RNA is the nucleic acid that is involved in the expression of genetic information, primarily in various aspects of protein synthesis.

1.10   a.   oxidizing agent – atom or group reduced during an oxidation/reduction reaction

b.   elimination – loss of an atom or group

c.   reducing agent – atom or group oxidized during an oxidation/reduction reaction

d.   isomerization – a shift of atoms or groups within a molecule

e.   nucleophilic substitution – displacement of an atom or group by an electron-rich species

1.11 Plants dispose of waste products either by degradation or by storage in vacuoles or cell walls.

1.13 a. modules – components or subsystems that perform specific functions within complex systems

b. noncoding RNA – RNA molecules that are not directly involved in protein synthesis

c. chemoautotrophs – organisms that transform energy contained in inorganic chemicals into chemical bond energy; compare with chemoheterotrophs, organisms that use preformed food molecules as their sole source of energy

d. biogeochemical cycles – global cycles of nutrients such as carbon, nitrogen, phosphorus, and sulfur

e. signal transduction – mechanism by which extracellular signals are received, amplified, and converted to a cellular response

1.14 The most important advantages of multicellular organisms over unicellular organisms are as follows: multicellular organisms have a stable internal environment; cells can be specialized (division of labor); there is a more efficient use of resources; sophisticated functions can be accomplished by such organisms.

1.16 a. metabolism – the sum of the chemical reactions carried out in a living cell

b. nucleophile – an atom or group with an unshared pair of electrons that is involved in a displacement (nucleophilic substitution) reaction

c. reductionism – the belief that complex processes can be understood by examining their simpler parts

d. electrophile – an electron-deficient species

e. energy – the ability to do work

1.17 Organelles are specialized subcellular structures found in eukaryotes. They permit the concentration of reactants and products at sites where they can be efficiently used.

1.19 a. system – multiple parts that work together to perform a specific function; e.g., the digestive system consists of interrelated organs that transform food into molecules that can be absorbed by cells (Note: the definition of a system in physical chemical/thermodynamics terms differs slightly – see Chapter 4.)

b. emergent properties – properties that emerge from interactions among parts; emergent properties have characteristics that cannot be predicted by analyzing its component parts

c. robustness – being able to maintain stability despite physical and/or chemical challenges

d. feedback – the process by which a product of a pathway serves to regulate that same pathway

e. degeneracy – the capacity of structurally different parts to perform the same or similar functions

1.20 Examples of the following reactions include:

a. nucleophilic substitution – the reaction of glucose with ATP to produce glucose-6-phosphate and ADP

b. elimination – the dehydration of 2-phosphoglycerate to form phosphoenolpyruvate

c. oxidation-reduction – the oxidation of ethyl alcohol to acetaldehyde

      d.   addition – the conversion of fumarate to malate

1.22   Both an airplane autopilot system and a biological system are robust, i.e., they have the ability to maintain stability despite changes in the environment or other events that threaten the continuation of system functions. They both have feedback control mechanisms, in which information regarding internal processes is used to adjust functions to maximize performance. The actual fail-safe mechanisms differ in that human-made systems have redundancy (duplicate parts) while biological systems have degeneracy, in which duplicate (or similar) functions may be carried out by different parts of the system.

1.23   In addition to being an important energy source, carbohydrates are important structural molecules in organisms and have a role in intracellular and intercellular communication.

1.25   a.   mutation – any change in the nucleotide sequence of a gene

      b.   extremozymes – an enzyme that functions under extreme conditions of temperature, pressure, pH, or ionic concentration

      c.   genome – the total genetic information possessed by an organism

      d.   autopoiesis – the concept of a living organism as autonomous, self-organizing, and self-maintaining via the action of thousands of biochemical reactions that are enzyme-catalyzed and self-regulated

1.26   The largest biomolecules in living organisms are the nucleic acids and the proteins. Nucleic acids store genetic information (DNA) and mediate the synthesis of proteins (RNA). The proteins are the tools that perform all of the tasks required to sustain living processes. Polysaccharides are also large biomolecules with structural and energy storage functions.

1.28   Order is maintained in living cells primarily by the synthesis of biomolecules, the selective transport of ions and molecules across cell membranes, the production of force and movement, and the removal of metabolic waste products.

1.29   Examples of waste products produced by animals include carbon dioxide, ammonia, urea, uric acid, and water.

1.31   The nucleotide base sequence of each type of mRNA molecule codes for the amino acid sequence of a specific polypeptide. Each tRNA molecule carries a specific amino acid which it subsequently delivers to the ribosome for incorporation into a polypeptide during protein synthesis. Ribosomal RNA molecules contribute to the structural and functional properties of ribosomes. Each polypeptide is manufactured as the base sequence information in the mRNA is translated by a ribosome. As base pairing occurs between the codon sequence of mRNA and the anticodon sequence of tRNA molecules, the amino acids are brought into close proximity and a peptide bond is formed.

1.32   Complex control mechanisms and protective systems allow living organisms to withstand various physical and/or chemical challenges, e.g., fluctuations in temperature, availability of nutrients, and energy needs.. As such, living organisms are robust, yet they are fragile in their vulnerability to unusual or rare events that cause irreparable damage. For example, a bleeding cut will clot and heal, but extended exposure to high levels of carbon monoxide causes death.

1.34   Both autopoietic systems and factories take in raw materials (nutrients or parts), use energy, manufacture products, and remove waste. Both maintain an inventory and have mechanisms in place to preserve system function and product integrity, via biological feedback and regulation mechanisms or factory quality control. In contrast to factories, autopoietic systems synthesize structural components, energy storage molecules, and molecules that regulate and/or modify system functions. This would be analogous to factories using raw materials to create their own employees, buildings, machines, computers, and fuel. Autopoietic systems are autonomous, self-organizing, and self-maintaining.

CHAPTER 1: SOLUTIONS TO THOUGHT QUESTIONS

1.35   The mechanisms of organic and biochemical reactions are the same. The means by which products are synthesized differ, as biochemical reactions are catalyzed by enzymes.

1.37   Prokaryotes are single-celled organisms that are smaller and less complicated than the eukaryotes and have short life cycles. Biochemists make the useful assumption that the basic elements of living processes in the two types of organisms are similar. Finally, some prokaryotes are easier to obtain, manipulate, and investigate than are multicellular eukaryotes.

1.38   The C–H bonds of fatty acids are the most reduced form of carbon found in organic molecules. Oxidation of these molecules to form carbon dioxide – the most oxidized form of carbon – has the highest energy yield. Also, fatty acids are stored without the need for water. They are therefore stored in smaller areas and with less mass than polysaccharides.

1.40   The new molecule forms three hydrogen bonds with guanine:

2-Amino-6-methoxypurine   Guanine (G)

1.41   The capacity of healthy bodies to adapt to high-cholesterol diets by inhibiting cholesterol synthesis is an example of the means by which living organisms regulate their metabolic processes.

1.43   Both normal and tumor cells are robust, i.e., they remain alive despite perturbations. The genetic instability of tumor cells facilitates the survival of the tumor. Out of the millions of cells in a tumor, there is a distinct possibility that one or more will express the P-glycoprotein. Under the selection pressure of the drug, these cells will survive. Cells not expressing P-glycoprotein or another means of detoxifying the drug will die.

1.44   There are 20 standard amino acids, so $X = 20$. The chain length is 10, so $n = 10$. The number of possible decapeptides is 20^{10} or 1.024×10^{13}. To draw all of the possibilities at the rate of one every 5 minutes would require about 97,300,000 years.

2 Living Cells

Brief Outline of Key Terms and Concepts

2.1 BASIC THEMES

WATER; HYDROPHILIC VS. HYDROPHOBIC

BIOLOGICAL MEMBRANES are lipid bilayers.
INTEGRAL AND PERIPHERAL PROTEINS, CHANNEL AND
CARRIER PROTEINS, RECEPTORS, ANCHOR PROTEINS

SELF-ASSEMBLY
In living organisms, biomolecules in supramolecular structures are able to assemble spontaneously because of the steric information they contain.

MOLECULAR MACHINES
Molecular machines are complexes in living organisms that function like mechanical devices, with moving parts that perform work. MOTOR PROTEIN

MACROMOLECULAR CROWDING
Cells are densely crowded with macromolecules of diverse types. Macromolecular crowding is a significant factor in a wide variety of cellular processes. EXCLUDED VOLUME

SIGNAL TRANSDUCTION
Organisms receive, interpret, and respond to environmental information by means of the process of signal transduction, which has three phases: reception, transduction, and response. SIGNALS, NEURO-TRANSMITTERS, HORMONES, CYTOKINES; LIGAND

2.2 STRUCTURE OF PROKARYOTIC CELLS
Prokaryotic cells are small and structurally simple, have a CELL WALL and a PLASMA MEMBRANE, lack a nucleus and other ORGANELLES, and have circular DNA in the NUCLEOID.

CELL WALL
PEPTIDOGLYCAN; GRAM-POSITIVE, GRAM-NEGATIVE;
LIPOPOLYSACCHARIDES (outer membrane); ENDOTOXINS;
PORINS, TRANSMEMBRANE PROTEIN COMPLEXES;
PERIPLASMIC SPACE; GLYCOCALYX, SLIME LAYERS, BIOFILMS

PLASMA MEMBRANE; PHOTOSYNTHESIS, RESPIRATION
CYTOPLASM; NUCLEOID, CHROMOSOME, PLASMIDS
PILI AND FLAGELLA: CELL MOVEMENT

2.3 STRUCTURE OF EUKARYOTIC CELLS
ENDOMEMBRANE SYSTEM, VESICLES

PLASMA MEMBRANE
Provides strength and shape to the cell, controls transport in and out of the cell, and contains RECEPTORS to allow the cell to respond to external stimuli. GLYCOCALYX, EXTRACELLULAR MATRIX, CELL CORTEX

ENDOPLASMIC RETICULUM (ER)
CISTERNAL SPACE, LUMEN; ROUGH ER (RER), SMOOTH ER (SER); ER STRESS, UNFOLDED PROTEIN RESPONSE (UPR), ER OVERLOAD RESPONSE (EOR), APOPTOSIS; BIOTRANS-FORMATION REACTIONS; SARCOPLASMIC RETICULUM (SR)

GOLGI APPARATUS (GOLGI COMPLEX)
Formed from relatively large, flattened, sac-like membranous VESICLES, the Golgi apparatus packages and secretes cell products. CISTERNA; CIS FACE, TRANS FACE; CISTERNAL MATURATION MODEL; SECRETORY VESICLES (SECRETORY GRANULES); EXOCYTOSIS; DICTYOSOMES (in plants)

NUCLEUS
The nucleus contains the cell's genetic information and the machinery for converting that information into protein molecules. NUCLEOPLASM; CHROMATIN FIBERS; HISTONES; NUCLEAR MATRIX; NUCLEOLUS (ribosomal RNA synthesis), SPECKLES; NUCLEAR ENVELOPE, NUCLEAR PORES, PERINUCLEAR SPACE, INNER MEMBRANE, NUCLEAR PORE PLUG

VESICULAR ORGANELLES
Examples include: acid hydrolase-rich LYSOSOMES and plant vacuoles, α-GRANULES (platelets), and MELANOSOMES (in melanocytes). ENDOCYTOSIS, PHAGO-CYTOSIS, AUTOPHAGY; LYSOSOMAL STORAGE DISEASES

MITOCHONDRIA: LOCATION OF AEROBIC RESPIRATION
AEROBIC RESPIRATION generates most of the energy needed by eukaryotes. ATP synthesis occurs in RESPIRATORY ASSEMBLIES embedded in the mitochondrion's INNER MEMBRANE. OUTER MEMBRANE, INTERMEMBRANE SPACE, CRISTAE, MATRIX

PEROXISOMES generate and break down peroxides.
GLYOXYSOMES (in plants)

PLASTIDS (in plants, algae, and some protists)
LEUCOPLASTS, CHROMOPLASTS, CHLOROPLASTS; PHOTO-SYNTHESIS; THYLAKOID MEMBRANE, GRANA, THYLAKOID LUMEN, STROMA, STROMA LAMELLAE

CYTOSKELETON
This highly structured network of proteinaceous filaments is responsible for cell shape, large- and small-scale cell movement, solid state biochemistry, and signal transduction. MICROTUBULES, MICROFILAMENTS, INTERMEDIATE FILAMENTS

RIBOSOMES are RNA/protein complexes that synthesize proteins.

BIOCHEMISTRY IN THE LAB: CELL TECHNOLOGY

OVERVIEW

Understanding more about *where* biomolecules react – within specific organelles, in the cytoplasm, or on the outside surface of the cell membrane, for example – places biochemical reactions in the context of their impact within the organism as a whole. Realizing that a specific reaction (and its pathway) occurs in a specific location, for a specific reason, and under a specific set of circumstances gives greater meaning to the functions of the biomolecules, their pathways, and even the overall systems.

STUDY HINTS AND STRATEGIES

Creating your own diagrams, tables, and/or images is a powerful learning tool. Make your learning active by organizing the material in a way that will help you the most, whether as pictures or diagrams of the organelles and cells, as lists or summaries that include functions, or as tables. Small sticky notes with the name of an organelle on one side and its function on the reverse may be helpful. Notes of different colors can be used to keep prokaryotic- or plant-specific features separate.

Create a table in which each row is devoted to a specific organelle. Check out the following example. You'll want to tailor the column headers to the emphasis given in your specific class or to challenges that you're facing in learning more about these great little packages.

Item*	Structure	Location	Function	Notes
What is it?	What's it made of?	Where is it?	What does it do?	Why is it important? How is it unique?

* Suggested items for the first column: Plasma Membrane, Endoplasmic Reticulum, RER, SER, Golgi Apparatus, Nucleus, Mitochondria, Vesicular Organelles, Lysosomes, Peroxisomes, Plastids, Cytoskeleton, Ribosomes

2.1 BASIC THEMES

WATER

Whether a biomolecule is HYDROPHILIC (polar and/or charged) or HYDROPHOBIC (nonpolar) – or both – affects its overall shape. Hydrophilic molecules readily interact with water molecules via dipole-dipole interactions or hydrogen bonds,[1] and hydrophobic molecules avoid water. Large biomolecules that have both hydrophilic and hydrophobic sections arrange themselves accordingly. Example: Proteins fold so that the hydrophobic sections are tucked inside, away from the aqueous environment, and their hydrophilic areas (with polar or charged amino acid R groups) face outward to interact with the surrounding water.

BIOLOGICAL MEMBRANES: SELECTIVE PHYSICAL BARRIERS

Importance of the control of transport across membranes
Basic structure: LIPID BILAYER (mostly phospholipids) with proteins
PHOSPHOLIPIDS: hydrophilic head, hydrophobic tails

---

[1] Remember that hydrogen bonds are relatively strong *interactions* between an unshared pair of electrons on one molecule and a polarized H on another molecule. That polarized H needs to be attached to an O or N atom that has a lone pair (an O atom in a biomolecule always will; N^+ won't). Hydrogen bonds are very weak, though, compared to a covalent bond.

MEMBRANE PROTEINS:
    INTEGRAL PROTEINS (embedded within) vs. **PERIPHERAL PROTEINS** (on the surface)
    CHANNEL PROTEINS: transport specific ions across the membrane
    CARRIER PROTEINS: transport specific molecules across the membrane
    RECEPTORS: signal transduction
    ANCHOR PROTEINS: attach the membrane to macromolecules

SELF-ASSEMBLY INTO SUPRAMOLECULAR STRUCTURES

Why does self-assembly occur? **STERIC INFORMATION** in the macromolecules: complementary shapes fit together to optimize hydrophilic interactions, hydrophobic interactions, and many, many weak interactions.

MOLECULAR CHAPERONES or templates may provide assistance with assembly (folding).

MOLECULAR MACHINES

- consist of proteins (and protein complexes) that perform work and have moving parts
- convert energy into directed motion via **ENERGY-TRANSDUCING MECHANISMS:**
    1. Nucleotide (e.g. ATP or GTP) binds to **MOTOR PROTEINS** (protein subunits).
    2. Nucleotide hydrolyzes and releases energy.
    3. This energy causes a precisely targeted change in the protein subunit's shape.
    4. This change is transmitted to nearby subunits.

Examples of molecular machines:
    RIBOSOMES – rapidly and accurately incorporate amino acids into polypeptides
    SARCOMERES – contractile units of skeletal muscle; actin and myosin are proteins

MACROMOLECULAR CROWDING
- Describes cell conditions better than "concentrated"
- **EXCLUDED VOLUME** = volume occupied by macromolecules
- Impacts many intracellular processes

SIGNAL TRANSDUCTION

- How cells receive, interpret, and respond to signals such as molecules or light
- Examples of eukaryotic signal molecules: neurotransmitters, hormones, cytokines

PHASES:
1. **RECEPTION:** signal molecule (**LIGAND**) binds to and activates a **RECEPTOR** on the membrane surface, causing transduction.
2. **TRANSDUCTION:** a change in the receptor's 3D-shape, which triggers:
3. **RESPONSE:** inside the cell, a cascade of events that involves covalent modification of proteins and results in changes such as enzyme activity, gene expression, and motion.

2.2 STRUCTURE OF PROKARYOTIC CELLS

Features: relatively small, able to move using pili or flagella, able to retain specific dyes

Identify prokaryotes based upon nutritional requirements, energy sources, chemical composition, and biochemical capacities.

Common features: Cell wall   Circular DNA molecules
                            Plasma membrane   No internal membrane-enclosed organelles

CELL WALL

GRAM-POSITIVE cells have a thick **PEPTIDOGLYCAN** layer outside the plasma membrane

GRAM-NEGATIVE cell walls are more complex. Layers from the outside in:

GLYCOCALYX (slime layer, biofilm, or capsule)

OUTER MEMBRANE consists of LPS = LIPOPOLYSACCHARIDES (lipid A + polysaccharide)
  —contains porins (channel proteins for transport through the membrane)

PERIPLASMIC SPACE with peptidoglycans and proteins

INNER (PLASMA) MEMBRANE

Archaea vary: some are Gram-positive, some Gram-negative, some have no cell wall

PLASMA MEMBRANE (CYTOPLASMIC MEMBRANE)
selectively permeable barrier
contains RECEPTOR PROTEINS; may contain proteins for PHOTOSYNTHESIS and RESPIRATION

CYTOPLASM

NUCLEOID – contains a chromosome (circular DNA molecule)

PLASMIDS – additional small circular DNA molecules
RIBOSOMES, INCLUSION BODIES

PILI AND FLAGELLA FOR MOTION AND CONJUGATION (singular: pilus and flagellum)

2.3 STRUCTURE OF EUKARYOTIC CELLS

ENDOMEMBRANE SYSTEM; VESICLES

PLASMA MEMBRANE
Controls transport of molecules into and out of the cell
Transport is facilitated by carrier and channel proteins
Glycocalyx, Receptors, Extracellular Matrix; Cell Cortex

ENDOPLASMIC RETICULUM (ER)
LUMEN (OR CISTERNAL SPACE)
ROUGH ER (RER): synthesis of membrane proteins and proteins for export from the
  cell; contains ribosomes
SMOOTH ER (SER): lipid synthesis, biotransformation
SARCOPLASMIC RETICULUM (SR) – SER in striated muscle

GOLGI APPARATUS (OR GOLGI COMPLEX)
Packages and distributes cell products to internal and external compartments
VESICLES: secretory vesicles (or secretory granules)
Cisterna (plate); *cis* and *trans* faces
EXOCYTOSIS (secretion, see Figure 2.19)
DICTYOSOMES = Golgi apparatus in plants

NUCLEUS
Contains cell's hereditary information, regulates metabolism by directing the
synthesis of protein cell components
NUCLEOPLASM; LAMINS, CHROMATIN FIBERS, HISTONES; NUCLEAR MATRIX (or NUCLEOSKELETON)
NUCLEAR ENVELOPE; PERINUCLEAR SPACE; INNER MEMBRANE
NUCLEAR PORE; NUCLEAR PORE PLUG
NUCLEOLUS: synthesis of ribosomal RNA

VESICULAR ORGANELLES
ENDOCYTOSIS; PHAGOCYTOSIS
LYSOSOMES: contain digestive enzymes (acid hydrolases)
Other examples: plant vacuoles, α-granules (in platelets), melanosomes (in melanocytes)

MITOCHONDRIA
Aerobic metabolism – oxygen-dependent synthesis of ATP
OUTER MEMBRANE / INTERMEMBRANE SPACE / INNER MEMBRANE with **CRISTAE** (folds)**/ MATRIX**
Respiratory assemblies: ATP synthesis
Regulation of **APOPTOSIS** (genetically programmed events that lead to cell death)

PEROXISOMES: CONTAIN OXIDATIVE ENZYMES

Most important function: generate and break down peroxides (R–O–O–R)

PLANTS ONLY:

VACUOLE (contains acid hydrolases, analogous to lysosomes in animals)
MICROFILAMENTS
CELL WALL (contains cellulose)
DICTYOSOMES (analogous to Golgi apparatus in animal cells)
PEROXISOMES: Types and function
1. in leaves; responsible for photorespiration, and
2. **GLYOXYSOMES**, in germinating seed; convert lipids to carbohydrates

PLASTIDS ARE ANALOGOUS TO MITOCHONDRIA IN ANIMAL CELLS
– Plant, algae, and some protist cells contain plastids.
– **PROPLASTIDS** are plastid precursors.

Types of plastids:
1. **LEUCOPLASTS (storage)**;
2. **CHROMOPLASTS** (accumulate plant pigments)
   – **CHLOROPLASTS**: photosynthesis (converts light energy into chemical energy)
   – **THYLAKOID MEMBRANE, GRANA, THYLAKOID LUMEN** (or channel)
   – **STROMA** (analogous to the mitochondrial matrix), **STROMA LAMELLAE**

CYTOSKELETON
MICROTUBULES: structural support for long, thin cells; protein = tubulin
MICROFILAMENTS: cytoplasmic streaming and amoeboid movement; protein = actin
INTERMEDIATE FILAMENTS: maintain cell shape under mechanical stress; various proteins
(Example: keratin filaments in outer skin cells)

FUNCTIONS:
1. maintains cell shape,
2. facilitates coherent cellular movement, both large- and small scale; provides supporting structure to guide organelle movement within the cell
3. solid state biochemistry: provides a platform for enzyme complexes, greatly increasing reaction rates
4. signal transduction: provides structural continuity for signal cascade protein

RIBOSOMES: RNA/protein complexes that function in protein synthesis

BIOCHEMISTRY IN THE LAB: CELL TECHNOLOGY

CELL FRACTIONATION
DIFFERENTIAL CENTRIFUGATION, MICROSOMES, DENSITY-GRADIENT CENTRIFUGATION; MARKER ENZYMES

ELECTRON MICROSCOPY; limit of resolution

AUTORADIOGRAPHY

AFTER STUDYING THIS CHAPTER, YOU SHOULD BE ABLE TO:

- Draw a diagram of a prokaryotic cell. Label and describe the function(s) of each component.

- Draw a diagram of a eukaryotic cell, with each organelle labeled.

- Describe the functions of eukaryotic organelles.

- Compare the storage of genetic information (DNA) in a prokaryotic cell with that of a eukaryotic cell.

- Given a diagram of a cell, identify the various components and describe their functions. Identify the cell as prokaryotic or eukaryotic.

- Identify organelles that occur only in plant cells. Describe their functions.

- Compare the features and functions of rough endoplasmic reticulum, smooth endoplasmic reticulum, sarcoplasmic reticulum, and SER in hepatocytes.

- Draw a diagram of a eukaryotic plasma membrane. Include (and label) the various types of proteins that occur in plasma membranes.

- Describe the features of a prokaryotic cell wall and plasma membrane. Compare these to a eukaryotic plasma membrane.

- Other biological membrane structures include those of the nucleus and of mitochondria. Compare their similarities and differences with the plasma membrane.

- Describe the structure and function(s) of the cytoskeleton.

- Demonstrate an understanding of important intermolecular interactions and their impact on cellular structure and function. Examples include: hydrophilicity vs. hydrophobicity and biomembrane self-assembly; cell surface receptors and signal transduction; endocytosis and exocytosis; and lysosomal storage diseases.

Use this space to note any additional objectives provided by your instructor.

CHAPTER 2: SOLUTIONS TO REVIEW QUESTIONS

2.1   The cell is the basic unit of life that is separated from its environment by a plasma membrane.

2.2   Refer to Figure 2.15 on page 47 in your text. Integral proteins extend through the lipid bilayer, from the external surface to the internal surface of the membrane. Peripheral proteins are located on the surface of the cell membrane.

2.4   Channel proteins form an opening through which specific ions may pass. Carrier proteins transport specific molecules through a membrane. Receptors are proteins with binding sites for extracellular ligands (signal molecules). Ligand-receptor binding triggers a cellular response. Anchor proteins attach the membrane to macromolecules on either side of the membrane.

2.5   The plasma membranes of both prokaryotic and eukaryotic cells control the flow of substances into and out of the cell. In addition, plasma membrane receptors bind to specific molecules in the cell's external environment. In prokaryotes, for example, some receptors allow the organism to respond to the presence of food molecules. In eukaryotes, numerous cell receptors bind specific hormone or growth factor molecules. The prokaryotic cell wall is sufficiently rigid that it maintains the organism's shape and protects against mechanical injury.

2.7   On the surfaces of biomolecules, functional groups that can form noncovalent interactions will facilitate the formation of supramolecular structures with biomolecules that have properties that are similar (e.g., hydrogen bonding) or complementary (e.g., oppositely charged ions). As these noncovalent interactions form, more of the molecules' surfaces are drawn closer to each other, making further interactions possible. Large numbers of these interactions stabilize the complexes formed from these molecules. These interactions are augmented when the biomolecules (such as proteins and nucleic acids) have intricate shapes that are complementary to each other.

2.8   a.   exocytosis – a cellular process that consists of the fusion of membrane-bound secretary organelles with the plasma membrane. The contents of the granules are then released into the extracellular space.

            b.   biotransformation – a biochemical process in which water-insoluble organic molecules are prepared for excretion, usually by increasing their water solubility

            c.   grana – tightly stacked portions of thylakoid membrane within chloroplasts

            d.   supramolecular – stable and functional complexes of molecules and polymers; examples include biological membranes, ribosomes, sarcomeres, and proteasomes

            e.   self-assembly – the spontaneous formation of supramolecular complexes made possible by the interactions of specific biomolecules, each of which has an intricately shaped surface

2.10   The components of the endomembrane system are the plasma membrane, endoplasmic reticulum, Golgi apparatus, nucleus, and lysosomes. All of these control transport of ions and molecules across its membrane. Each membrane encloses an internal space that requires such control to function properly, i.e., for key biochemical reactions to take place. The compartments of the endomembrane system are connected via membranous vesicles that bud off from a donor membrane in one component in the system and fuse with the membrane of another component. For example, proteins synthesized in the RER are transferred via vesicles to the Golgi apparatus for further processing reactions.

2.11 The cytoskeleton provides structural continuity for intracellular signal transduction by providing a solid support for signal cascade proteins. Protein-protein interactions trigger sequential protein structure changes, resulting in the flow of information within the cell.

2.13 Plant cells; leucoplasts; chromoplasts.

2.14 Examples of diseases linked to organelles and their underlying causes are as follows:

(1) Cystic fibrosis (CF) is caused by the misfolding of the regulator protein CFTR, which functions as a plasma membrane chloride channel. The misfolded CFTR becomes trapped within the ER and is degraded. Without this Cl⁻ channel, thick mucus accumulates and compromises the lungs, pancreas, and other organs.

(2) Congenital disorders of glycosylation (CDG) are caused by mutations in genes that code for glycosylation enzymes or glycosylation-linked transport proteins in the Golgi apparatus.

(3) Progeria is caused by a specific mutation in the lamin A gene that codes for lamin, a component of the lamina of the nuclear envelope.

(4) Emery-Dreifuss muscular dystrophy is caused by the absence or mutation of the gene that codes for emerin, a protein of the inner membrane of the nuclear envelope.

(5) Tay-Sach's disease and Gaucher's disease are lipid storage diseases that are caused by the absence of a lysosomal enzyme.

(6) Pompe's disease (glycogen storage disease type II) is also caused by the absence of a lysosomal enzyme.

(7) I-cell disease is caused by the defective import of enzymes into lysosomes. In addition, ER stress is an important feature of Alzheimer's, Huntington's, and Parkinson's diseases, as well as heart disease and diabetes.

2.16
a. biofilm – a protective, adhesive barrier layer secreted by bacteria and consisting of substances such as polysaccharides and proteins
b. vesicle – a membranous sac that buds off from a donor membrane and subsequently fuses with the membrane of another compartment or with the plasma membrane
c. extracellular matrix – a gelatinous material, containing proteins and carbohydrates, that binds cells and tissue together
d. sarcoplasmic reticulum – specialized smooth endoplasmic reticulum that is located in striated muscle and serves as a reservoir for calcium ions, the signal that triggers muscle contraction
e. thylakoid - an intricately folded membrane system that is responsible for several chloroplast metabolic functions

2.17 Cell technology has contributed to modern medicine by improving the ability to study disease at the cellular and molecular levels. Most notable from an historical perspective are the discovery of DNA and its structure, and advances in microscopy and centrifugation techniques. Specific examples are: cell fractionation techniques to study organelles outside of cells, electron microscopy (both TEM and SEM), and autoradiography to study the intracellular location and behavior of cellular components.

2.19 The nucleus is the repository of the cell's hereditary information. The nucleus also exerts a profound influence over all the cell's metabolic activities through the expression of that information.

2.20 Among the roles of plasma membrane proteins are transport, response to stimuli, cell-cell contact, and catalytic functions.

2.22   The Golgi apparatus processes, sorts, and packages protein and lipid molecules for distribution to other regions of the cell or for export.

2.23   All living cells have similar chemical compositions (i.e., they are all composed of molecules such as carbohydrates, proteins, and lipids) and they all utilize DNA as genetic material.

2.25   Specific examples of vesicular organelles include lysosomes, which contain digestive enzymes; the glyoxysomes of fat and oil storing cells in seeds that are involved in gluconeogenesis from these fats and oils; the melanin-containing melanosomes, which migrate from the basal layer to the epithelial layer of the skin, and plant vacuoles, which that contain numerous enzymes and biomolecules needed for plant growth and development. (For additional examples of secretory lysosomes, see solution to Review Question 2.35.)

2.26   Peroxisomes are small spherical membranous organelles that contain oxidative enzymes. Primary functions of peroxisomes are the generation and degradation of peroxides and the oxidation of toxic molecules. Additional functions include the synthesis of certain membrane lipids and the degradation of fatty acids and purine bases. To form peroxisomes, nuclear genes code for the enzymes and membrane proteins, which are synthesized on cytoplasmic ribosomes and then imported into preperoxisomes. The ER provides the peroxisomal membrane, and peroxins (a group of proteins) assemble the peroxisomes. [Peroxisomes are also involved in photorespiration in plants (Chapter 13 of your text.)]

2.28   a.   membrane potential – potential difference across the membrane of living cells; usually measured in millivolts [Note: Membrane potential is described more fully in Chapter 3. "Periplasmic space" is offered as an alternate term to define.]
periplasmic space – the region between the outer membrane and the inner (plasma) membrane in a prokaryotic cell wall

   b.   transmembrane protein – a protein that extends from the inner surface to the outer surface of the cell membrane

   c.   peripheral protein – a protein that is attached to the cell membrane surface and is not embedded in the cell membrane

   d.   receptor protein – a protein on the cell surface that binds to a specific extracellular nutrient molecule and facilitates its entry into the cell; other receptors bind chemical signals and direct the cell to respond appropriately

   e.   anchor protein – a protein that attaches the cell membrane to macromolecules on either side of the membrane

2.29   a.   proplastids – small nearly colorless plant cell structures that develop into the plastids of differentiated cells

   b.   motor protein – a component of a biological machine that binds nucleotides; nucleotide hydrolysis drives precise changes in the protein's shape

   c.   hydrophobic – refers to molecules that possess few, if any, electronegative charges; do not dissolve in water

   d.   hydrophilic – refers to molecules that possess positive or negative charges or contain relatively large numbers of electronegative oxygen or nitrogen atoms; dissolves easily in water

2.31   a.   signal transduction – mechanisms by which extracellular signals are received, amplified, and converted to a cellular response

b. neurotransmitter – a molecule released at a nerve terminal that binds to and influences the function of other nerve cells or muscle cells

c. hormone – a molecule produced by specific cells that influences the function of distant target cells

d. ligand – a molecule that binds to a specific site on a larger molecule. In the context of signal transduction, a ligand is an external signal molecule that binds to and activates a cell receptor.

e. endotoxin – toxins that are released when a cell disintegrates

2.32 Hepatocyte SER functions include synthesis of the lipid components of very-low-density lipoproteins (VLDL), and biotransformation reactions, which convert water-insoluble metabolites and xenobiotics into more soluble products for excretion. Striated muscle SER is called the sarcoplasmic reticulum (SR) and is a reservoir for calcium ions, the signal that triggers muscle contraction.

2.34 a. endomembrane system – interconnecting internal membranes that divide the cell into functional compartments; its components are the plasma membrane, endoplasmic reticulum, Golgi apparatus, nucleus, and lysosomes

b. extracellular matrix (ECM) – a gelatinous material consisting of structural proteins and complex carbohydrates that binds cells together within animal tissues

c. excluded volume – the cellular volume that is occupied by macromolecules

d. solid state biochemistry – biochemical reactions that occur via enzymes that are bound to cytoskeletal filaments

e. cell cortex – a three-dimensional meshwork of proteins that reinforces the inner surface of eukaryotic plasma membranes

2.35 Secretory lysosomes can fuse with the plasma membrane and release substances onto the cell surface. Specific examples include osteoclast lysosomes that release enzymes to aid bone resorption; platelet lysosomes that release α-granules that contain adhesive protein ligands; and melanosomes that release melanin.

CHAPTER 2: SOLUTIONS TO THOUGHT QUESTIONS

2.37 Specialized cells can perform very sophisticated functions that make multicellular organisms possible. Cell specialization can be considered a disadvantage because such cells cannot exist independently; that is, they can exist only as part of a multicellular organism where their metabolic needs (e.g., energy requirements and waste product removal) are met.

2.38 The immobilization of enzymes and organelles on the cytoskeleton facilitates the highly organized set of living processes required to sustain the living state. For example, the close proximity of immobilized enzymes in a biochemical pathway allows the rapid delivery of the product of one enzyme to the active site of the next. This circumstance requires lower concentrations of reactant molecules than the time-consuming diffusion process.

2.40 The diphtheria toxin first binds to receptors on the cell surface. It is then transported into the cell, where it dissociates into the A and B subunit. The A subunit then moves to the ribosome, where it binds and blocks protein synthesis. Since protein synthesis is stalled, general cell processes are impeded and the cell dies.

2.41   The presence of DNA or possibly RNA in the organelle would strongly suggest that it may once have been free living.

2.43   The volume of a ribosome is calculated as follows:

$$\pi r^2 h = (3.14)(0.007 \ \mu m)^2(0.02 \ \mu m) = 3 \times 10^{-6} \ \mu m^3$$

The volume of a bacterial cell (from question 2.42) is 1.6 μm^3. The number of ribosomes that can fit in a bacterial cell is $1.6/3 \times 10^{-6} = 5 \times 10^5$, but because they occupy only 20% of the cell's volume, divide by 5 to give 1×10^5 ribosomes per bacterial cell.

2.44   The volume of the *E. coli* cell is given by $V = \pi r^2 h = (3.14)(0.5 \mu m)^2(2 \mu m) = 1.57 \ \mu m^3$

The surface area is: $A = 2\pi r^2 + 2\pi r h$

$$= (2)(3.14)(0.5 \mu m)^2 + (2)(3.14)(0.5 \mu m)(2 \mu m) = 1.57 \ \mu m^2 + 6.28 \ \mu m^2 = 7.85 \ \mu m^2$$

The *E. coli* surface-to-volume ratio = $7.85 \ \mu m^2/1.57 \mu m^3 = 5.0 \ \mu m^{-1}$

The volume of the eukaryotic cell is $V = (4/3)(3.14)(10)^3 = 4189 \ \mu m^3$

The surface area is $4\pi r^2 = 4(3.14)(10)^2 = 1256 \ \mu m^2$

The eukaryotic cell surface-to-volume ratio = $1256 \mu m^2/4189 \mu m^3 = 3.0 \ \mu m^{-1}$

The eukaryotic cell has a much smaller surface to volume ratio than does the *E. coli*. In order to import enough material to sustain the functions of the cell, the membrane must become more efficient. Eukaryotes have significantly greater membrane transport capacity because of membrane transport proteins that are more sophisticated and present in exceptionally large numbers, and extensive membrane folding, which increases the surface to volume ratio. Note that the loss of the prokaryote cell wall preceded the membrane folding.

Use this space to draw the structure of a prokaryotic or a eukaryotic cell. Label all of the organelles and the various membrane components.

3 Water: The Matrix of Life

Brief Outline of Key Terms and Concepts

3.1 MOLECULAR STRUCTURE OF WATER
POLAR BOND; POLAR MOLECULE; DIPOLE
ELECTROSTATIC INTERACTIONS

3.2 NONCOVALENT BONDING
Noncovalent bonds are important in determining the physical and chemical properties of living systems.

IONIC INTERACTIONS (SALT BRIDGES)

HYDROGEN BONDS,
with both dipole-dipole and covalent character, play a critical role in the properties of water and its place in the structure and function of cells.

VAN DER WAALS FORCES
dipole-dipole interactions; dipole-induced dipole; induced dipole-induced dipole

3.3 THERMAL PROPERTIES OF WATER
Hydrogen bonding is responsible for water's unusually high freezing and boiling points. Because water has a high HEAT CAPACITY, it can absorb and release heat slowly. Water plays an important role in regulating heat in living organisms.

3.4 SOLVENT PROPERTIES OF WATER
Water's dipolar structure and its capacity to form hydrogen bonds enable water to dissolve many ionic and polar substances. SOLVATION SPHERES

HYDROPHILIC MOLECULES, CELL WATER STRUCTURING, and SOL-GEL TRANSITIONS

HYDROPHOBIC MOLECULES AND THE HYDROPHOBIC EFFECT (HYDROPHOBIC INTERACTIONS)
Nonpolar molecules cannot form hydrogen bonds with water and are excluded via clathrate formation.

AMPHIPATHIC MOLECULES, such as fatty acid salts, spontaneously rearrange themselves in water to form MICELLES.

OSMOSIS
Osmosis is the movement of water across a semipermeable membrane from a dilute solution to a more concentrated solution. HYPERTONIC, HYPOTONIC, AND ISOTONIC SOLUTIONS; DIALYSIS HEMOLYSIS, CRENATION
OSMOTIC PRESSURE is the pressure exerted by water on a semipermeable membrane as a result of a difference in the concentration of solutes on either side of the membrane.
OSMOLARITY $= i$M   M $=$ molarity
$i =$ degree of ionization (van't Hoff factor)
OSMOTIC PRESSURE $= \pi = i$MRT
R $= 0.082$ L·atm/K·mol T$=$ Kelvin
CELL MEMBRANE POTENTIAL

3.5 IONIZATION OF WATER

ACIDS, BASES, AND pH
STRONG VS. WEAK ACIDS AND BASES; CONJUGATE BASE
Liquid water molecules have a limited capacity to ionize to form H^+ and OH^- ions. pH SCALE
Hydrogen ion concentration is a crucial feature of biological systems primarily because of their effects on biochemical reaction rates and protein structure. ACIDOSIS, ALKALOSIS

BUFFERS
A buffer is a mixture of a weak acid and its conjugate base. Buffers prevent changes in pH.
LE CHATELIER'S PRINCIPLE, BUFFERING CAPACITY
HENDERSON-HASSELBALCH EQUATION

$$pH = pK_a + \log \frac{[A^-]}{[HA]}$$

Buffer ranges $= pK_a \pm 1$

PHYSIOLOGICAL BUFFERS
BICARBONATE BUFFER: CO_2/HCO_3^-
– regulation by lungs and kidneys
– carbonic anhydrase
PHOSPHATE BUFFER $H_2PO_4^-/HPO_4^{2-}$
PROTEIN BUFFERS: Buffering capacity provided by proteins' amino acid side chains.

OVERVIEW

Without the unique properties of water, life as we know it could not exist. This chapter serves as a refresher for a number of concepts that you've seen in general and organic chemistry (electronegativity, polarity, hydrogen bonds, heat capacity, osmotic pressure, acid-base chemistry, buffers) and places them solidly in the context of biological systems. If some of these concepts happened to slip by you in previous courses, chances are that you'll find them more interesting and accessible here. You'll be seeing much of this material applied to amino acids, proteins, enzymes, and in the chapters beyond, so be sure to practice (and learn) this material early. This chapter includes examples of buffer problems with detailed solutions, and extra practice problems.

3.1 MOLECULAR STRUCTURE OF WATER: H_2O

Water is **POLAR**.

*   O–H is a **POLAR BOND** because O is much more **ELECTRONEGATIVE** than H. The O has a partial negative charge, and the H has a partial positive charge.

* H_2O is a **POLAR MOLECULE** because it has a **DIPOLE**. Whether or not a molecule is polar also depends on its geometry (or overall shape). Because the H_2O molecule is bent at an angle of 104.5°, it has a dipole – the "side" of the molecule with the oxygen's unshared pairs of electrons is more negative than the "side" with the hydrogen atoms. [Compare H_2O with CO_2. Both have polar bonds, but CO_2 is nonpolar. The polarity of the two C=O bonds cancels out because CO_2 is linear: O=C=O.]

The polar nature of water allows it to interact with a variety of other molecules. For example, table salt (NaCl) dissolves completely in water because of ion-dipole interactions. Alcohol dissolves in water via hydrogen bonding and dipole-dipole interactions.

*   **ELECTROSTATIC INTERACTIONS** occur between opposite charges (including the partial charges on atoms in a polar bond).

3.2 NONCOVALENT BONDING

IONIC INTERACTIONS

> Example: In proteins, **SALT BRIDGES** are ionic interactions between oppositely charged amino acid side groups. Example: $R–CO_2^-$ ..... $^+H_3N–R$

HYDROGEN BONDS

*   Occur between a hydrogen that's attached to an oxygen (or nitrogen) and a lone pair of electrons (on O, N, or S).

*   Each H_2O molecule can form hydrogen bonds with up to 4 other H_2O molecules.

*   Hydrogen bonding explains water's relatively high melting and boiling points, heat of vaporization, heat capacity, surface tension, and viscosity.

VAN DER WAALS FORCES

*   The more easily an atom can be polarized, the stronger the van der Waals forces.

*   Types of van der Waals forces:
    *   **DIPOLE-DIPOLE** interactions
    *   **DIPOLE-INDUCED DIPOLE** interactions
    *   **INDUCED DIPOLE-INDUCED DIPOLE** interactions (London dispersion forces); individually, these are the weakest, but they become significant when a great many of these interactions are present

3.3 THERMAL PROPERTIES OF WATER

*   Higher-than-expected melting and boiling points due to hydrogen bonding
*   High heat of vaporization – water doesn't boil easily

- High heat capacity – water can absorb and store heat and release it slowly.
- Living organisms use these properties to regulate temperature: high water content helps to retain heat because of water's high heat capacity; evaporation is used as a cooling mechanism.

3.4 SOLVENT PROPERTIES OF WATER

Water is polar and can form hydrogen bonds.

HYDROPHILIC MOLECULES, CELL WATER STRUCTURING, AND SOL-GEL TRANSITIONS
- Hydrophilic = "water loving"; soluble in water
- Polar molecules and ionic substances are hydrophilic
- Solvation spheres (shells of H_2O molecules formed by water around solutes):
  - depend on charge density: the smaller and more highly charged the ion, the larger the solvation sphere
  - an ion with a larger solvation sphere moves more slowly (example: Na^+ moves more slowly than K^+)

STRUCTURED WATER
- Layers of H_2O, constantly rearranging, between adjacent macromolecules
- H_2O molecules that are closer to polar surfaces move more slowly
- Helps to stabilize macromolecular structure, yet its dynamics allow the macromolecule to function

SOL-GEL TRANSITIONS
- Cytoplasm has gel-like properties because the polar surfaces of biopolymers have highly structured solvation layers of water.
- Sol-gel transitions may be caused by temperature changes, matrix architecture, or inclusion of solutes.
- Contributes to cell movement (and other cell functions); actin-binding proteins; amoeboid motion

HYDROPHOBIC MOLECULES AND THE HYDROPHOBIC EFFECT
- Hydrophobic molecules are **NONPOLAR**.
- Hydrophobic effect or hydrophobic interactions: Why do hydrophobic molecules group together in water?

  Water molecules maximize the formation of hydrogen bonds with other water molecules, and minimize any association with the nonpolar molecules. H_2O does this by forming a large "cage" around a group of nonpolar molecules, as opposed to many small "cages" around individual nonpolar molecules.

  Those weak van der Waals forces may contribute a little, but it's the *water* that *excludes* nonpolar molecules, *not* the attraction between nonpolar molecules, that causes them to separate.

  So – water hydrogen-bonds with itself as much as possible, *excluding* the hydrophobic molecules.[1]

---

[1] Also, in Chapter 4 you'll learn that the overall disorder of the water increases when a lipid micelle or membrane is formed (or when a protein folds), and that the water's disorder outweighs the resulting increased order of the micelle.

AMPHIPATHIC MOLECULES (CONTAIN BOTH HYDROPHOBIC AND HYDROPHILIC ENDS)

Molecules with a hydrophilic "head" and a long hydrophobic "tail" form micelles or bilayers (example: phospholipids form bilayers to create biological membranes).

OSMOTIC PRESSURE (π)

- **OSMOSIS, OSMOMETER**
- Osmotic pressure $= \pi = iMRT$ i = degree of ionization (van't Hoff factor)
  M = molarity, R = 0.082 L·atm/K·mol
  T = temperature in Kelvin (Recall K = °C + 273)
- **OSMOLARITY** $= iM$
- **ISOTONIC** vs. **HYPERTONIC** vs. **HYPOTONIC**; **HEMOLYSIS** (cell swells and may burst) vs. **CRENATION** (cell shrinks)
- **CELL MEMBRANE POTENTIAL** – cytoplasmic side of the cell membrane is negatively charged due to charged amino acid R groups on proteins.
- Cells regulate their osmolarity, usually by pumping ions across the membrane.
- Calculation of molecular weight using mass and osmotic pressure data: Use the equation (above) to solve for moles, then divide: #g / #moles = molecular weight

3.5 IONIZATION OF WATER

ACIDS, BASES, AND pH

$$HA \rightleftharpoons H^+ + A^-$$

- Strong vs. weak acids and bases
- weak acid (HA) and conjugate base (A^-)
- K_a is the equilibrium constant for the loss of an H^+

$$K_a = \frac{[H^+][A^-]}{[HA]}$$

- pK_a: the lower the pK_a, the stronger the acid (see below)
- The stronger the acid, the more dissociated the acid. That means there's more H^+ in solution, so the K_a will be larger. But remember that the $pK_a = -\log K_a$. If the K_a is 10^{-4}, the pK_a is 4. For a K_a that's a hundred times greater at 10^{-2}, the pK_a is 2. So, the lower the pK_a, the stronger the acid.

BUFFERS

Buffer = solution of a weak acid (HA) and its conjugate base (A^-)

Buffers resist pH changes when H^+ or OH^- is added (Le Chatelier's principle)

ACIDOSIS vs. **ALKALOSIS**

BUFFERING CAPACITY depends on total buffer concentration and ratio of $[A^-]/[HA]$

  (total buffer concentration = [HA] + [A$^-$])

HENDERSON-HASSELBALCH EQUATION: $pH = pK_a + \log \dfrac{[A^-]}{[HA]}$

Buffers are most effective when $[A^-] = [HA]$ or in the pH range of $pK_a \pm 1$. Titration curves show relatively flat areas when the pH = pK_a. These flat areas indicate good buffer ranges because a relatively large amount of OH^- may be added with very little change in pH.

WEAK ACIDS WITH MORE THAN ONE IONIZABLE GROUP

Examples: amino acids, H_3PO_4 (phosphoric acid)

PHYSIOLOGICAL BUFFERS

BICARBONATE BUFFER

The bicarbonate buffer system is slightly more complicated that one would expect because CO_2 reacts with H_2O to form H_2CO_3:

$$CO_2 + H_2O \rightleftharpoons H_2CO_3 \rightleftharpoons H^+ + HCO_3^-$$

This can be simplified to:

$$CO_2 + H_2O \rightleftharpoons H^+ + HCO_3^- \qquad pK_a = 6.37$$

Carbonic anhydrase is the enzyme in the blood that catalyzes this reaction (otherwise it'd be much to slow). Note: Of course you know that CO_2 itself isn't an acid, but this equation is a shortcut to take into account the chemistry of carbonic acid, the action of carbonic anhydrase, and the resulting capability of the blood to use CO_2 to control $[HCO_3^-]$ and $[H^+]$, i.e., to control pH.

Given that its pK_a differs from blood pH (7.4) by more than one pH unit, how can bicarbonate serve as an important buffer in the blood?

> Both CO_2 and HCO_3^- can be regulated by the lungs and kidneys. CO_2 can be exhaled, and the kidneys can remove H^+ from the blood. Both of these actions are needed to keep the ratio $[HCO_3^-]/[CO_2]$ high.

Why does the ratio $[HCO_3^-]/[CO_2]$ need to be high in order to be an effective buffer at a pH of 7.4?

> Compare the buffer pH with the pK_a. Because the buffer pH is more basic than the pK_a, the buffer needs more conjugate base than acid. Alternatively, we could use the Henderson-Hasselbalch equation, which gives a ratio of about 11:1.

PHOSPHATE BUFFER

$$H_2PO_4^- \rightleftharpoons H^+ + HPO_4^{2-} \qquad pK_a = 7.2$$

Given that the pK_a of dihydrogen phosphate is so close to blood pH (7.4), why *isn't* $H_2PO_4^-/HPO_4^{2-}$ an important buffer system in the blood?

> Their concentrations are too low. Phosphate buffers *are* important in intracellular fluids, where $[H_2PO_4^-]$ and $[HPO_4^{2-}]$ are higher.

PROTEIN BUFFER

> Some amino acid side groups are weak acids or bases, so many proteins – like serum albumins – can act as buffers to help regulate the pH of the blood.

SOLVING BUFFER PROBLEMS: USE THE HENDERSON-HASSELBALCH EQUATION

	One way to remember this equation is
$$pH = pK_a + \log \frac{[A^-]}{[HA]}$$ $$\frac{[A^-]}{[HA]} = \frac{\text{conjugate base}}{\text{weak acid}}$$ Total buffer concentration $= [A^-] + [HA]$	**H comes before K, and *AHA*!!** pH is equal to pK_a plus the log of *AHA*!! Yell the "Aha!" with enthusiasm and you'll always remember this equation!

A buffer is a mixture of a weak acid and its conjugate base. The toughest part of solving buffer problems is often identifying the weak acid (HA) and its conjugate base (A^-). Remember that an acid donates an H^+ and a base accepts the H^+. So, the weak acid will always have an extra H^+ when compared to its conjugate base, and the conjugate base will have *one* extra negative charge.

$$HA \rightleftharpoons H^+ + A^-$$

EXAMPLES:

HA (WEAK ACID)	**A⁻ (CONJUGATE BASE)**
H_2CO_3	HCO_3^-
$H_2PO_4^-$	HPO_4^{2-}

These examples were chosen for two reasons: (1) they are present in living systems as physiological buffers, and (2) they can be confusing because one of the forms listed in each example (above) can serve as either an acid or a conjugate base, depending what else is present (and the pH). Pay close attention to the problem, and be sure that you identify the acid as the one with the extra H^+. For example, HCO_3^- is the conjugate base here, but if CO_3^{2-} had been present, then HCO_3^- would have been the weak acid.

HA (WEAK ACID)	**A⁻ (CONJUGATE BASE)**
HCO_3^-	CO_3^{2-}
H_3PO_4	$H_2PO_4^-$

Hint: It helps to write out the acid-base reaction in this form. Be sure that both the number of H's and the charges balance.	$HA \rightleftharpoons H^+ + A^-$
Check out these examples. Writing reactions like this helps to make it more clear which is HA and which is A^-. Note that $H_2PO_4^-$ is the conjugate base in the first equation, but it's the weak acid in the second equation.	$H_3PO_4 \rightleftharpoons H^+ + H_2PO_4^-$ $H_2PO_4^- \rightleftharpoons H^+ + HPO_4^{2-}$

This carboxylic acid group is the weak acid, and its carboxylate is the conjugate base.	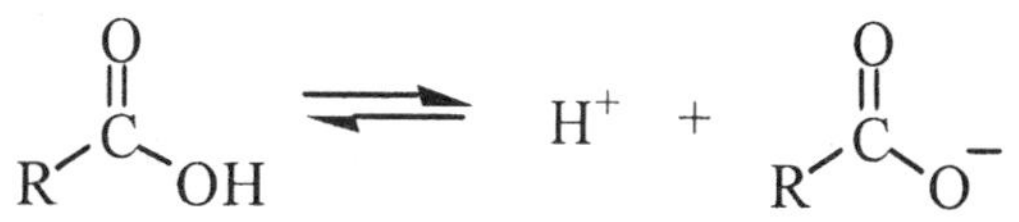

$$R-\underset{\overset{\parallel}{O}}{C}-OH \rightleftharpoons H^+ + R-\underset{\overset{\parallel}{O}}{C}-O^-$$

An amine group is a conjugate base, and its conjugate acid is shown here on the left.	

$$R-NH_3^+ \rightleftharpoons H^+ + R-NH_2$$

HOW TO PREPARE A BUFFER GIVEN A TARGET pH AND TOTAL CONCENTRATION

Use the Henderson-Hasselbalch equation to solve for $[A^-]/[HA]$. Rearrange your answer in the form: $[HA] = x[A^-]$ With the equation for the total concentration of a buffer, you can solve for $[A^-]$ by substituting "$x[A^-]$" for the value of $[HA]$. Once you have $[A^-]$, you can use either equation to solve for $[HA]$. See Example Problem # 3, below.

Also, note that you can't buy "A^-." It typically comes as a sodium or potassium salt (NaA or KA). For example, the buffer system $HPO_4^{2-}/H_2PO_4^-$ would be prepared by dissolving the appropriate amount of NaH_2PO_4 and Na_2HPO_4 in water, then diluting to the final volume. So, the final step often involves converting moles to grams of salt, using the molar mass.

WHAT HAPPENS WHEN YOU ADD A STRONG ACID TO A CONJUGATE BASE,

either in a buffer solution or alone? Write out the acid-base equation and think about what's happening. [*Hint*: Be sure to work in moles (or millimoles).] For example, think about what would happen if you added 0.02 moles of strong acid (H^+) to 0.05 moles of conjugate base (A^-). The 0.02 moles of H^+ would react completely with 0.02 moles of A^-, forming 0.02 moles of HA, and leaving $(0.05 - 0.02) = 0.03$ moles of A^- left over.

EXAMPLES OF BUFFER PROBLEMS *(DETAILED SOLUTIONS FOLLOW)*

1. What is the pH of a phosphate buffer if $[H_2PO_4^-] = 20$ mM and $[HPO_4^{2-}] = 15$ mM? The $pK_a = 7.20$ *(from Table 3.4, page 90 in your text).*

2. Calculate the ratio $\dfrac{[HPO_4^{2-}]}{[H_2PO_4^-]}$ at pH 6.2, 7.2, and 8.2. The pK_a is 7.2.

3. Describe how you would prepare one liter of 0.20 M lactate buffer with a pH of 4.2. What ratio of lactate salt to lactic acid would you use? The pK_a of lactic acid is 3.86.

4. a. What is the pH of a solution prepared by mixing 100 mL of 1.00 M HCl with 300 mL of 0.500 M sodium succinate? The pK_{a1} of succinic acid is 4.21.

    b. Is this buffer system effective? If so, why? If not, how could you correct it?

    c. Oops! Somebody just sploshed 80 mL of 1.0 M NaOH into your carefully-made solution! What's the final pH? Did the buffer work? How do you know? Is the solution still a good buffer?

EXAMPLES OF BUFFER PROBLEMS: *SOLUTIONS*

1. *What is the pH of a phosphate buffer when [$H_2PO_4^-$] = 20 mM and [HPO_4^{2-}] = 15 mM? The pK_a = 7.20.*

   First, identify the acid and the conjugate base: $H_2PO_4^- \rightarrow H^+ + HPO_4^{2-}$
   Since $H_2PO_4^-$ donates an H^+, it is the acid and HPO_4^{2-} is the conjugate base.
   Use the Henderson-Hasselbalch equation to determine the pH of the buffer solution.

   $$pH = pK_a + \log \frac{[HPO_4^{2-}]}{[H_2PO_4^-]} = 7.20 + \log 0.75 = 7.20 - 0.12 = 7.08$$

   Always think about your answer to be sure that it makes sense. Compare the amount of acid vs. base present with the pH vs. pK_a. For example, in this problem we have more acid than base, and at 7.08, the final pH is more acidic than the pK_a. Yay!

2. *Calculate the ratio of* $\dfrac{[HPO_4^{2-}]}{[H_2PO_4^-]}$ *at pH 6.2, 7.2, and 8.2. The pK_a is 7.2.*

   Identify the acid and the conjugate base: $H_2PO_4^- \rightleftharpoons HPO_4^{2-} + H^+$

   As before, $H_2PO_4^-$ is the acid and HPO_4^{2-} is the conjugate base. Plug the pH and pK_a values into the Henderson-Hasselbalch equation and solve for the ratio $\dfrac{[A^-]}{[HA]}$.

   $$pH = pK_a + \log \frac{[A^-]}{[HA]}$$

   At pH 6.2: $\log \dfrac{[A^-]}{[HA]} = pH - pK_a = 6.2 - 7.2$

   $$\frac{[A^-]}{[HA]} = 10^{(6.2-7.2)} = 10^{-1} = 0.1 \text{ or } \frac{1}{10}$$

   At pH 7.2: $\dfrac{[A^-]}{[HA]} = 10^{(7.2-7.2)} = 10^0 = 1 \text{ or } \dfrac{1}{1}$

   At pH 8.2 $\dfrac{[A^-]}{[HA]} = 10^{(8.2-7.2)} = 10^1 = 10 \text{ or } \dfrac{10}{1}$

3. *Describe how you would prepare a 0.20 M lactate buffer with a pH of 4.2. What ratio of lactate salt to lactic acid would you use? The pK_a of lactic acid is 3.86.*

   Use the Henderson-Hasselbalch equation to calculate the ratios of the salt and acid, where the salt is "A^-" and the acid is "HA".

   $$pH = pK_a + \log \frac{[A^-]}{[HA]}$$

   Substituting these values into the equation gives:

$$4.2 = 3.86 + \log [A^-]/[HA]$$
$$0.34 = \log [A^-]/[HA]$$
$$100.34 = 2.19 = [A^-]/[HA]$$
$$[A^-] = (2.19)[HA] \qquad \text{(equation 1)}$$

Since the total concentration of the lactate buffer is 0.20 M, we know that:

$$[A^-] + [HA] = 0.2 \text{ M} \qquad \text{(equation 2)}$$

Use simultaneous equations (i.e., substituting the value for $[A^-]$ from equation 1 into equation 2) to determine the concentrations of the salt and acid for this particular buffer solution to give:

$$(2.19)[HA] + [HA] = 0.20 \text{ M}$$
$$(3.19)[HA] = 0.20 \text{ M}$$
$$[HA] = 0.063 \text{ M}$$

Take this value and inserting it into (equation 1) to solve for $[A^-]$:

$$[A^-] = (2.19)[HA] = (2.19)(0.063) = 0.14 \text{ M}$$

To prepare this buffer, place 0.14 moles of lactate (or sodium lactate) and 0.063 moles of lactic acid in a 1 L volumetric flask and dilute with water to the 1 L mark.

And now to double-check: We're adding more base than acid to make this buffer. Since our final pH, 4.2, is more basic than our pK_a, our answer makes sense. Yay!

4. a. *What is the pH of a solution prepared by mixing 100 mL of 1.00 M HCl with 300 mL of 0.500 M sodium succinate? The pK_{a1} of succinic acid is 4.21.*

First, identify the weak acid and its conjugate base. Sodium succinate would give Na^+ and (succinate)$^-$ in solution. HCl is a strong acid that would react with the (succinate)$^-$ to form H-succinate, or succinic acid. So, the weak acid would be succinic acid and the conjugate base would be (succinate)$^-$.

The number of moles of (succinate)$^-$ initially is (300 mL)(0.500 M) = 150 mmol.

The number of moles of HCl added is (100 mL)(1.00 M) = 100 mmol.

It's a good assumption that HCl will react completely with the (succinate)$^-$, so that would give: 150 mmol (succinate)$^-$ – 100 mmol reacted with HCl

= 50 mmol (succinate)$^-$ left over and 100 mmol of succinic acid formed.

Now, use the Henderson-Hasselbalch equation:

$$\boxed{pH = pK_a + \log \frac{[\text{succinate}^-]}{[\text{succinic acid}]}}$$

$$pH = 4.21 + \log \frac{[50 \text{ mmol}]}{[100 \text{ mmol}]} = 4.21 - 0.30 = \textbf{3.91}$$

A pH of 3.91 makes sense because more acid than base is present, and 3.91 is more acidic than 4.21, the pK_a.

Note that this method included a shortcut: millimoles rather than molarity was used. This is valid because the total volume is the same and would cancel out. To use molarity, divide both the numerator (0.0045 mol) and the denominator (0.0030 mol) by the total volume, 0.450 L (300 mL + 150 mL).

b. *Is this buffer system effective? If so, why? If not, how could you correct it?*

This is a good buffer because its pH is within the range of the $pK_a \pm$ one pH unit, or 3.21 to 5.21. For a pH lower than 3.21, more succinate should be added. For a pH higher than 5.21, more acid should be added.

c. *Oops! Someone just splooshed 80 mL of 1.0 M NaOH into your carefully-made solution! What's the final pH? Did the buffer work? How do you know? Is the solution still a good buffer?*

pH = 4.61, yes (see below), yes

From (a), we know that before the addition of NaOH, we have 50 mmol of succinate and 100 mmol of succinic acid. OH^-, a strong base, will react completely with the succinic acid to form succinate. We have (80 mL)(1.0 M) = 80 mmol OH^-. So,

100 mmol succinic acid – 80 mmol reacted (with OH^-) = 20 mmol succinic acid

50 mmol succinate + 80 mmol formed = 130 mmol succinate

Let's plug these new values into the Henderson-Hasselbalch equation:

pH = 4.21 + log(130 mmol/20 mmol) = 4.21+0.81 = **5.02** (which makes sense!)

The buffer worked. If 80 mmol of OH^- were present in the same volume of pure water, the final pH would be 12.5. It's still a pretty good buffer since 5.02 is less than one pH unit from the pK_a. However, it will have a greater buffering capacity for acids than for bases.

PRACTICE PROBLEMS TO PROMOTE PERFECTION

Solutions are at the end of this chapter, after the solutions to the Review and Thought Questions.

1. Describe how you would prepare 1.0 L of a 0.30 M acetate buffer with a pH of 5.4. What ratio of acetate salt to acetic acid would you use? The pK_a of acetic acid is 4.76.

2. An acetate buffer is prepared by adding 40.0 mL of 1.00 M acetic acid to 200 mL of an aqueous solution containing 2.00 grams of sodium acetate. The final solution is diluted to 300 mL. (The molecular weight of sodium acetate is 82.0 g/mol, and the pK_a of acetic acid is 4.76.)

    a. What is the pH?

    b. What is the total buffer concentration?

    c. What would the final pH be if 20 mL of 1.0 M of hydrochloric acid were added to this buffer solution? Is this still a good buffer? What would the pH have been if the buffer hadn't been present? (Assume the same total volume of water.)

3. For laboratory experiments that are extremely sensitive to pH, why is it often recommended to use *freshly* distilled water?

4. An 8oz. serving (240 mL) of a popular cola contains 27 grams of sugars, listed as "high fructose corn syrup and/or sugar." What is the osmotic pressure that would be exerted by 27 grams of fructose in 240 mL of water at 37°C? If the sugars were 27 grams of sucrose, how would that affect the osmotic pressure? The molecular weight of fructose is 180 g/mol, and the molecular weight of sucrose is 342 g/mol.[2]

5. That same 240 mL serving of cola contains 25 mg sodium. If we assume that the sodium is present as sodium chloride,[3] that corresponds to 4.53×10^{-3} M NaCl. What would be the osmolarity and osmotic pressure of an aqueous solution of NaCl at this concentration at 37°C? Assume 100% ionization of the NaCl.

6. A solution of 0.200 g of an unknown molecule in 100 mL of water exerts an osmotic pressure of 0.465 atm at 25°C. Calculate the molecular weight of this nonelectrolyte.

---

[2] Obviously, this isn't a very good approximation for the cola, since the sugars are a mixture and there are other ingredients, such as caffeine, phosphoric acid, citric acid, carbonation, and "natural flavors."

[3] O.k., so it's probably a terrible assumption, especially since "salt" isn't listed in the ingredients.

AFTER STUDYING THIS CHAPTER, YOU SHOULD BE ABLE TO:

NONCOVALENT INTERACTIONS

- Identify the types of noncovalent interactions that can occur between given molecules (or between a given molecule and water).

- State whether a given molecule has a dipole moment.

- Will a micelle be able to form in a solution of a given molecule?

ACIDS AND BASES

- Identify weak acid – conjugate base pairs.

- Calculate pH from a given $[H^+]$, or $[H^+]$ from a given pH.

BUFFERS: USE THE HENDERSON-HASSELBALCH EQUATION

- Identify mixtures that can form buffer systems, and give the pH range where each would be most effective.

- Calculate the pH of a buffer given the pK_a and amounts of a weak acid and its conjugate base. (Calculate the amounts of a weak acid and its conjugate base given the amounts of conjugate base and HCl, *or* given the amounts of weak acid and NaOH.)

- Solve for the ratio of conjugate base to acid ($[A^-]/[HA]$).

- Determine how to prepare a specific buffer given the target pH, total buffer concentration, and total volume.

- Calculate the pH of a buffer solution after HCl or NaOH has been added.

- Titration curves: Estimate the pK_a and the effective buffer range.

OSMOTIC PRESSURE

- Calculate osmotic pressure given the amount of solute and volume using $\pi = iMRT$

- Calculate osmolarity (iM).

- Given osmotic pressure data, calculate the molar mass.

- Predict the direction that water will flow during dialysis. Predict whether a cell will shrink or swell, given various changes in osmotic pressure.

Use this space to note additional objectives provided by your instructor.

CHAPTER 3: SOLUTIONS TO REVIEW QUESTIONS

3.1 Both c and d are weak acid – conjugate base pairs.

3.2 $pH = -\log [H^+]$

$8.3 = -\log [H^+]$

$[H^+] = 10^{-8.3} = 5.0 \times 10^{-9} \text{ M}$

3.4 To prepare a 0.1 M phosphate buffer with a pH of 7.2, use the Henderson-Hasselbalch equation to calculate the ratio of the conjugate base to acid, where the conjugate base is "A⁻" and the acid is "HA":

$$pH = pK_a + \log \frac{[A^-]}{[HA]}$$

From a table of ionization constants, choose the phosphate conjugate acid base pair that has a pK_a closest to 7.2:

$$H_2PO_4^- \rightleftharpoons H^+ + HPO_4^{2-}$$

      acid   conjugate base

$$pK_a = 7.2$$

Substituting these values into the equation gives:

$7.2 = 7.2 + \log [A^-]/[HA]$

$0 = \log [A^-]/[HA]$

$10^0 = 1 = [A^-]/[HA]$

$[A^-] = [HA]$ (equation 1)

The concentrations of the conjugate base and acid must be equal. We also know that the total concentration of the phosphate buffer is 0.1 M. Therefore,

$[A^-] + [HA] = 0.1 \text{ M}$ (equation 2)

Using simultaneous equations (i.e., substituting the value for [HA] from equation 1 into equation 2) to determine the concentrations of the conjugate base and acid for this particular buffer solution gives:

$[A^-] + [A^-] = 0.1 \text{ M}$

$2[A^-] = 0.1 \text{ M}$

$[A^-] = 0.05 \text{ M}$

Taking this value and inserting it into (equation 1) gives: $[HA] = 0.05$ M

To prepare this buffer, place 0.05 mol of the acid and 0.05 mol of the conjugate base in a 1-L volumetric flask and dilute with water to the 1-L mark.

3.5 Molecules b and d can form hydrogen bonds with like molecules and with water. Molecules a and c can form weak hydrogen bonds between their lone pairs and the hydrogen in water.

3.7 In a solution of 1 M sodium lactate, water flows into the dialysis bag. In solutions of 3 M or 4.5 M sodium lactate, water flows out of the dialysis bag.

3.8 a. water and ammonia – hydrogen bonds
b. lactate and ammonium ion – ionic interactions
c. benzene and octane – van der Waals forces
d. carbon tetrachloride and chloroform – van der Waals forces
e. chloroform and diethyl ether – van der Waals forces

3.10 Arrows indicate atoms that would be involved in hydrogen bonding:

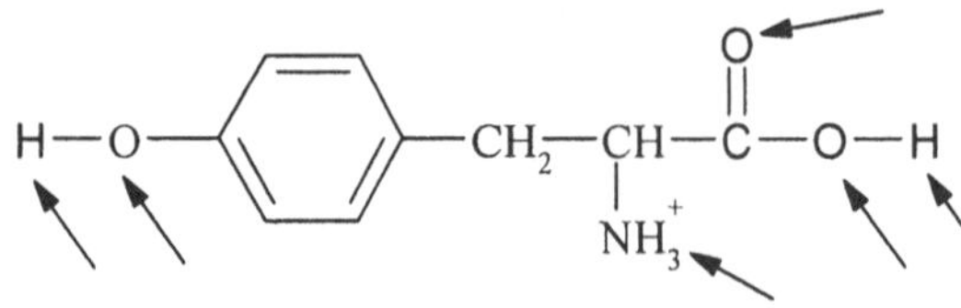

34

3.11 a. Hydrogen bonds are electrostatic interactions between a hydrogen covalently bonded to an oxygen, nitrogen, or sulfur (the O, N, or S must have a lone pair to polarize the H), and the lone pairs on nearby oxygen, nitrogen, or sulfur atoms.

      b. $pH = -\log[H^+]$

      c. A buffer is a solution that resists changes in pH, and is a mixture of a weak acid and its conjugate base.

      d. Osmotic pressure is the pressure needed to stop the net flow of water across a membrane.

      e. Osmolytes are osmotically active substances that cells produce to restore osmotic balance.

3.13 When they are very close together, molecules of d are capable of forming micelles because one end of the molecule is polar and the other end is nonpolar.

3.14 a. Isotonic solutions have identical osmotic pressures, i.e., they have the same particle concentration.

      b. Amphipathic molecules contain both polar and nonpolar domains.

      c. Hydrophobic interactions are the association of nonpolar molecules when they are placed in water.

      d. A dipole is a separation of partial charge in a molecule (one part of the molecule will have a partial positive charge, and one part of the molecule will have a partial negative charge) that results when a molecule has an unsymmetrical arrangement of polar bonds.

      e. An induced dipole is a dipole that is created in a molecule by a nearby charge (ion) or partial charge (dipole or polar molecule).

3.16 The buffering capacity of a system is increased by raising the concentrations of both buffer components but not changing their ratio ($[A^-]/[HA]$). Increasing the concentration of only the weak acid, for example, would increase the buffer capacity for added base, but would lower the buffer capacity for added acid.

3.17 Molecules b, c, and e are all weak acids because they are only partially ionized; a and d are strong acids (a is hydrochloric acid and d is nitric acid).

3.19 A buffer is composed of a weak acid and its conjugate base. Only c is a buffer.

3.20 Hyperventilation drives the transfer of carbon dioxide from the blood. This process, which shifts the following equilibrium to the left, consumes protons, thereby making the blood more alkaline.

$$CO_2 + H_2O \rightleftharpoons H_2CO_3 \rightleftharpoons HCO_3^- + H^+$$

$$\longleftarrow$$

3.22 a. A solvation sphere refers to the "shell" or cagelike arrangement of hydrogen-bonded water molecules that surround individual cations and anions in solution.

      b. A hypertonic solution has a higher osmotic pressure (a greater particle concentration) than the solution with which it's being compared.

      c. Alkalosis is a condition in which the blood pH is above 7.45 for a prolonged time.

      d. A dipole-dipole interaction is the noncovalent attraction between the partial negative charge on one polar molecule and the partial positive charge of another polar molecule.

e. The term *structured water* refers to the hydrogen-bonded, layered network of water molecules associated with macromolecules such as proteins. (See Figure 3.10) Structured water molecules bridge the space between adjacent macromolecules.

3.23   No. The carbonic acid and carbonate react to produce bicarbonate. It is possible to have either a buffer system of carbonic acid and bicarbonate or a buffer system of bicarbonate and carbonate.

3.25   pH = 4.22

Ascorbic acid is the weak acid (HA) and sodium hydrogen ascorbate is its conjugate base (A^-). Note that a more accurate representation is H_2A and HA^-.

The number of moles of hydrogen ascorbate initially:
$$(300 \text{ mL})(0.25 \text{ M}) = 75 \text{ mmol } HA^-$$
The number of moles of HCl added: $(150 \text{ mL})(0.2 \text{ M}) = 30 \text{ mmol HCl}$

Since HCl reacts completely with the hydrogen ascorbate,
    30 mmol HCl added = 30 mmol HA^- that react with HCl to form H_2A

The number of moles of H_2A remaining after addition is:
    75 mmol HA^- – 30 mmol that react with HCl = 45 mmol H_2A left over.
    30 mmol of ascorbic acid were formed.

Next, use the Henderson-Hasselbalch equation (pK_{a1} = 4.04):

$$pH = pK_a + \log\frac{[HAscorbate^-]}{[H_2 Ascorbate]} \quad \text{or} \quad pH = pK_a + \log\frac{[HA^-]}{[H_2A]}$$

$$pH = 4.04 + \log\frac{[45 \text{ mmol}]}{[30 \text{ mmol}]}$$
$$= 4.04 + 0.18 = \mathbf{4.22}$$

Does a pH of 4.22 make sense? Yes, because we have more base than acid present, and 4.22 is more basic than 4.04, the pK_a. Note that this method included a shortcut: millimoles rather than molarity was used. This is valid because the total volume is the same and would cancel out. To use molarity, divide both the numerator (0.0045 mol) and the denominator (0.0030 mol) by the total volume (0.450 L).

3.26   The contribution from the ionization of water must be considered. The hydrogen ion concentration is 1×10^{-8} M from acid and 10^{-7} M from water for a total acid concentration of 1.1×10^{-7} M acid. Therefore, the pH = $-\log(1.1 \times 10^{-7})$ = 6.96.

Note: Had the problem stated that $[H^+] = 1 \times 10^{-8}$ M, then the pH would be 8. However, without information to the contrary, only HCl and H_2O are present, which could be true only if the HCl had been added to pure water. (If NaOH were added to a more concentrated solution of HCl, then NaCl would also be in solution.) If acid is added to pure water, the pH will be acidic – perhaps only very slightly acidic – but definitely *not* basic.

3.28   In a mixture of one mole of benzoic acid ("HA") and one mole of sodium benzoate ("A^-"), [HA] = [A^-], and so [A^-]/[HA] = 1. Since log(1) = 0, the Henderson-Hasselbalch equation simplifies to pH = pK_a = 4.2.

$$pH = pK_a + \log\frac{[A^-]}{[HA]} \qquad pH = 4.2 + \log(1) \qquad pH = 4.2$$

3.29   Detergents that readily form micelles in water have structures similar to those of the amphipathic lipid molecules that make up a cell membrane. These detergents, therefore, should be able to associate with – and form micelles with – the lipid components of the cell membrane, resulting in membrane disruption and cell death. (Bacteria that are resistant to the action of detergents have a more complex membrane structure, outer coating(s), and/or additional structures that protect its inner cell membrane.)

3.31 When H^+ (HCl) is added to an acetic acid/acetate buffer, the H^+ will react with the acetate (A^-) to form more acetic acid (HA).

moles H^+ (HCl) added = $(1 \times 10^{-3} \text{ L})(1 \text{ M}) = 1 \times 10^{-3}$ mol H^+ added

moles of HA and A^- initially present: $(1 \text{ L})(1 \text{ M HA}) = 1$ mol HA = 1 mol A^-

$$H^+ + A^- \rightleftharpoons HA$$

The number of moles of HA will increase:
1 mol HA (initial) + 0.001 mol HA (from the added HCl) = 1.001 mol HA

The number of moles of A^- will decrease:
1 mol A^- (initial) – 0.001 mol A^- (that reacted with HCl) = 0.999 mol A^-

(Note that it's not necessary to calculate the total volume and the molarity, since the volume would be in both the numerator and the denominator, and would cancel.)

$$pH = pK_a + \log\frac{[A^-]}{[HA]}$$

pH = 4.75 + log [(0.999 mol A^-)/(1.001 mol HA)]
pH = 4.75 + (-0.0009) = 4.7491 ≈ 4.75

Compare this insignificant pH change, upon the addition of HCl to a buffer, with the addition of the same amount of HCl to water (Review Problem 3.32).

3.32 When 1 mL of 1 M HCl is added to 1 L of water, the new $[H^+]$ becomes:
(0.001 L) × (1 M H^+) = 0.001 moles H^+
(0.001 moles H^+)/(1.001 L total) = 9.99×10^{-4} M H^+
pH = $-\log[H^+]$ = $-\log(9.99 \times 10^{-4}$ M $H^+)$
pH = 3
Compare this pH change (from 7 to 3) to the addition of the same amount of HCl to a buffer solution (Review Problem 3.31).

CHAPTER 3: SOLUTIONS TO THOUGHT QUESTIONS

3.34 The highly concentrated sugar solution pulls water out of any bacterial cells present, which kills them, thereby preserving the fruit.

3.35 The regular crystal lattice of the ice crystal is more open than the tightly hydrogen-bound liquid water. If ice were more dense than water, ice formed in lakes and oceans would sink to the bottom. Eventually, only a narrow layer at the surface would be liquid. This environmental condition is incompatible with life (for most aquatic life, and they would not be able to survive).

3.37 The blood is so highly buffered by the bicarbonate buffer and the large amounts of blood proteins that under normal physiological conditions the transport of weak acids in the blood does not appreciably change its pH. For example, in the presence of bicarbonate, any acid that ionizes produces carbon dioxide (which is exhaled). The pH of the blood then remains virtually unchanged.

$$HCO_3^- + H^+ \rightleftharpoons H_2CO_3 \rightleftharpoons CO_2 + H_2O$$

3.38 The pH scale is derived using the ionization constant of water. To establish the pH scale for another solvent, the ionization constant of that solvent would have to be used, and the pH scale would be different from the pH scale for water.

3.40 No. The structure of cells is based on the phase separation of hydrophobic and hydrophilic substances. The function of the cell membrane is possible only because lipids are insoluble in water. If water dissolved every molecule, living organisms would not be able to create a barrier (membranes!) between themselves and their surroundings, and living organisms would not be possible.

3.41 The small water molecules can crowd closely around the ions and effectively disperse the charge, thereby facilitating solution. The bulky R group of the alcohol prevents this close interaction of solvent and solute. As a result the ionic compound does not dissolve as easily.

3.43 Hydration tends to make ionization easier. The hydrated acid group on the protein surface would have a higher K_a than one in the anhydrous interior of the protein.

3.44 Water weakens ionic interactions by forming a solvation sphere around each ion. As the distance increases between the cation and the anion, the attractive force decreases between them. (See Figure 3.9.) In other words, the polar water molecules crowd around the ions, interacting with the ions and weakening the interactions between oppositely charged ions.

3.46 Sugars contain many –OH (alcohol) groups per molecule. Alcohols are structurally similar to water (ROH vs. HOH) and have similar chemical properties. The OH groups of the alcohols "hydrate" the proteins and prevent them from aggregating.

3.47 Magnesium, with a double positive charge, forms a strong hydration sphere. For Mg^{2+} to move into the structured water of macromolecules, its hydration sphere must be removed in a process that requires a large amount of energy. Chloride ion, on the other hand, has only a single negative charge and is a larger ion. Its hydration sphere is not as tightly held and it would take less energy to remove. As a result, chloride would be more easily incorporated.

3.49 The syrup is actually a mixture of the sugar and small amounts of water that remain tightly held by hydrogen bonding to the OH groups of the sugars. Such tightly held water molecules prevent the direct hydrogen bonding between –OH groups in the sugar molecules required for crystallization.

3.50 The titration curve for tyrosine is as follows:

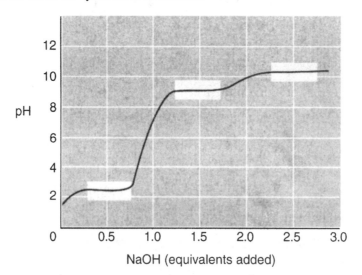

3.52 The equation is set up exactly the same as question 3.51.

69.9 J/g(1g) + 1.03 J/g(1g)(85.5-59.955) + 549 J/g(1g) = 646.6 J

The energy required to vaporize water is 4.7 times that of hydrogen sulfide. Most of this difference is in the heat of vaporization. In water, strong hydrogen bonds must be broken. Hydrogen sulfide does not have strong hydrogen bonds to break, hence the energy requirement is less.

3.53   The solvation sphere of the potassium ion in methyl alcohol is as follows:

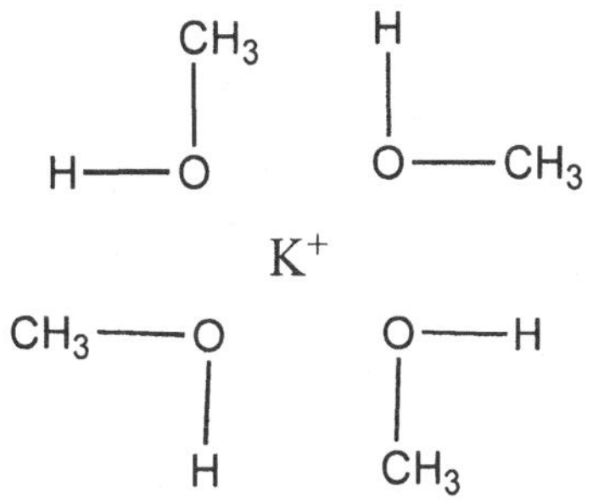

3.55   Oxygen is a smaller atom than nitrogen. As a result, the hydrogens of the water molecule can approach the oxygen more closely to form stronger bonds. Oxygen is also more electronegative than nitrogen, resulting in an O–H bond being more polar than a N–H bond.

3.56   The energy of solvation stabilizes both the departing proton and the acetate ion. The extra energy released helps drive the process and makes ionization easier. As a result, the pK_a of acetic acid in water should be lower than in the absence of water.

SOLUTIONS TO *PRACTICE PROBLEMS TO PROMOTE PERFECTION*

1.   *Describe how to prepare 1.0 L of a 0.30 M acetate buffer with a pH of 5.4. What ratio of acetate salt to acetic acid would you use? (The pK_a of acetic acid is 4.76.)*

$$pH = pK_a + \log[A^-]/[HA]$$

$$5.4 = 4.76 + \log[A^-]/[HA]$$

$$0.64 = \log[A^-]/[HA]$$

[A⁻]/[HA] = 4.4 = ratio of acetate salt to acetic acid

$$[A^-] = (4.4)[HA]$$

Substitute this equation for [A⁻] into the total buffer concentration equation:

Total buffer concentration = $[A^-] + [HA] = 0.30$ M

$$(4.4)[HA] + [HA] = 0.30 \text{ M}$$

$$(5.4)[HA] = 0.30 \text{ M}$$

[HA] = 0.056 M

Substitute this value for [HA] into the total buffer concentration equation:

$$[A^-] + 0.056 \text{ M} = 0.30 \text{ M}$$

$$[A^-] = 0.24 \text{ M}$$

To prepare the buffer, mix 0.24 mol of acetate salt with 0.074 mol of acetic acid in a volumetric flask and dilute to the mark with water. (To be more specific, dissolve 20 grams of sodium acetate in water in a 1 L volumetric flask. Add 74 mL of 1 M acetic acid. Dilute to the 1 L mark with freshly distilled water.)

2. a. millimoles of acetic acid = (40.0 mL)(1.00M) = 40.0 mmol

    millimoles of acetate = (2.00 g)/(82.0 g/mol) = 24.4 mmol

    pH = pK_a + log[A$^-$]/[HA] = 4.76 + log(24.4/40.0) = **4.55**

    (Note that we don't need to use the total volumes to get the correct answer. If you calculated molarity, the equation would be:

    pH = 4.76 + log([0.0813 M A$^-$]/[0.133 M HA]) = 4.55

    b. Total buffer concentration = (40.0 mmol + 24.4 mmol)/300 mL = 0.215 M

    c. HCl: (10 mL)(1 M) = 10 mmol HCl, which would react with the acetate to form acetic acid

    acetate: 24.4 mmol – 10 mmol (reacted with HCl) = 14.4 mmol acetate

    acetic acid: 40.0 mmol + 10 mmol (formed) = 50 mmol acetic acid

    pH = 4.76 + log(14.4/50) = 4.76 – .54 = **4.22**

    It's still a good buffer! And, if 10 mmol HCl were in a total volume of 310 mL, the pH would be **1.5**. Buffers work!

3. Distilled water that has been exposed to air for any length of time will probably contain dissolved CO_2, which would lower the pH.

    $$CO_2 + H_2O \rightleftharpoons H_2CO_3 \rightleftharpoons H^+ + HCO_3^-$$

4. First, calculate molarity of the fructose: 27 grams/180 g/mol = 0.15 moles
    0.15 moles/0.240 L = 0.625 M

    T = 37 + 273 = 310 K

    $\pi = iMRT$ = (1)(0.625 M)(0.082)(310 K) = 16 atm

    The molarity of sucrose would be 27 grams/342 g/mol = 0.0790 mol
    0.0790 mol/0.240 L = 0.329 M

    $\pi = iMRT$ = (1)(0.329 M)(0.082)(310 K) = 8.4 atm

5. osmolarity = iM = (2)(4.53 x 10^{-3} M) = 9.06 x 10^{-3} M
    osmotic pressure = $\pi = iMRT$ = (2)(4.53 x 10^{-3} M)(0.082)(310 K) = 0.23 atm

6. $\pi = iMRT$, R = 0.082 L·atm/K·mol   T = 25°C + 273 = 298 K
    i = 1 (for a nonelectrolyte)   M = moles/liter = # mol/(0.100 L)

    First, use $\pi = iMRT$ to solve for the number of moles:

    0.465 atm = (1)(moles/0.100L)(0.082 L·atm/K·mol)(298 K)
    1.90 x 10^{-3} mol

    molecular weight = grams/mol = (0.200 g)/(1.90 x 10^{-3} mol) = 105 g/mol

4 Energy

Brief Outline of Key Terms and Concepts

OVERVIEW OF THERMODYNAMICS

All living organisms unrelentingly require energy. BIOENERGETICS, the study of energy transformations, can be used to determine the direction and extent to which biochemical reactions proceed. ENTHALPY (a measure of heat content) and ENTROPY (a measure of disorder) are related to the first and second laws of thermodynamics, respectively. FREE ENERGY is the portion of total energy that is available to do work, and is related to enthalpy and entropy.

4.1 THERMODYNAMICS

FIRST LAW OF THERMODYNAMICS

INTERNAL ENERGY CHANGE = HEAT + WORK

At constant pressure, a system's ENTHALPY CHANGE ΔH is equal to the flow of HEAT ENERGY. EXOTHERMIC processes release heat energy, and have a negative $(-)$ ΔH. ENDOTHERMIC reactions have a positive $(+)$ ΔH, and require heat energy from the surroundings. In ISOTHERMIC processes no heat is exchanged with the surroundings. Total energy, free energy, enthalpy, and entropy are state functions.

$$\Delta H^{\circ}_{(reaction)} = \Sigma \Delta H_f^{\circ}_{(products)} - \Sigma \Delta H_f^{\circ}_{(reactants)}$$

SECOND LAW OF THERMODYNAMICS

The second law of thermodynamics states that the universe tends to become more disorganized. ENTROPY increases may take place anywhere in the system's universe. For bioprocesses, entropy increases take place in the surroundings.

4.2 FREE ENERGY

FREE ENERGY is a thermodynamic function that can be used to predict the spontaneity of a process. FREE ENERGY, a state function that relates the first and second laws of thermodynamics, represents the maximum useful work obtainable from a process.

$$\Delta G = \Delta H - T\Delta S_{sys}$$

SPONTANEOUS reactions are EXERGONIC $(-\Delta G)$; they release energy. Nonspontaneous reactions are ENDERGONIC $(+\Delta G)$; they need energy input to proceed.

STANDARD FREE ENERGY CHANGES

ΔG° is the standard free energy change (ΔG) at standard conditions: 25°C, 1 atm, and 1 M solute. In bioenergetics, $\Delta G^{\circ\prime}$ = the standard free energy change at pH 7. For: $aA + bB \rightleftharpoons cC + dD$,

$$\Delta G = \Delta G^{\circ} + RT \, ln\frac{[C]^c[D]^d}{[A]^a[B]^b} \qquad R = 8.315 \, \frac{J}{mol \cdot K}$$

(Remember to convert temperature into Kelvin, and to convert J to kJ if necessary.)

$$K_{eq} = \frac{[C]^c[D]^d}{[A]^a[B]^b}$$

When the system is at equilibrium, $\Delta G = 0$, so:

$$\Delta G^{\circ} = -RT \, ln \, K_{eq}$$

COUPLED REACTIONS

The hydrolysis of ATP immediately and directly provides the free energy to drive an immense variety of endergonic biochemical reactions.

$$\Delta G^{\circ\prime}_{(total)} = \Delta G^{\circ\prime}_{(reaction\ 1)} + \Delta G^{\circ\prime}_{(reaction\ 2)}$$

THE HYDROPHOBIC EFFECT REVISITED

To understand why biomolecules spontaneously self-assemble into complex ordered systems (e.g., lipid bilayers), consider the total entropy (disorder) of the *entire system*, including the huge increase in entropy of the surrounding water molecules.

BIOCHEMISTRY IN PERSPECTIVE: NONEQUILIBRIUM THERMODYNAMICS

Living organisms are far-from-equilibrium DISSIPATIVE STRUCTURES. They create internal organization via a continuous flow of energy.

4.3 THE ROLE OF ATP

ATP hydrolysis provides most of the free energy required for living processes. ATP is ideally suited to its role as universal energy currency, as ATP is easily hydrolyzed into products that have the following stabilizing advantages over ATP: (1) less electrostatic repulsion between adjacent $(-)$ charges, (2) more RESONANCE HYBRIDIZATION, (3) more easily solvated, (4) greater entropy (more molecules). Since ATP has an INTERMEDIATE PHOSPHORYL GROUP TRANSFER POTENTIAL, it can carry phosphoryl groups from high-energy compounds to low-energy compounds.

BIOCHEMISTRY IN PERSPECTIVE: EXTREMOPHILES

utilize an array of redox reactions to generate the energy required to sustain life in hostile environments. ELECTRON DONOR, ELECTRON ACCEPTOR, CHEMO-LITHOTROPHS, CHEMOORGANOTROPHS, FERMENTATION, AEROBIC RESPIRATION, ANAEROBIC RESPIRATION

4.1 THERMODYNAMICS

ENTHALPY (*H*), ENTROPY (*S*), FREE ENERGY (*G*)

Open vs. closed systems

State functions – values are independent of path

Enthalpy, entropy, and free energy are state functions, but work and heat are not.

Energy is the capacity to do work.

Energy is exchanged (between system and surroundings) as work and/or heat.

FIRST LAW OF THERMODYNAMICS

Energy can neither be created nor destroyed, but it can be transformed from one form into another.

ENTHALPY
$$\Delta H_{reaction} = H_{products} - H_{reactants}$$

EXOTHERMIC *vs.* ENDOTHERMIC
$\Delta H = (-)$ $\qquad$ $\Delta H = (+)$

Since enthalpy is a state function, the reaction mechanism (i.e., how you get from the reactants to the products) doesn't affect ΔH. That means that we can calculate the standard enthalpy of any reaction by summing up the ΔH_f° (the standard enthalpy of formation per mole) of the products and subtracting the sum of the ΔH_f° of the reactants:

$$\Delta H^\circ{}_{(reaction)} = \Sigma \Delta H_f^\circ{}_{(products)} - \Sigma \Delta H_f^\circ{}_{(reactants)}$$

Don't forget to take the stoichiometry of the reaction into consideration. You'll need to multiply the standard enthalpy of formation of each molecule by its coefficient in the chemical reaction.

SECOND LAW OF THERMODYNAMICS

The disorder (entropy, ΔS) of the universe always increases.

Spontaneous reactions or processes are **EXERGONIC** – energy is released.

Systems *can* spontaneously become more ordered (decrease in entropy) IF the surroundings become more disordered (increase in entropy) and the overall disorder (entropy) of the universe *increases*.

4.2 FREE ENERGY = Δ*G* = GIBBS FREE ENERGY CHANGE

$$\Delta G = \Delta H - T\Delta S \quad \text{(at constant temperature and pressure)}$$

EXERGONIC (**SPONTANEOUS**, releases energy)
$\qquad$ vs. **ENDERGONIC** (nonspontaneous, requires energy)

The sign of ΔG allows us to predict whether or not a chemical reaction can occur.

IF	THEN THE REACTION IS:
$\Delta G = -$	spontaneous (exergonic, favorable)
$\Delta G = +$	nonspontaneous (endergonic, not favorable)
$\Delta G = 0$	at equilibrium (no change)

Consider a general reaction, A + B → C + D. If the ΔG for this reaction is negative (energy is released and it's spontaneous), then the reverse reaction, C + D → A + B, has to be nonspontaneous (ΔG is positive). Since ΔG is a state function, the magnitude of ΔG in the forward reaction is the same as the reverse reaction. So, when the direction of a chemical reaction is reversed, the sign of ΔG is also reversed.

When ΔG = zero, there is no net change in the chemical reaction. The rate forward (A+B $\rightarrow$ C+D) equals the rate of the reverse reaction (C+D $\rightarrow$ A+B), and the reaction is at equilibrium.

STANDARD FREE ENERGY CHANGES: $\Delta G°$

Since ΔG depends on temperature, pressures, and concentrations, there needs to be some reference point. The *standard free energy*, $\Delta G°$, is defined as ΔG at standard state conditions: 25°C, 1 atm pressure, and 1 M reactant concentration. However, nearly all biochemical reactions occur in dilute, aqueous mixtures, so biochemists have developed their own reference point, $\Delta G°'$, which is $\Delta G°$ at pH = 7.[1]

ΔG	=	free energy change of a reaction under actual conditions
$\Delta G°$	=	free energy change at standard state conditions: 25°C, 1.0 atm, 1.0 M
$\Delta G°'$	=	free energy change at biochemical standard state conditions: $\Delta G°$ at pH 7.

So, the symbol (°) indicates standard state conditions, and a prime (') indicates biochemical standard state conditions.

ΔG is related to the reference $\Delta G°$ (or $\Delta G°'$) by the following equation[2]:

$$\Delta G = \Delta G° + RT \, ln\frac{[C]^c[D]^d}{[A]^a[B]^b}$$

for the reaction: $aA + bB \rightarrow cC + dD$

$$R = 8.315 \text{ J/mol·K}$$
$$T \text{ (in Kelvin)} = °C + 273$$

At equilibrium, $\Delta G = 0$, so: $\quad \boldsymbol{\Delta G° = -RT \, ln \, K_{eq}}$

K_{eq} is the equilibrium constant: $K_{eq} = \dfrac{[\text{products}]_{\text{at equilibrium}}}{[\text{reactants}]_{\text{at equilibrium}}} = \dfrac{[C]^c[D]^d}{[A]^a[B]^b}$

The sign and value of $\Delta G°$ indicate the direction and magnitude of a particular reaction at equilibrium (and under standard state conditions). For a given $\Delta G°$, the K_{eq} can be calculated.

EXAMPLE:

Calculate K_{eq} for the hydrolysis of ATP, given the following:

$$ATP + H_2O \rightarrow ADP + P_i \qquad \Delta G° = -30.5 \text{ kJ/mol} \qquad K_{eq} = \frac{[ADP][P_i]}{[ATP]}$$

(Since water is the solvent, $[H_2O]$ is omitted from K_{eq}.)

SOLUTION: We can solve for K_{eq} using $\quad \Delta G° = -RT \, ln \, K_{eq}$

$$-30.5 \text{ kJ/mol} = -(8.315 \text{ J/mol·K})(298K) \, ln \, K_{eq}$$

$$12.31 = ln \, K_{eq}$$

$$K_{eq} = 2.22 \times 10^5 \text{ or } 222,000/1$$

---

[1] Note that if $[H^+]$ does not affect the chemical reaction, then $\Delta G° = \Delta G°'$.

[2] Equilibrium constants are really defined for "activities" rather than "concentrations". However, concentrations are simpler to use and adequate for our purposes.

In this example, you can see that a negative $\Delta G°$ produces a large K_{eq}. The large K_{eq} indicates that at equilibrium (at 25°C), there is 222,000 times more product than reactant, and the reaction will proceed in the forward direction. Another way to look at this: At equilibrium, for each molecule of ATP present, there will be 471 ADP molecules and 471 P_i molecules (since [ADP][P_i] = 222,000, and the square root of 222,000 is 471).

What if the $\Delta G°$ had been +30.5 kJ/mol? Then the K_{eq} would be 4.5 x 10^{-6} 1/222,000.

$$ATP + H_2O \quad \rightarrow \quad ADP + P_i \quad \quad \Delta G° = -30.5 \text{ kJ/mol, spontaneous}$$
$$ADP + P_i \quad \rightarrow \quad ATP + H_2O \quad \Delta G° = +30.5 \text{ kJ/mol, nonspontaneous}$$

Although a reaction with a positive $\Delta G°$ is nonspontaneous, that same reaction may have a negative ΔG under different (nonstandard state) conditions, e.g., in a living cell. The actual direction of a reaction can be altered by changing the concentration of products or reactants. This is described by the general equation:

$$\Delta G = \Delta G° + RT \, ln\frac{[\text{products}]}{[\text{reactants}]} \quad \quad \text{or} \quad \quad \Delta G = \Delta G° + RT \, ln\frac{[C]^c[D]^d}{[A]^a[B]^b}$$

EXAMPLE:

glucose-1-phosphate $\rightarrow$ glucose-6-phosphate $\quad \Delta G° = -7.1$ kJ/mol

From the negative $\Delta G°$, we already know that the reaction proceeds forward to form glucose-6-phosphate under biochemical standard state conditions. But, can this reaction be reversed? Intuitively, if we were to add a high concentration of glucose-6-phosphate, we would think that this reaction could be "pushed" backwards. What if the concentration of glucose-6-phosphate was 100 mM and the concentration of glucose-1-phosphate was 1.0 mM? In which direction would the reaction proceed?

Let's use the equation: $\quad \Delta G = \Delta G° + RT \, ln\frac{[\text{glucose-6-phosphate}]}{[\text{glucose-1-phosphate}]}$

[glucose-6-phosphate] = 100 mM
[glucose-1-phosphate] = 1.0 mM

$$\Delta G = -7100 \text{ J/mol} + (8.315 \text{ J/mol K})(298K) \, ln \, \frac{[100 \text{ mM}]}{[1 \text{ mM}]}$$

$$\Delta G = +4.3 \text{ kJ/mol}$$

The positive ΔG tells us that this reaction will proceed in the opposite direction, as written. So, by increasing the concentration of the product, we can reverse the direction of the reaction.

As you can see, the direction of a reaction under actual cellular conditions depends not only on $\Delta G°$, but also on the concentrations of reactants and products.

COUPLED REACTIONS

A nonspontaneous reaction ($+\Delta G$) can be driven forward by coupling it with a spontaneous reaction ($-\Delta G$) to produce a negative ΔG overall.
Coupled reactions allow the cell to harness the energy produced by catabolism.

EXAMPLE:

Compare the reactions for PEP (phosphoenolpyruvate) hydrolysis vs. ATP synthesis:

$$PEP + H_2O \quad \rightarrow \quad \text{pyruvate} + P_i \quad \quad \Delta G°' = -61.9 \text{ kJ/mol}$$
$$ADP + P_i \quad \rightarrow \quad ATP + H_2O \quad \quad \Delta G°' = +30.5 \text{ kJ/mol}$$

Note that the hydrolysis of PEP is spontaneous ($\Delta G°$ is negative, and the reaction proceeds in the forward direction as written), but the reaction to form ATP is not ($\Delta G°$ is positive). To capture some of the energy of this PEP hydrolysis, consider coupling (adding) these two reactions:

PEP + H_2O	$\rightarrow$	pyruvate + P_i	$\Delta G°' = -61.9$ kJ/mol
ADP + P_i	$\rightarrow$	ATP + H_2O	$\Delta G°' = +30.5$ kJ/mol
PEP + ADP	$\rightarrow$	pyruvate + ATP	$\Delta G°' = -31.4$ kJ/mol

When we add these reactions, we also add the individual $\Delta G°'$ values to obtain the $\Delta G°'$ of the overall reaction. (This is valid because ΔG is a state function.)

$$\Delta G°'{}_{(total)} = \Delta G°'{}_{(reaction\ 1)} + \Delta G°'{}_{(reaction\ 2)}$$

The overall reaction is still thermodynamically favorable ($-\Delta G°'$), but we used some of the free energy to "drive" an unfavorable chemical reaction forward. Some of the energy released from the hydrolysis of PEP (phosphoenolpyruvate) was captured by ADP to form ATP. THIS CAPTURED ENERGY CAN BE RELEASED LATER (WHEN IT'S NEEDED) BY HYDROLYZING ATP TO REGENERATE ADP.

For example, consider the phosphorylation of glucose:

Glucose + P_i	$\rightarrow$	Glucose-6-phosphate + H_2O	$\Delta G°' = +13.8$ kJ/mol

When this reaction is coupled to ATP hydrolysis, the overall $\Delta G°'$ is negative. The energy from ATP hydrolysis drives this endergonic reaction forward.

Glucose + P_i	$\rightarrow$	Glucose-6-phosphate + H_2O	$\Delta G°' = +13.8$ kJ/mol
ATP + H_2O	$\rightarrow$	ADP + P_i	$\Delta G°' = -30.5$ kJ/mol
Glucose + ATP	$\rightarrow$	Glucose-6-phosphate + ADP	$\Delta G°' = -16.7$ kJ/mol

Since cells use ATP synthesis and hydrolysis to store and use chemical energy, ATP is often referred to as the "energy currency" of the cell.

In summary, a reaction that is not spontaneous can be driven forward if it's coupled with a spontaneous reaction and the overall ΔG is negative. In other words, the nonspontaneous reaction needs the energy from another reaction to drive it forward.

How can we get a nonspontaneous reaction to go forward (assuming constant temperature)?

- Couple it with a spontaneous reaction that will supply enough energy to give an overall negative ΔG, or
- Change the relative concentrations of reactant and product so that the actual ΔG will be negative.

THE HYDROPHOBIC EFFECT REVISITED

The hydrophobic effect explains why nonpolar molecules aggregate in water, why micelles and bilayers form, and why proteins fold. Although these processes seem to result in a decrease in entropy (i.e., an increase in order), the overall entropy (including that of the surrounding water molecules) is higher.[3]

---

[3] This is an oversimplification. Surface tension effects, van der Waals interactions, and the ordered arrangement of the water molecules immediately adjacent to a nonpolar molecule (i.e., the nature of structured water) should also be

4.3 THE ROLE OF ATP

ATP is produced by using the energy released by breaking down nutrient molecules (catabolism) and by the light reactions of photosynthesis. Hydrolysis of ATP releases 30.5 kJ/mole of energy that is used to drive endergonic processes such as:

1. Biosynthesis (anabolic pathways)
2. Active transport of substances across cell membranes
3. Mechanical work (e.g., muscle contraction)

Why is ATP hydrolysis so exergonic? The products are more stable than the reactants, because the final products:

1. have less electrostatic repulsion (of negative charges)
2. have more resonance structures than the reactants[4]
3. are more easily solvated than the reactants

PHOSPHORYL GROUP TRANSFER POTENTIAL is the tendency of a phosphoryl-containing molecule to hydrolyze, resulting in its phosphoryl group being released as HPO_4^{2-} or transferred to another molecule.

The greater the phosphoryl group transfer potential, the more energy is released when a phosphoryl group is hydrolyzed (and the more stable a molecule would be without its phosphoryl group).

Since ATP has an *intermediate* phosphoryl group transfer potential, it can take a phosphoryl group from a higher-energy compound and transfer it to a lower-energy compound. For instance, recall the two examples given earlier for coupled reactions:

PEP + ADP $\rightarrow$ Pyruvate + ATP $\quad \Delta G° = -31.4$ kJ/mol

Glucose + ATP $\rightarrow$ Glucose-6-phosphate + ADP $\quad \Delta G° = -16.7$ kJ/mol

Essentially, a phosphoryl group could be transferred from PEP to ATP, then from ATP to glucose, because both of these coupled reactions are spontaneous. PEP has a higher phosphoryl group transfer potential than ATP, and ATP has a higher phosphoryl group transfer potential than glucose-6-phosphate.

THERMODYNAMICS VS. KINETICS

THERMODYNAMICS tell us whether or not a reaction will be **SPONTANEOUS**.

If a reaction is spontaneous, there will be more products than reactants when the reaction is in equilibrium. That is, the reaction proceeds forward until it reaches equilibrium. Spontaneous reactions are **EXERGONIC** – they release energy.

KINETICS tell us **HOW *FAST*** the reaction will go.

Consider the combustion of glucose: $\quad C_6H_{12}O_6 + 6O_2 \rightarrow 6CO_2 + 6H_2O$

Glucose reacts with oxygen to form carbon dioxide and water. Thermodynamics tell us that this reaction is spontaneous and highly exothermic, but kinetics tell us that the rate at room temperature is incredibly slow. That's why glucose in contact with air doesn't simply burst into flames.

So, the chemical term *spontaneous* has a different meaning from the common dictionary definition. Glucose will *spontaneously* combust, but it will not *suddenly* combust (unless enough energy is applied to overcome its activation energy, and/or the activation energy is

---

taken into consideration. In general, at biological temperatures, these processes are driven by an increase in entropy (disorder) of the surroundings.

[4] Of course, this assumes that the H^+ of ADP dissociates.

lowered with the help of a catalyst or enzyme).[5]

THERMODYNAMICS	KINETICS
SPONTANEITY: *Can* a reaction happen? Is a reaction *spontaneous*?	RATE: *How fast* will a reaction happen?
Gibbs free energy change, ΔG	E_a, activation energy
exergonic vs. endergonic	k (rate constant)
enthalpy, ΔH, and entropy, ΔS	order of a reaction
$\Delta G = \Delta H - T\Delta S$	depends on mechanism (path)
equilibrium	*Enzymes* are biochemical catalysts that reduce the activation energy and cause a reaction to go *faster*.
K_{eq} (equilibrium constant)	
Chapter 4: **ENERGY**	Enzymes affect the rate – they do not affect the thermodynamics of a reaction.
	Chapter 6: **ENZYMES**

FOOD FOR THOUGHT: What's wrong with the following statement?
ATP releases energy when its high-energy phosphoryl bond is broken.

ANSWER:

1. The "high-energy phosphoryl bond" is not simply "broken." It's hydrolyzed (cleaved with the addition of water). Some bonds are broken, and others are formed.
2. The term "high-energy bond" suggests that the bond is unstable. Stability really refers to its ability to participate in reactions as opposed to the magnitude of the bond energy. The phosphoanhydride bond of ATP is actually more stable relative to compounds that have a higher phosphoryl group transfer potential.

EQUATIONS

$\Delta H^{\circ}_{(\text{reaction})} = \Sigma \Delta H_f^{\circ}{}_{(\text{products})} - \Sigma \Delta H_f^{\circ}{}_{(\text{reactants})}$

$\Delta G = \Delta H - T\Delta S$

$K_{eq} = \dfrac{[\text{products}]_{\text{at equilibrium}}}{[\text{reactants}]_{\text{at equilibrium}}} = \dfrac{[C]^c[D]^d}{[A]^a[B]^b}$ (for the reaction $a\text{A} + b\text{B} \rightleftharpoons c\text{C} + d\text{D}$)

(For the K_{eq} term, remember to use exponents if the coefficients in the chemical equation are greater than one.)

$\Delta G = \Delta G^{\circ} + RT \, ln\dfrac{[\text{products}]}{[\text{reactants}]}$ $R = 8.315$ J/mol K; T (in Kelvin) = $^{\circ}$C + 273

$\Delta G = \Delta G^{\circ} + RT \, ln\dfrac{[C]^c[D]^d}{[A]^a[B]^b}$

Hint: Be careful to use *either* **J or kJ** in your final equation, since R contains joules(J) and ΔG is typically in kilojoules(kJ). (1000 J = 1 kJ)

$\Delta G^{\circ} = -RT \, ln \, K_{eq}$ (at equilibrium)

$\Delta G^{\circ}{}'_{(\text{total})} = \Delta G^{\circ}{}'_{(\text{reaction 1})} + \Delta G^{\circ}{}'_{(\text{reaction 2})}$ (for coupled reactions)

---

[5] "Spontaneous combustion" as defined by popular usage is beyond the scope of this discussion.

AFTER STUDYING THIS CHAPTER, YOU SHOULD BE ABLE TO:

- Calculate $\Delta H_{reaction}$ given ΔH_f° data.

- Determine whether a reaction is: endothermic or exothermic, exergonic or endergonic, spontaneous or nonspontaneous.

- Use $\Delta G = \Delta G^\circ + RT \, ln\dfrac{[\text{products}]}{[\text{reactants}]}$ to determine whether or not a reaction will proceed forward at particular concentrations. That is, will it be spontaneous under different cellular conditions?

- Coupled reactions: Determine whether the overall reaction will be spontaneous using the equation: $\Delta G^{\circ}{}'_{(total)} = \Delta G^{\circ}{}'_{(reaction\ 1)} + \Delta G^{\circ}{}'_{(reaction\ 2)}$

- Coupled reactions: Choose a reaction that will provide enough energy to drive a nonspontaneous reaction.

- Use $\Delta G^\circ = -RT \, ln \, K_{eq}$ to calculate either ΔG° or K_{eq}.

Use this space to note any additional objectives provided by your instructor.

CHAPTER 4: SOLUTIONS TO REVIEW QUESTIONS

4.1   a. thermodynamics – the study of the heat and energy transformations in a chemical reaction

        b. endergonic – chemical reactions that absorb energy (i.e., have a negative free energy charge)

        c. enthalpy – a measure of the heat evolved during a reaction

        d. free energy – a measure of the tendency of a reaction to occur

        e. high-energy bond – a bond that liberates large amounts of free energy when it is broken

4.2   State functions are those that are independent of path. Entropy, enthalpy, and free energy are all path independent.

4.4   Given that the ionization constant for formic acid is 1.8×10^{-4}, the $\Delta G°$ for the reaction would be calculated as follows:

$$\Delta G° = -RT \ln K_{eq}$$

$$\Delta G° = -(8.315 \text{ J/mol·K}) (298 \text{ K}) \ln(1.8 \times 10^{-4})$$

$$\Delta G° = -(8.315) (298) (-8.62) \text{ J/mol}$$

$$\Delta G° = +21366 \text{ J/mol or } 21.4 \text{ kJ/mol}$$

4.5   a. dissipative structure – a far-from-equilibrium system that can form ordered structures under the influence of an energy gradient, such as that which occurs between the sun and the Earth; the maintenance of dissipative systems requires continuous work

        b. respiration – a biochemical process whereby fuel molecules are oxidized and their electrons are used to generate ATP

        c. electron donor – a molecule, atom, or ion that contains a lone pair of electrons that can react with an electron-deficient species to form a covalent bond; a Lewis base

        d. high-quality energy – a form of energy that has the capacity to do significant work; examples included electrical energy and chemical energy

        e. resonance hybrid – the combination of structures for a molecule with two or more alternative structures that differ only in the position of electrons

4.7   Statements (a) and (b) are undetermined, since no rate information was provided.

        Statement (c) is false. Both reactions have a negative $\Delta G°'$, so both reactions are spontaneous.

        Statement (d) is true.

4.8   Work is defined as a change in energy that produces a physical change. Physiological examples include the maintenance of concentration gradients across membranes, biomolecule synthesis, active transport across membranes, and muscle contraction.

4.10   a. fermentation – the anaerobic degradation of sugars; an energy-yielding process in which organic molecules serve as both donors and acceptors of electrons

        b. anaerobic respiration – a biochemical process whereby fuel molecules are oxidized and their electrons are used to generate ATP *in the absence of O_2*; molecules, atoms, or ions other than oxygen are the terminal electron acceptors

   c. redox reaction – reactions that involve changes in the oxidation number of the reactants

   d. chemolithotroph – an organism that derives chemical energy from minerals

   e. phosphoryl group transfer potential – the tendency of a phosphate bond to undergo hydrolysis; the $\Delta G^{\circ\prime}$ of hydrolysis of phosphorylated compounds

4.11   AMP hydrolysis involves cleavage of an ester bond and therefore releases the least energy. Hydrolysis of the other phosphate linkages involves either the hydrolysis of an anhydride or enol bond.

4.13   Under standard conditions the following statements are true: b, d, and e.

4.14 $\Delta G^{\circ\prime} = -RT \ln K_{eq}$

$-7100 \text{ J/mol} = -(8.315 \text{ J/molK})(298K)(\ln K_{eq})$

$\ln K_{eq} = 2.865$

$K_{eq} = 17.56$

4.16   The reduction of CO_2 to methane by *Methanococcus janaschii*, a hyperthermophile, is possible due to the high temperatures and pressures of the hydrothermal vent environments in which the organism thrives. This reduction is thermodynamically favorable under these conditions, but not at ambient temperature and pressure. Note that using a coupled reaction to drive CO_2 reduction forward would result in a favorable ΔG for the overall process, but would not change the ΔG for the reduction reaction alone. In other words, a coupled reaction can provide the energy to drive CO_2 reduction, but cannot convert an energy-requiring reaction into an energy-producing reaction. This is the inherent difference between an organism using ATP hydrolysis, for example, to drive CO_2 reduction forward, and an organism that obtains energy from CO_2 reduction to synthesize ATP.

4.17   glucose-1-phosphate $\rightarrow$ glucose + P_i $\Delta G^{\circ\prime} = -20.9 \text{ kJ/mol}$

[glucose] = [P_i] = 4.8 mM = 4.8×10^{-3} M

$\Delta G^{\circ\prime} = -RT \ln K_{eq}$ $K_{eq} = \dfrac{[\text{glucose}][P_i]}{[\text{glucose-1-phosphate}]}$

$-20.9 \text{ kJ/mol} = -(8.315 \times 10^{-3} \text{ kJ/molK})(298K)(\ln K_{eq})$

$\ln K_{eq} = 8.43$

$K_{eq} = 4.60 \times 10^3 = \dfrac{[\text{glucose}][P_i]}{[\text{glucose-1-phosphate}]}$

$4.60 \times 10^3 = \dfrac{[4.8 \times 10^{-3} \text{ M}][4.8 \times 10^{-3} \text{ M}]}{[\text{glucose-1-phosphate}]}$

$[\text{glucose-1-phosphate}] = \dfrac{[2.3 \times 10^{-5}]}{[4.6 \times 10^3]} = 5.0 \times 10^{-9}$ M

4.19   In the absence of Mg^{2+}, increased repulsion between adjacent negative charges of ATP would cause it to have less stability than ATP with Mg^{2+} present.

4.20   Yes, the value of $\Delta G^{\circ\prime}$ for the hydrolysis of pyrophosphate (PP_i) would change as the pH changes from 7 to 9. Although H^+ is neither a reactant nor a product in the hydrolysis reaction, the pK_a values for PP_i must be considered to determine its degree of ionization. With pK_a values of 6.70 ($H_2P_2O_7^{2-}$) and 9.32 ($HP_2O_7^{3-}$), the predominant charge at both pH 7 and pH 9 would be –3. At pH 7, PP_i exists as a 2:1 mixture of $HP_2O_7^{3-}$ and $H_2P_2O_7^{2-}$, whereas at pH 9, PP_i exists as a 2:1 mixture of $HP_2O_7^{3-}$ and $P_2O_7^{4-}$. Since these reactant ions differ in the number of negative charges, their stabilities would differ due to the varying amount of repulsion between adjacent negative charges.

Their hydrolysis products differ in stability as well:

$$H_2P_2O_7^{2-} + H_2O \rightarrow 2H_2PO_4^{-}$$

$$HP_2O_7^{3-} + H_2O \rightarrow H_2PO_4^{-} + HPO_4^{2-}$$

$$P_2O_7^{4-} + H_2O \rightarrow 2HPO_4^{2-}$$

These ratios were calculated using the Henderson-Hasselbalch equation. However, the same conclusion may be obtained using a more qualitative approach. At pH 8.01, the midpoint between the two pK_a values, 100% $HP_2O_7^{3-}$ exists. Above and below pH 8.01, different reactant mixtures exist that differ in charge, stability, and hydrolysis products.

With different degrees of stability in both the reactants and in the hydrolysis products, the $\Delta G°'$ values would be expected to differ as well.

4.22 [glucose] = [P_i] = 4.8 mM = 4.8 x 10^{-3} M

[glucose-6-phosphate] = 0.25 mM

$$K_{eq} = \frac{[glucose][P_i]}{[glucose\text{-}6\text{-}phosphate]}$$

$$K_{eq} = \frac{(4.8\times10^{-3}M)(4.8\times10^{-3}M)}{(2.5\times10^{-4}M)}$$

K_{eq} = 9.2 x 10^{-2} M

4.23 $$\Delta G = \Delta G°' + RT \ln \frac{[glucose][P_i]}{[glucose\text{-}6\text{-}phosphate]}$$

$\Delta G°' = -13.8$ kJ/mol

$$\frac{[glucose][P_i]}{[glucose\text{-}6\text{-}phosphate]} = 9.2 \times 10^{-2} \text{ (From Problem 4.22)}$$

$\Delta G = -13.8$ kJ/mol + (8.315 x 10^{-3} kJ/molK)(298)(ln 9.2 x 10^{-2})

$\Delta G = -13.8$ kJ/mol + (2.48)(–2.39) kJ/mol

$\Delta G = -19.7$ kJ/mol

4.25

ATP $\rightarrow$ AMP + PP_i	$\Delta G°' = -32.2$ kJ/mol
$PP_i \rightarrow 2 P_i$	$\Delta G°' = -33.5$ kJ/mol
ATP $\rightarrow$ AMP + 2P_i	$\Delta G°' = -65.7$ kJ/mol

$\Delta G°' = -RT \ln K_{eq}$

–65.7 kJ/mol = –(8.315 x 10^{-3})(298K)($\ln K_{eq}$)

$\ln K_{eq}$ = 26.52

K_{eq} = 3.29 x 10^{11}

CHAPTER 4: SOLUTIONS TO THOUGHT QUESTIONS

4.26   The energy liberated by the hydrolysis of 12.5 mol of ATP is

(12.5 mol)(–30.5 kJ/mol) = – 381.3 kJ

The energy required to produce 12.5 mol of ATP is 1142.2 kJ. The apparent efficiency of the process is:

(381.3/1142.2) x 100 = 33.4%

4.28   Although only a few molecules are involved, the laws of thermodynamics are still obeyed.

4.29   In the case of endothermic solutions, the enthalpy may be negative but the entropy is sufficiently positive to make the overall $\Delta G°$ favorable.

4.31   ATP has an intermediate phosphoryl group transfer potential. This makes it possible for ATP to serve as a carrier of phosphoryl groups from high energy compounds to those of lower energy.

4.32   The following information is needed to determine the $\Delta G°'$ of a reaction: temperature, concentrations of reactants and products, and ΔG.

4.34   Phosphoenolpyruvate has the greatest phosphoryl group transfer potential. On hydrolysis, the enol which has a resonance restricted structure, rapidly converts to the more stable, resonance-hybridized keto form and drives the equilibrium to the right.

4.35   When magnesium ions coordinate with the phosphate groups of ATP, the repulsion between adjacent oxygen anions is decreased. Consequently, ATP is stabilized and the free energy of hydrolysis is reduced. [Conditions that stabilize the reactant (ATP) would make the ΔG less negative.]

4.37   Both acetate and phosphate are resonance stabilized. However phosphate has more resonance forms than acetate. Hydrolysis reduces the repulsions between nearby oxygen anions in phosphate. There is no such repulsion in acetate.

4.38   (Note that this question should refer to problem #33. The data provided is as follows:

$\Delta H° = -88$ kJ/mol $\Delta S° = 0.3$ kJ/mol·K $\Delta G° = 0$

$\Delta G° = \Delta H° - T\Delta S°$

$0 = -88 \text{kJ/mol} - T(0.3 \text{kJ/mol})$

$88 \text{kJ/mol} = -T(0.3 \text{ kJ/mol})$

$T = -88/0.3 = -293.3 \text{K}$

This is *below* absolute zero, an absolute impossibility. There is no temperature at which the denaturation of this protein can be in equilibrium.

4.40   Glucose-1-phosphate contains a high-energy anhydride bond. Glucose-6-phosphate, on the other hand, has a lower-energy ester bond.

4.41   Phosphoenolpyruvate is an enol, and hydrolysis liberates the enol form, which rapidly tautomerizes to the keto form. Equilibrium is driven to the right. Isopropyl phosphate and allyl phosphate do not have this secondary reaction to drive the transfer of the phosphate group.

5 Amino Acids, Peptides, and Proteins

Brief Outline of Key Terms and Concepts

OVERVIEW

POLYPEPTIDES are polymers composed of AMINO ACIDS linked by PEPTIDE BONDS. The AMINO ACID SEQUENCE is the order of the amino acids in a polypeptide. An AMINO ACID RESIDUE is an amino acid within a polypeptide chain. PEPTIDES have less than 50 amino acid residues; PROTEINS have more than 50. Proteins may contain more than one chain.

5.1 AMINO ACIDS

AMPHOTERIC, ZWITTERION

AMINO ACID CLASSES: NONPOLAR, POLAR, ACIDIC, and BASIC. Amino acids are classified according to their capacity to interact with water.

BIOLOGICALLY ACTIVE AMINO ACIDS

HORMONES, NEUROTRANSMITTERS, nucleotide (and other) precursors, metabolic intermediates

MODIFIED AMINO ACIDS IN PROTEINS

AMINO ACID STEREOISOMERS

All amino acids – except glycine – have a chiral carbon (asymmetric carbon) atom, and as a result, can exist as STEREOISOMERS. ENANTIOMERS are mirror-image forms of a molecule. Most asymmetric molecules in living organisms occur in only one stereoisomeric form. With few exceptions, only L-amino acids (not D-) are found in proteins.

TITRATION OF AMINO ACIDS

Titration is useful in determining the relative ionization potential of acidic and basic groups in an amino acid or peptide. The ISOELECTRIC POINT is the pH at which an amino acid has no net charge.

AMINO ACID REACTIONS

PEPTIDE BOND FORMATION

Peptide bonds have partial double-bond character, making them rigid and flat. Amino acid sequences are written from N-terminal to C-terminal.

CYSTEINE OXIDATION FORMS DISULFIDE BRIDGES that help stabilize polypeptide and protein structure.

SCHIFF BASE FORMATION

When amine groups react reversibly with carbonyl groups, they form Schiff bases (ALDIMINES).

5.2 PEPTIDES

Peptides have significant biological activity that includes a variety of signal transduction processes.

5.3 PROTEINS

Functions include: catalysis, structure, movement, defense, regulation, transport, storage, and stress response (HEAT SHOCK PROTEINS). MULTIFUNCTION PROTEINS, PROTEIN FAMILIES AND SUPERFAMILIES FIBROUS PROTEINS; GLOBULAR PROTEINS

CONJUGATED PROTEINS contain PROSTHETIC GROUPS (non-protein components). HOLOPROTEINS Have their prosthetic groups; in APOPROTEINS, prosthetic groups are ABSENT.

PROTEIN STRUCTURE

INVARIANT amino acid residues are essential to its function. HOMOLOGOUS polypeptides have similar amino acid sequences and a common origin. PRIMARY STRUCTURE is the sequence of amino acid residues connected by peptide bonds. Molecular diseases arise from mutations that result in changes in key protein's primary structure. SECONDARY STRUCTURE describes hydrogen-bond stabilized polypeptide segments with 3D order, such as α-helices and β-pleated sheets. MOTIF, SUPERSECONDARY STRUCTURE

TERTIARY STRUCTURE is the unique three dimensional conformation that a protein assumes (PROTEIN FOLDING) because of the interactions between amino acid side chains. (FOLD – the core 3D structure) Interactions that stabilize tertiary structure: the hydrophobic effect, electrostatic interactions (SALT BRIDGES), hydrogen bonds, covalent bonds, and hydration. MODULAR (or MOSAIC) PROTEINS, LIGANDS

QUATERNARY STRUCTURE describes proteins with several separate polypeptide SUBUNITS held together by noncovalent and covalent bonds. OLIGOMER, PROTOMER; ALLOSTERY, ALLOSTERIC TRANSITION, EFFECTOR, MODULATOR

UNSTRUCTURED PROTEINS: INTRINSICALLY UNSTRUCTURED PROTEINS (IUPs), NATIVELY UNFOLDED PROTEINS

LOSS OF PROTEIN STRUCTURE – DENATURATION

THE FOLDING PROBLEM: The primary structure contains all the information required for each newly synthesized polypeptide to fold into its biologically active conformation Some relatively simple polypeptides fold spontaneously; other larger molecules require the assistance of MOLECULAR CHAPERONES such as the hsp70S, hsp60S (CHAPERONINS). MOLTEN GLOBULE

FIBROUS PROTEIN function usually involves structural support.

GLOBULAR PROTEIN function usually involves binding to small LIGANDS or to other macromolecules. Examples: myoglobin, hemoglobin. COOPERATIVE BINDING

STUDY STRATEGY:

I *highly recommend* learning the amino acids A.S.A.P. *Knowing* the structures without having to think twice will be a tremendous advantage as you build understanding and work the protein problems later in the chapter. You'll continue to see the positive effects of knowing these structures as you move forward. Specific amino acid structures are critical to enzyme function (Chapter 6). It follows, then, that every pathway that includes discussion of the mode of action or the regulation of an enzyme mentions specific amino acid structures (Chapters 8, 9, 10, 12, etc.). Check out the HINTS FOR LEARNING THE 20 STANDARD AMINO ACIDS on the next page.

5.1 AMINO ACIDS

20 standard amino acid structures and abbreviations

STANDARD vs. NONSTANDARD amino acids (standard: commonly found in proteins; nonstandard: chemically modified after a peptide or protein was synthesized)[1]

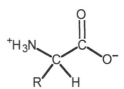

AMPHOTERIC (can act as an acid or a base)

at pH=7, amino acids exist as ZWITTERIONS (two ions on two different atoms in the same molecule)

AMINO ACID CLASSES: DETERMINED BY HOW THE R GROUP INTERACTS WITH H_2O

All amino acids have an acidic carboxyl group $-CO_2H$ and a basic amino group $-NH_2$, so amino acids are classified by the *one thing* that makes *each* amino acid *unique*: its R group.

NONPOLAR	R contains aromatic, aliphatic, and/or sulfur groups
POLAR, neutral	R contains $-OH$ or an amide
ACIDIC	R contains $-CO_2H$ that can donate an H^+
BASIC	R contains an N that can accept an H^+

Be aware that amino acid R groups that are acidic or basic are almost always written as they exist at pH 7. So, the R group of an acidic amino acid such as aspartate, with $R = -CH_2CO_2H$, will have lost its H^+ at pH 7, so it's written as $-CH_2CO_2^-$. Even though it's written in this form (i.e., its conjugate base), its official amino acid classification is still, and will always be, ACIDIC.

THREE-LETTER ABBREVIATIONS = FIRST THREE LETTERS *EXCEPT FOR THESE:*

Ile	=	Isoleucine	Asn =	Asparagine
Trp	=	Tryptophan	Gln =	Glutamine

---

[1] Note that the definitions of *standard* vs. *nonstandard* amino acids differ from the dietary *essential* vs. *nonessential* amino acids.

HINTS FOR LEARNING THE 20 STANDARD AMINO ACIDS

Set out to LEARN the amino acids, not just to memorize them. Flash cards help. Write the name on one side *and the structure on the other side* (to make the learning active rather than passive – it makes a difference..).

Write out the structures for yourself. As you learn the structures, make a shorter list of the ones that continue to be troublesome for you.

Group similar amino acids together and compare them.

Group confusing amino acids together and compare them. For example, leucine and isoleucine are troublemakers, since leucine has the isobutyl group, not isoleucine. We're all stuck with the chore of keeping those two straight. Also, there's aspartate and glutamate, but it's helpful that aspartate has the shorter chain and also comes before glutamate in the dictionary.

Try drawing the structures differently. For example, comparing tryptophan and histidine drawn as shown below may help you to learn these two problem children.

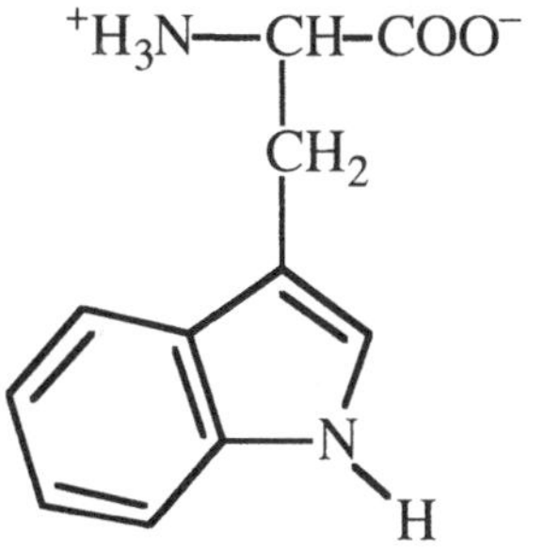

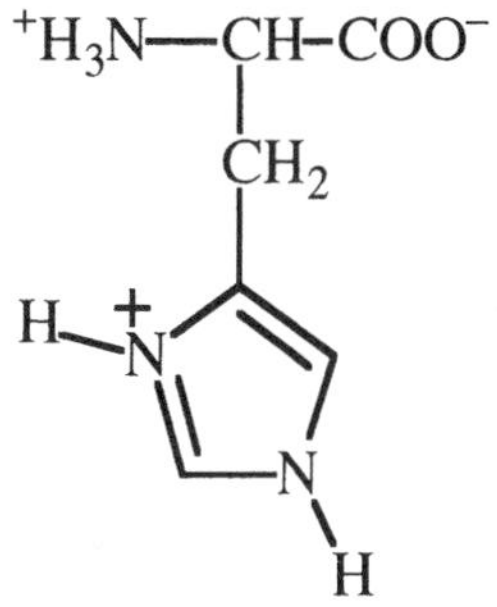

BIOLOGICALLY ACTIVE AMINO ACIDS

Chemical messengers; Examples:
NEUROTRANSMITTERS: glycine, glutamate, GABA, serotonin, melatonin;
HORMONES: thyroxine, indole acetic acid (derived from amino acids)

Precursors to complex N-containing molecules (nucleotides, heme, chlorophyll)

Metabolic intermediates

MODIFIED AMINO ACIDS IN PROTEINS

Examples of how amino acids can be modified to form **AMINO ACID DERIVATIVES** include carboxylation, hydroxylation, and phosphorylation. Phosphorylation of Ser, Thr, and/or Tyr residues in proteins is used to regulate entire metabolic pathways.

What do the R groups of serine, threonine, and tyrosine all have in common?
—an –OH group that can be phosphorylated

AMINO ACID STEREOISOMERS

AMINO ACID STEREOCHEMISTRY affects PROTEIN STRUCTURE, which affects FUNCTION

All of the α-amino acids – except for glycine – has at least one chiral carbon. (Glycine's R group is just an H, so the α-carbon is –CH$_2$– instead of –CHR–.)

Proteins always contain **L**-amino acids, not D-. *(Well, almost always…)*

CONNECTION TO ORGANIC CHEMISTRY:
> Remember these terms?
>> **CHIRAL CARBON** (or chiral center), **STEREOISOMERS, ENANTIOMERS** (R vs. S), **OPTICAL ISOMERS** [dextrorotary(+) vs. levorotary(–) (how a sample rotated a plane of polarized light)]
>
> **How Biochem is the Same:** Biochemistry uses D- and L- too, BUT it's not *quite* the same…
>
> **How Biochem is Different:** "L-" refers to a similarity with L-glyceraldehyde at the chiral carbon furthest from the C=O. So, L-alanine shares a structural similarity with L-glyceraldehyde – that's it.

TITRATION OF AMINO ACIDS

1. Draw the amino acid in its most acidic form.

2. Draw the amino acid structure that occurs when it reacts with one OH–. Then, draw structures that react with further OH– until the amino acid is in its most basic form. (Groups with the lowest pK_a values donate their H$^+$ ions first.)

3. **ISOELECTRIC POINT (pI)** = the pH where the overall charge is zero

   ### TO DETERMINE THE ISOELECTRIC POINT:

   1. Identify the acid(s) and base(s) on the amino acid or peptide. Which functional groups can happily lose or gain an H$^+$? (*'Happily'* means within the pH range of 1–14.)

   2. Rank the functional groups in the order in which they will lose an H$^+$. Remember that the lower the pK_a, the stronger the acid, so the acid with the lowest pK_a will lose its H$^+$ first.

   3. Average the pK_a values on either side of the isoelectric (neutral) molecule. Note: Be aware that if an acidic or basic side group is present, you *never* average *all* of the pK_as. Only average these two pK_as: the pK_a to go from a net charge of +1 to 0, and the pK_a to go from a net charge of 0 to –1.

   If it's not clear which two pK_a values to average, first draw the amino acid or peptide at the lowest (acidic) pH and label each ionizable group with its pK_a. Then remove one H$^+$ at a time (in order of pK_a) and draw each structure. It helps to label the reaction arrows with their pK_a values and each structure with its net charge.

   Remember that the lower the pK_a, the stronger the acid. Also: when pH = pK_a, [A$^-$] = [HA]. Below the pK_a, there will be more HA. Above the pK_a, there will be more A$^-$.

A detailed example of lysine is located after the table of amino acids with ionizable R groups on the next page. Also, the examples of alanine and glutamic acid are in your text (pp. 132-133), along with a tetrapeptide that includes lysine and aspartic acid.

In addition to the Review and Thought Questions at the end of Chapter 5, **more pI practice problems are located at the end of this study guide chapter.**

AMINO ACIDS WITH IONIZABLE R GROUPS

Amino Acid	Structure at pH=7 and R group pK_a	Amino Acid	Structure at pH=7 and R group pK_a
Aspartate Asp	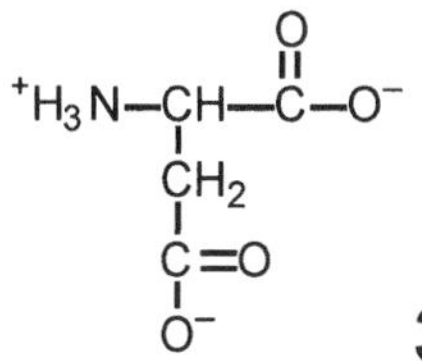 3.86	**Glutamate** Glu	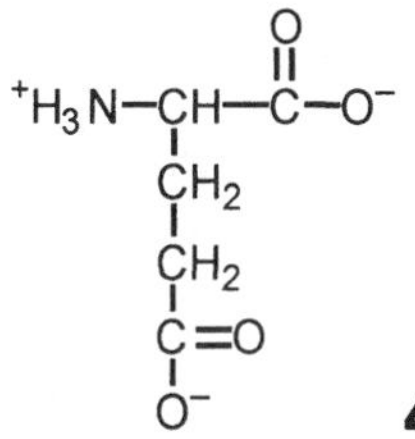 4.25
Tyrosine Tyr	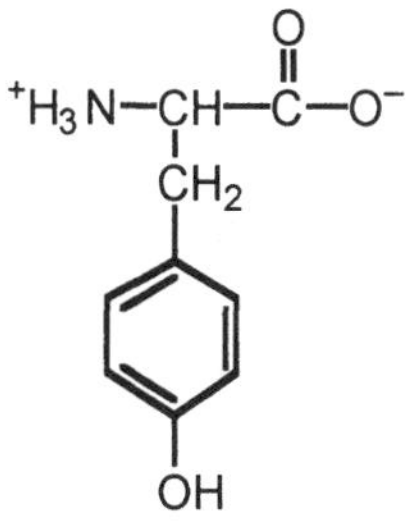 10.07	**Lysine** Lys	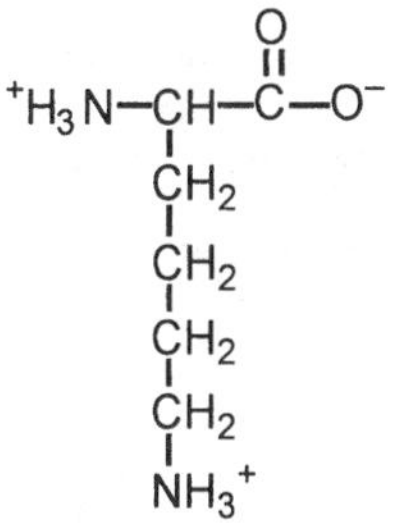 10.79
Histidine His	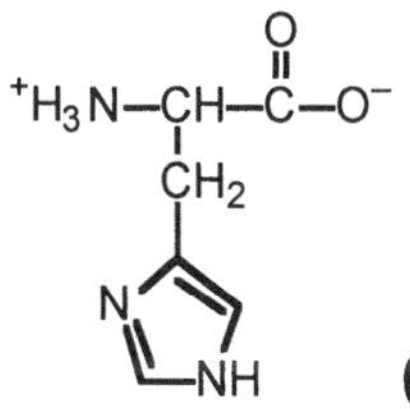 6.0	**Cysteine** Cys	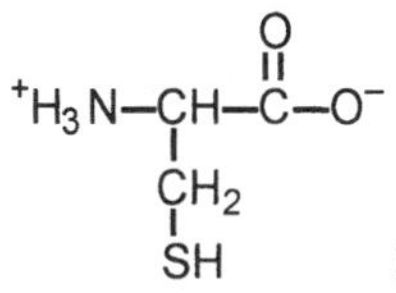 8.33
Arginine Arg	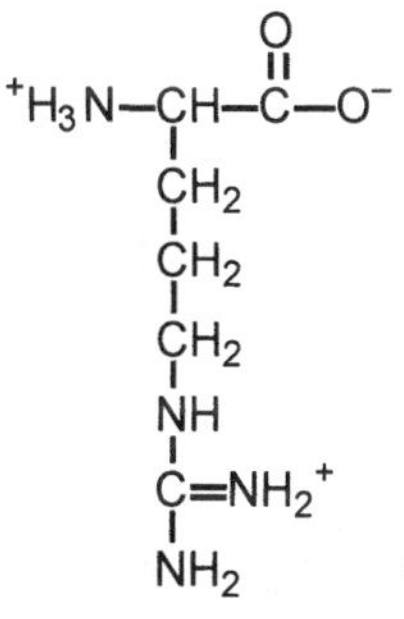 12.48		

Unusual Functional Groups

- Guanidino: in arginine's R group
- Imidazole: in histidine's R group
- Phenol: in tyrosine's R group
- Sulfhydryl: in cysteine's R group (thiol)

CALCULATING pI: LET'S LOOK AT THE EXAMPLE OF LYSINE IN MORE DETAIL.

1. Identify acidic and basic functional groups.

–COOH is a weak acid and can donate its H^+ to form –COO⁻. (pK_a = 2.18)

–NH₂ is a weak base and can accept an H^+ to form –NH₃⁺. (pK_a = 8.95)

The R group also has an –NH₂ that can accept an H^+ to form –NH₃⁺. (pK_a = 10.07)

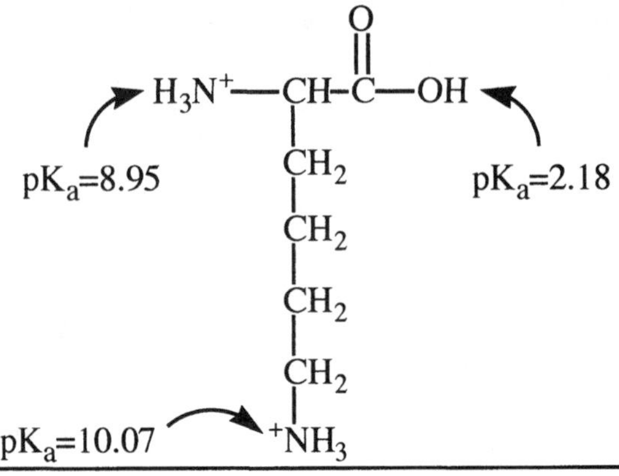

pK_a=8.95

pK_a=2.18

pK_a=10.07

2. Rank functional groups according to pK_a.

What happens to each acid/base group at very low pHs (acidic conditions)?	Which group will donate its H^+ first (that is, at the lower pH values)?	The next group to give up its H^+ is the one with the next higher pK_a value, in this case, the amino group with a pK_a of 8.95.	The third and last group to lose its H^+ is the R group, with a pK_a of 10.07.
All acid and base groups will have their H^+s: –COOH, –NH₃⁺.	The one with the lowest pK_a, in this case, the –COOH with a pK_a of 2.18.		

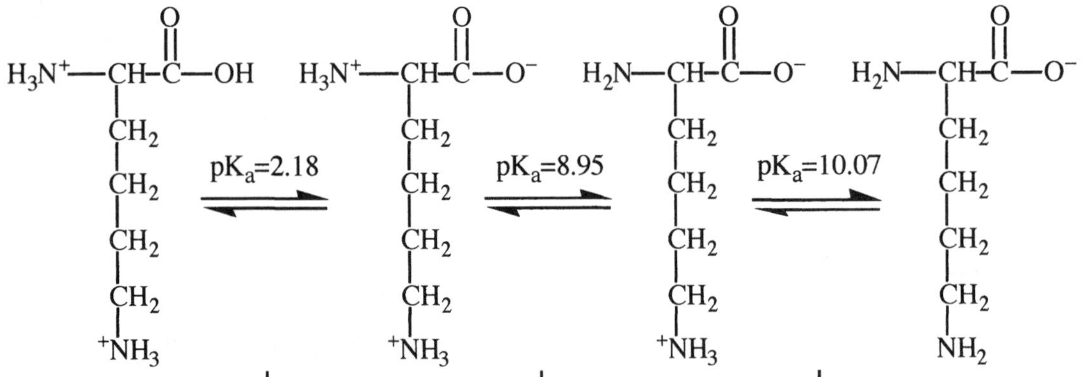

+2	+1	0	−1
Net Charge	Net Charge	Net Charge	Net Charge
pH < 2.18	2.18 < pH < 8.95	8.95 < pH < 10.07	pH > 10.07

3. To calculate the pI, average the pK_as on either side of the neutral molecule. For lysine, this would be the average of 8.95 and 10.07 = (8.95+10.07)/2 = **9.52**.

AMINO ACID REACTIONS

PEPTIDE BOND FORMATION

- peptide bond has partial double-bond character due to resonance, so the peptide bond is rigid and planar
  - α-carbon (Cα) is next to the peptide-bonded C=O
  - ψ = rotation around the Cα–N bond; φ = rotation around the Cα–C bond
- amino acid residues are amino acids that are part of a peptide or protein chain. Once an amino acid reacts to form a peptide or protein, 's more accurate to refer to a glycine residue rather than a glycine, since it is that part of the amino acid that has become part of a peptide or protein
- draw peptides from N-terminal to C-terminal
- name peptides (amino acid sequence from N-terminal to C-terminal)

PEPTIDE BOND = AN AMIDE BOND LINKING TWO AMINO ACIDS

Amino acids are linked by a covalent bond between the α-amino (–NH$_2$) of one amino acid and the α-carboxyl (–COOH) of another amino acid.

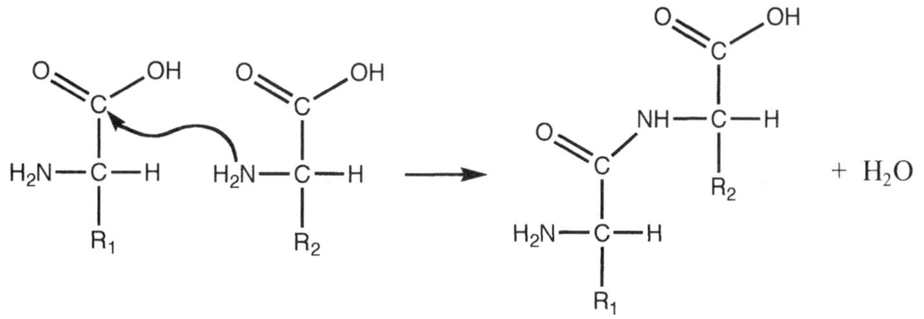

The curved arrow above shows where the new peptide bond will form. The amino acids are drawn in their neutral forms to illustrate the loss of an –OH from the carboxyl group and an –H from the amino group to produce a water molecule.
[Note: Of course, "real life" is much more complex – the amino acids would be ionized at pH 7, the general mechanism is multistep, and protein synthesis in living organisms is quite complicated (see Chapter 19).]

The peptide C–N bond is shorter and much more rigid than a typical carbon-nitrogen single bond. This can be explained by drawing the resonance structure for an amide bond:

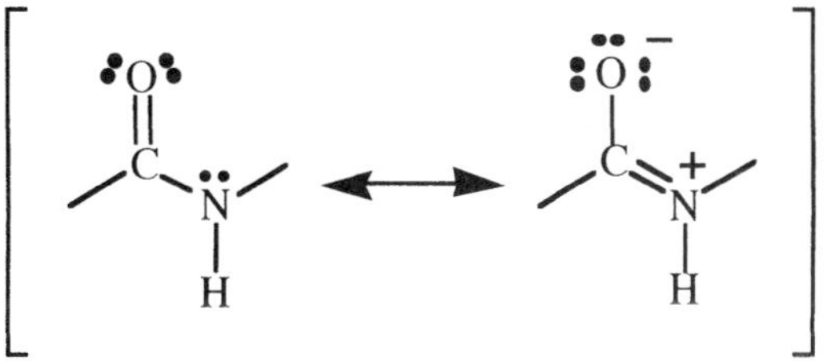

Although this resonance structure may seem unlikely because of the charge separation, it is indeed significant. Resonance explains the observed similarities of the peptide bond to a carbon-carbon double bond, namely, its rigidity and its shorter-than-expected bond length.

A carbon-nitrogen double bond is planar, or flat, and this makes the peptide bond rigid. The planar, rigid peptide bond has important consequences for protein structure.

The effect that this rigidity has on possible protein shapes is to limit somewhat the number of possibilities. This introduces an element of control. Picture a heavy chain with rather long links. Between the links there is free rotation, but the links themselves are rigid. Compared to a rope, there are less possible ways that the chain could be contorted.

However, for the biological functions that proteins perform, having specific ways (rather than infinite ways) that proteins can fold and form shapes results in greater control and specificity in protein function.

CYSTEINE OXIDATION AND THE FORMATION OF DISULFIDE BRIDGES
- covalent S–S bond = DISULFIDE BOND, called a DISULFIDE BRIDGE in polypeptides
- formed by the oxidation of two cysteine R groups:

$$R–SH + HS–R \rightleftharpoons R–S–S–R \text{ (cystine)}$$

- can occur within a chain or between two separate chains

SCHIFF BASE FORMATION

Schiff Base: IMINE (C=N) produced when $R–NH_2$ reacts reversibly with a carbonyl

Schiff bases are also called aldimines when the carbonyl is an aldehyde

Schiff bases are intermediates in transamination reactions (Chapter 14).

IMINE functional group: C=N

5.2 PEPTIDES

PEPTIDE FUNCTIONS	SPECIFIC PEPTIDES THAT PERFORM THESE FUNCTIONS:
Reducing agent	glutathione (GSH, where "SH" indicates a cysteine R group) $$2\,GSH + H_2O_2 \rightarrow GSSG + 2\,H_2O$$ Note that GSH forms a disulfide bond with another GSH.
Appetite control	α-melanocyte stimulating hormone, cholecystokinin, galanin, neuropeptide Y
Blood pressure control	vasopressin, atrial natriuretic factor
Pain perception	opioid peptides (Met-enkephalin, Leu-enkephalin) vs. Substance P, bradykinin

WRITING PEPTIDE AND PROTEIN SEQUENCES: N-TERMINUS TO C-TERMINUS

Peptide and protein sequences are written from left to right as amino, or N-terminus, to carboxyl, or C-terminus. The amino/carboxyl refers to free amino or free carboxyl groups, meaning that they are not part of a peptide bond.

Example:

Draw the dipeptides Phe–Asp and Asp–Phe. They're different! Not only do they have different structures, but their pI values will also be different. Calculate their pI values.

Solution:

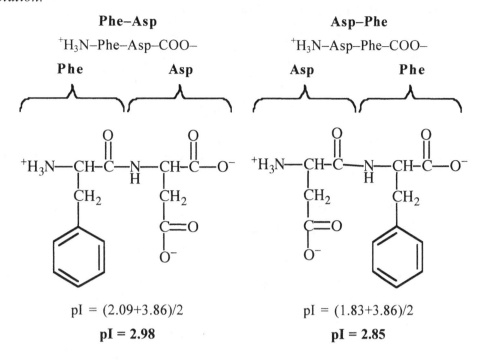

Phe–Asp	Asp–Phe
^+H_3N–Phe–Asp–COO–	^+H_3N–Asp–Phe–COO–

pI = (2.09+3.86)/2   pI = (1.83+3.86)/2

pI = 2.98   **pI = 2.85**

Aspartame™

Asp–Phe is an important commercial dipeptide. Its methyl ester (on the C-terminus) is Aspartame™, an artificial sweetener that is 200 times sweeter than sugar. Both amino acids have the L-configuration around the α-carbon. If either is in the D-configuration, then the peptide is bitter rather than sweet.

5.3 PROTEINS: CLASSIFIED BY FUNCTION, SHAPE, OR COMPOSITION

PROTEIN FUNCTIONS: catalysis, structure, movement, defense, regulation, transport, storage, stress response

PROTEIN SHAPES: **FIBROUS** proteins vs. **GLOBULAR** proteins

PROTEIN COMPOSITIONS: **SIMPLE** vs. **CONJUGATED**

Conjugated protein = simple protein + prosthetic group.

*H*oloproteins *H*AVE THEIR PROSTHETIC GROUPS, BUT

PROSTHETIC GROUPS ARE *A*BSENT IN *A*POPROTEINS.

Examples of conjugated proteins: **GLYCOPROTEINS, LIPOPROTEINS, METALLOPROTEINS, PHOSPHOPROTEINS, HEMOPROTEINS**

PROTEIN STRUCTURE

PRIMARY STRUCTURE = AMINO ACID SEQUENCE

HOMOLOGOUS POLYPEPTIDES have similar amino acid sequences and have arisen from the same ancestral gene

INVARIANT vs. **VARIABLE RESIDUES**; mutations that change the amino acid sequence and their connection to evolution and molecular diseases

HOMOZYGOUS vs. **HETEROZYGOUS**

MOLECULAR DISEASES result from amino acid substitutions (caused by DNA mutations) at invariant residues of key proteins; example: sickle-cell anemia caused by mutant hemoglobin

SECONDARY STRUCTURE = REPEATING PATTERNS OF LOCALIZED STRUCTURE

α-HELIX: right-handed helix, 3.6 residues per turn, pitch = 54 nm
Hydrogen bonds between N–H and C=O are four residues apart, and R groups extend *outward* from the helix.
Amino acids that are incompatible with the α-helix: Gly, Pro, and sequences with large numbers of charged and/or bulky R groups.

β-PLEATED SHEET; β-strand; parallel vs. antiparallel
Stabilized by H-bonds between N–H and C=O of adjacent chains
Each β-strand is fully extended. Antiparallel is more stable than parallel because the hydrogen bonds between chains are more direct (*colinear* and shorter).

SUPERSECONDARY STRUCTURES or **MOTIFS**

βαβ unit, β-turn, β-meander, αα-units, β-barrel, Greek key

TERTIARY STRUCTURE AND PROTEIN FOLDING OF GLOBULAR PROTEINS

Features: 1. Amino acids that are far apart in the primary structure may be close together once folded.

2. Globular proteins are compact.

3. Large globular proteins often contain **DOMAINS** – compact units with specific functions. (Examples: *EF hand* binds Ca^{2+}; *leucine zipper* and *zinc finger* domains found in DNA-binding proteins); **FOLD** = core 3-D structure of a domain

4. **MODULAR** or **MOSAIC PROTEINS** in eukaryotes contain numerous duplicate or imperfect copies of one or more domains that are linked in series

Stabilizing factors:

HYDROPHOBIC INTERACTIONS

ELECTROSTATIC INTERACTIONS (SALT BRIDGES)

HYDROGEN BONDING

COVALENT BONDS (DISULFIDE BRIDGES)

HYDRATION (structured water forms a dynamic hydration shell that stabilizes tertiary protein structure)

QUATERNARY STRUCTURE – SEVERAL SUBUNITS (POLYPEPTIDE CHAINS)

Oligomers and protomers; why multisubunit proteins are common

Noncovalent and covalent interactions hold the subunits in place. The most important interaction is the hydrophobic effect.

Covalent crosslinks: disulfide bridges, desmosine and lysinonorleucine

Allostery, ligand binding, allosteric transitions, effectors or modulators

UNSTRUCTURED PROTEINS

IUPs are **INTRINSICALLY UNSTRUCTURED PROTEINS.**

NATIVELY UNFOLDED PROTEINS have no ordered structure.

IUP functions include the regulation of signal transduction, transcription, translation, and cell proliferation. The disorder of the IUPs allow them to be more flexible and "search" for binding partners; IUPs tend to become more ordered upon binding with a target molecule.

LOSS OF PROTEIN STRUCTURE: PROTEIN DENATURATION – loss of 3-D structure.
Interactions between amino acid residues are disrupted, but peptide bonds are NOT broken.

STUDY HINT: Review the bonding, interactions, and forces that stabilize 3-D structure. Anything that can disrupt these forces can cause denaturation.

Denaturing agents include strong acids or bases, organic solvents, detergents, reducing agents (reduce disulfide bonds), salt concentration (salting out), heavy metal ions, temperature changes, and mechanical stress.

PROTEIN DYNAMICS AND FLEXIBILITY IN A PROTEIN'S 3-D STRUCTURE; why flexibility in a protein's 3-D structure is essential to most protein functions

BIOCHEMISTRY IN PERSPECTIVE: MOLECULAR MACHINES
Molecular machine function is made possible by conformational changes triggered by the hydrolysis of nucleotides bound to protein subunits called motor proteins.

MOTOR PROTEINS
1. Classical motors: myosins, kinesins, dyneins
2. Timing devices
3. Microprocessing switching devices
4. Assembly and disassembly factors

THE FOLDING PROBLEM

THE TRADITIONAL FOLDING MODEL: Interactions between amino acids side chains alone force the molecule to fold into its final shape.

LIMITATIONS OF THE TRADITIONAL FOLDING MODEL
1. Time constraints: Folding happens on the order of seconds (or a few minutes), not in years, as calculated by the traditional folding model.
2. Complexity: Think about the number of bonds that can rotate, not only in the backbone, but in the side groups. Phew!

RECENT ADVANCES IN PROTEIN FOLDING RESEARCH HAVE REVEALED:
• secondary structure (α-helix, β-sheet) forms early in the process
• hydrophobic interactions are very important

- larger polypeptides have partially-folded intermediate structures; molten globules are partially-organized globular structures that resemble the final protein structure

  - **MOLECULAR CHAPERONES** assist in protein folding:
    - bind to denatured or unfolded proteins
    - protect unfolded proteins from inappropriate interactions that would lead to incorrect structures
    - assist proteins in folding rapidly, precisely, and correctly
    - promote protein degradation when refolding isn't possible

- most molecular chaperones are hsps (heat shock proteins); two major classes are:

  - **Hsp70s** – bind to short hydrophobic segments in unfolded polypeptides to prevent their aggregation. ATP hydrolysis releases the polypeptide, which is then passed on to an hsp60.

  - **Hsp60s** (**chaparonins** or Cpn 60s) – family of chaperones that mediate protein folding and release the polypeptide upon ATP hydrolysis

FIBROUS PROTEINS TYPICALLY SERVE STRUCTURAL FUNCTIONS

– have high proportions of regular secondary structures; rodlike or sheetlike shapes

Examples: α-keratin, collagen, silk fibroin
Take a close look at the descriptions and figures of these examples in your text. The structures are rigid and well suited to their functions. Where are each of these located, and what are their specific functions?

Note that the helical structure in collagen is not an α-helix, but a triple left-handed helix. Earlier in the chapter, Gly and Pro were described as being likely to disrupt an α-helix. Review that section, and compare the α-helix with the description of collagen on p. 165. Note how the odd features of Gly and Pro work together well in the structure of collagen.

GLOBULAR PROTEINS TYPICALLY SERVE DYNAMIC FUNCTIONS

- Function: usually involves binding ligands or large biomolecules that induce a conformational change "linked to a biochemical event."

- Myoglobin (in skeletal and cardiac muscle), hemoglobin (in red blood cells)

- Heme protein decreases heme's affinity for O_2 and protects Fe^{2+} from irreversibly oxidizing to Fe^{3+} (hematin), so O_2 can bind reversibly (like a sticky note that can be placed or removed when needed).

- Fetal hemoglobin (HbF) has greater affinity for O_2 than HbA, maternal hemoglobin. (How else could a growing baby receive O_2 from the mother?)

- Myoglobin has a greater affinity for oxygen than hemoglobin. So, oxygen moves from blood to muscle, and myoglobin only gives up its oxygen when the muscle cell's O_2 concentration is very low.

- **TAUT STATE** vs. **RELAXED STATE** (T, R states)

- **COOPERATIVE BINDING**: The binding of the first ligand (example – O_2 to hemoglobin) causes a conformational change that facilitates binding of more ligands (3 more O_2 molecules to hemoglobin).

- **BOHR EFFECT**: Dissociation of O_2 from hemoglobin is enhanced at lower pH. Why? Higher levels of CO_2 cause higher $[H^+]$, since $CO_2 + H_2O \rightarrow HCO_3^-$ and H^+. H^+ stabilizes the deoxy form of hemoglobin, so it's formed faster.

THE IMPORTANCE OF ALLOSTERY

Allostery pulls together the complexity and importance of protein 3-D structure and ties them directly to function. It all boils down to those intra- and intermolecular forces (as well as covalent bonds) that determine a protein's overall shape. The shape affects (some might say determines) the protein's function.

The main idea is that a small molecule (or ion) – a **LIGAND** – binds to a protein and induces a conformational change (an **ALLOSTERIC TRANSITION)**, which changes its shape, which changes the protein's affinity for other ligands. So, binding one ligand results in the protein having an increased or decreased affinity for binding more ligands. Ligands that trigger allosteric transitions are called **EFFECTORS** or **MODULATORS**. Allosteric enzymes can be regulated by ligand binding. (We'll see more of this in Chapter 6.)

AMINO ACID SEQUENCING TECHNIQUES

- Carboxypeptidase: C-terminus
- Sanger's method: N-terminus: DNFB reacts with the N-terminal amino acid to form a DNP derivative; also called the chain-terminating method.
- Trypsin: hydrolyzes the C-side of Lys and Arg
- Chymotrypsin: hydrolyzes the C-side of Phe, Tyr, and Trp
- Cyanogen bromide: hydrolyzes the C-side of Met
- Edman degradation: sequentially determines the N-terminal residue of each fragment

PRACTICE PROBLEMS TO PROMOTE PERFECTION

Answers follow the Solutions to the Review and Thought Questions for this chapter.

1. Given the peptide: Val-Arg-Ala-Tyr-Gly
    a. Draw the structure of the peptide in its most acidic form.
    b. Name it.
    c. Would you expect this peptide to have a high or a low pI? Calculate the pI. (Refer to Table 5.2 in your text for pK_a values.)
    d. Draw the titration curve for this peptide.
    e. What is the net charge of this peptide at pH = 7? In gel electrophoresis, would it move towards the positive anode or the negative cathode?
    f. If this peptide was produced as a fragment during the determination of the amino acid sequence of a protein, which of the following methods could have been used to produce it? Would this peptide fragment give you any further information regarding the primary structure? Methods: carboxypeptidase, Sanger's method (DNFB), trypsin, chymotrypsin, Edman degradation

2. Given the peptide: Pro-His-Met-Ser-Phe
    a. Draw the structure of the peptide in its most acidic form.
    b. Calculate the pI. (Refer to Table 5.2 in your text for pK_a values.)
    c. What is the net charge of this peptide at pH = 12? In gel electrophoresis, would it move towards the positive anode or the negative cathode?
    d. Answer (1f), above, for this peptide.

3. Given the following peptide, write out the proper sequence of amino acids using their three-letter abbreviations.

(Here's a bonus question: There's an error in this structure that doesn't interfere with the problem. Can you find the error? What's wrong with this structure?)

4. Determine the amino acid sequence of a polypeptide given the following data:

   Treatment with carboxypeptidase liberates Val.

   Treatment with DNFB liberates DNP-Ser.

   Treatment with trypsin results in the following fragments:
   Glu-His-Phe-Arg, Pro-Val, Ser-Tyr-Ser-Lys, Val-Trp-Gly-Lys

   Treatment with chymotrypsin results in the following fragments:
   Arg-Val-Trp, Gly-Lys-Pro-Val, Ser-Lys-Glu-His-Phe, Ser-Tyr

   Total hydrolysis produces the following amino acids:
   Arg, Glu, Gly, His, Lys(2), Phe, Pro, Ser(2), Trp, Tyr, Val(2)

5. Determine the amino acid sequence of a polypeptide given the following data:

   Treatment with carboxypeptidase liberates Val.

   Treatment with DNFB liberates DNP-Ser.

   Treatment with trypsin results in the following fragments:
   Phe-Glu-His-Lys, Phe-Gly-Arg, Pro-Val, Ser-Tyr-Ser-Lys, Trp-Gly-Lys

   Treatment with chymotrypsin results in the following fragments:
   Gly-Arg-Trp, Gly-Lys-Pro-Val, Ser-Lys-Phe, Glu-His-Lys-Phe, Ser-Tyr

   Total hydrolysis produces the following amino acids:
   Arg, Glu, Gly(2), His, Lys(3), Phe(2), Pro, Ser(2), Trp, Tyr, Val

   (Hint: If you get stuck working from the C-terminal end, try working from the N-terminal end, and vice-versa.)

AFTER STUDYING THIS CHAPTER, YOU SHOULD BE ABLE TO:

- Classify amino acids as polar (neutral), nonpolar, acidic, or basic

- Draw the structure and give the net charge of an amino acid or peptide at a specific pH. Will it move towards the anode or the cathode in gel electrophoresis?

- Draw (or use) titration curves; determine (or use) pK_a and pI values.

- Name and draw peptides, given the three-letter abbreviations. Given a structure of a peptide, write it out using the three-letter abbreviations.

- Calculate the pI values of peptides.

- Determine the amino acid sequence of a peptide.

- Predict the secondary structure of a given polypeptide: which amino acids tend to stabilize or disrupt each type of secondary structure?

- Describe interactions, bonding, and other forces that contribute to stabilizing protein tertiary structure. Identify amino acids which would participate in a given interaction.

CHAPTER 5: SOLUTIONS TO REVIEW QUESTIONS

5.1   A polypeptide is a polymer containing more than 50 amino acid residues. A protein is composed of one or more polypeptide chains. A peptide is a polymer with fewer than 50 amino acid residues.

5.2   a. nonpolar   b. polar   c. acidic   d. basic   e. nonpolar

f. basic   g. nonpolar   h. polar   i. nonpolar   j. nonpolar.

5.4   First, draw His at the lowest (acidic) pH. Then, remove one H^+ at a time (in order of pK_a) and draw each structure. Remember (when you learned about buffers) that at each of the plateaus in the titration curve, $pH = pK_a$, and $[HA] = [A^-]$. The pK_a values, then, are equal to the pH at each of the plateaus. It helps to go ahead and label the reaction arrows with the pK_a values and each structure with its net charge. Going through this exercise helps to visualize the rest of the problem.

NET CHARGE: +2   **+1**   **0**   **−1**

a.   So, at the first plateau, the pH = 1.82 and the first two species above are present. At the second plateau, the pH = 6.0, and the second and third species are present. At the third plateau, the pH = 9.17, and the third and fourth species are present.

b.   The pK_a values for each species are approximately 1.8, 6, and 9.2, respectively.

c.   The isoelectric point for histidine is (6.0 + 9.2)/2 = 7.6

5.5   The name of the molecule is cysteinylglycyltyrosine. Its abbreviated structure is:

$$^+H_3N–Cys–Gly–Tyr–COO^-$$

5.7   Six examples of the major functions of protein in the body are catalysis, structure, movement, defense, regulation, and transport.

5.8   a.   Fibrous proteins, which possess water-insoluble sheetlike or ropelike shapes, typically have structural roles in living organisms. Globular proteins are compact spherical molecules (usually water-soluble) that typically have dynamic functions.

b.   Simple proteins contain only amino acids. A conjugated protein is a simple protein combined with a nonprotein component, such as a lipid or a sugar.

c.   An apoprotein is a protein without its prosthetic group. An apoprotein molecule combined with its prosthetic group is a holoprotein.

5.10   a.   The amino acid sequence is a polypeptide's primary structure.

b.   A β-pleated sheet is one type of secondary structure.

c.   Inter- and intra-chain hydrogen bonds between N–H groups and carbonyl groups of peptide bonds are the principal feature of secondary structure. Hydrogen bonds formed between polar side chains are important in tertiary and quaternary structure.

d.   Disulfide bonds are strong covalent bonds that contribute to tertiary and quaternary structure.

5.11 a. Polyproline – left-handed helix

b. Polyglycine – β-pleated sheet

c. Ala-Val-Ala-Val-Ala-Val – α-helix

d. Gly-Ser-Gly-Ala-Gly-Ala – β-pleated sheet

5.13 a. heat – hydrogen bonding (secondary and tertiary structure)

b. strong acid – hydrogen bonding (secondary and tertiary structure) and salt bridges (secondary and tertiary structure)

c. saturated salt solution – salt bridges (tertiary structure)

d. organic solvents – hydrophobic interactions (tertiary structure)

5.14 Amino acids with basic side groups such as lysine, arginine, or tyrosine would contribute to a high pI value.

5.16 The first step in the isolation of a specific protein is the development of an assay, which allows the investigator to detect it during the purification protocol. Next, the protein, as well as other substances, are released from source tissue by cell disruption and homogenization. Preliminary purification techniques include salting out, in which large amounts of salt are used to induce protein precipitation, and dialysis, in which salts and other low molecular weight material are removed. Further purification methods, which are adapted to each research effort at the discretion of the investigator, include various types of chromatography and electrophoresis. Three chromatographic methods are: ion-exchange chromatography, gel-filtration chromatography, and affinity chromatography. Gel electrophoresis may be used to purify a protein and/or to assess the purity of a protein.

5.17 Refer to p. 172 for a discussion of chromatographic separation techniques.

5.19 a. mosaic proteins – eukaryotic proteins that contain numerous duplicate or imperfect copies of one or more domains that are linked in series

b. homologous polypeptide – a protein molecule whose amino acid sequences and functions are similar to those of another protein

c. cooperative binding – a mechanism in which binding one ligand to a target molecule promotes the binding of other ligands

d. aldol condensation – an aldol addition involving the elimination of a water molecule

e. globular protein – a protein that adopts a rounded or globular shape.

5.20 With overlapping fragments, the segments can be fitted together because fragments that fit together have common sequences at their ends. If the segments are not overlapping, the order of the amino acids in the protein cannot be determined.

5.22 The following fragments are produced when bradykinin is treated with the indicated reagents:

a. carboxypeptidase: Arg and Arg-Pro-Pro-Gly-Phe-Ser-Pro-Phe

b. chymotrypsin: Arg-Pro-Pro-Gly-Phe, Ser-Pro-Phe, and Arg

c. trypsin: Arg and Pro-Pro-Gly-Phe-Ser-Pro-Phe-Arg

d. DNFB (followed by HCl): DNP-Arg and the following amino acid residues: 3 Pro, 1 Gly, 2 Phe, 1 Ser, and 1 Arg

5.23 The primary structure (the amino acid sequence) of a polypeptide provides opportunities for an exceedingly large number of interactions between side groups, in addition to hydrogen bonding that can take place between amide groups along the backbone.

As a result, the number of three-dimensional shapes that are theoretically possible is too high to be able to predict the one specific conformation the polypeptide will adopt after synthesis.

5.25 As components of molecular machines, motor proteins undergo conformational changes that, in turn, trigger additional conformational changes in adjacent subunits. Motor proteins bind nucleotides that are hydrolyzed in order to drive the initial conformational changes. Organisms use motor proteins as classical motors to move a load along a protein filament, as timing devices to provide a delay period for complex processes, as on-off molecular switches in signal transduction pathways, and as assembly/disassembly factors to reversibly form larger molecular complexes from protein subunits.

5.26 In addition to mediating protein folding, molecular chaperones help protect newly synthesized proteins from inappropriate, premature protein-protein interactions by binding to and stabilizing proteins during early stages of folding. Molecular chaperones also direct the refolding of proteins that have partly unfolded; if refolding isn't possible, they promote protein degradation.

CHAPTER 5: SOLUTIONS TO THOUGHT QUESTIONS

5.28 Hydrophobic amino acids such as valine, leucine, isoleucine, methionine, and phenylalanine usually occur within the anhydrous core of proteins because of the hydrophobic effect. Hydrophilic amino acids such as arginine, lysine, aspartic acid, and glutamic acid occur most often on or near the surface of proteins, where they interact with water molecules. Glycine and alanine are hydrophobic amino acids and so tend to occur in the interior of proteins. Glutamine has a polar side chain that can form hydrogen bonds and, therefore, is often on the surface of proteins.

5.29 Living cells possess complex mechanisms for assisting the proper folding of nascent polypeptides. These mechanisms are poorly understood and cannot yet be duplicated in the laboratory.

5.31 The immobilized water of protein molecules is locked into position by hydrogen bonding between polar and ionic groups and water molecules. This gives rise to a three-dimensional structure in which the water molecules have very little freedom of motion, i.e., they are frozen in place.

5.32 The peptide bond is stronger than the ester bond for two reasons. The N is closer to the size of C than the O is, which makes for greater covalency in the bond. Also, because the O and N differ in electronegativity, and both have lone pair(s) of electrons, the amide has resonance hybridization. The amide bond, therefore, has partial double bond character.

5.34 The primary sequence of β-endorphin is:

Tyr–Gly–Gly–Phe–Met–Thr–Ser–Glu–Lys–Ser–Gln–Thr–Pro–Leu–Val–Thr–Leu–

–Phe–Lys–Asn–Ala–Ile–Val–Lys–Asn–Ala–His–Lys–Lys–Gly–Gln

5.35 a. The isoelectric point is calculated by taking the average pK_a values for the amino group of glycine (9.6) and the carboxyl group of valine (2.32). The pI is 5.96.

 b. At pH = 1 the tripeptide is positively charged and will move toward the negative electrode. At pH = 5 the peptide has zero net charge and will not migrate. At pH 10 and 12 it will have a –1 charge and will move toward the positive electrode.

69

5.37 Proline and hydroxyproline are both imino acids and do not lose their nitrogen atom when they react with ninhydrin. As a result, when proline reacts with ninhydrin, the compound shown below is formed.

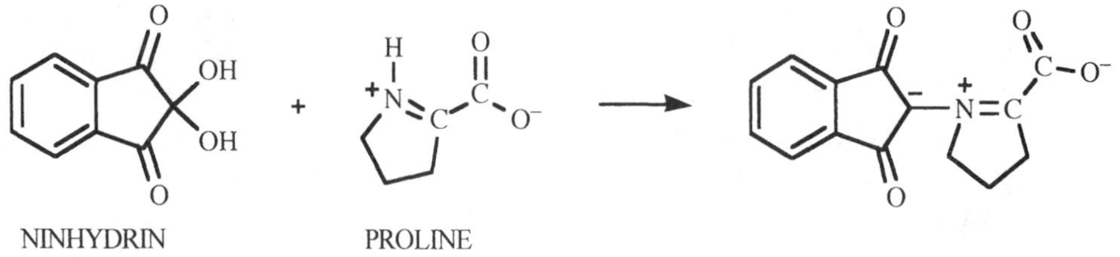

NINHYDRIN   PROLINE

Note: The reaction of the free amino groups of the amino acids with ninhydrin produces blue dye molecules. When ninhydrin reacts with either proline or hydroxyproline, both of which have nitrogen locked into a ring, a yellow dye molecule is formed.

5.38 The properties that may be responsible for the different functions of a multifunctional protein include different ligand binding sites and the capacity to form homo- or heterocomplexes. In addition, the structural properties of such a protein may be affected by its cellular location.

5.40 Glyceraldehyde-3-phosphate dehydrogenase and the crystallins are two examples of multifunction proteins. The process of normal genetic mutation will produce changes in protein structure. Most of these changes will be neutral or detrimental. In rare instances mutations facilitate a new function.

5.41 The synthesis of protein requires large amounts of energy. Proteins capable of several functions reduce this energy cost.

5.43 The hydrophobic amino acids glycine, phenylalanine, methionine, valine, and leucine should all seek the relatively water-free interior of the decapeptide.

5.44 The products are glycine, phenylalanine, methionine, valine, leucine, serine, histidine, and aspartic acid. The asparagine would be changed to aspartic acid by the hydrolysis conditions.

5.46 The extended α-helix probably contains amino acids with nonpolar side chains such as lysine, leucine, and phenylalanine. These need to be replaced by amino acids such as glycine and proline. Glycine's small R group permits a contiguous proline to assume a *cis* orientation, resulting in a tight turn in the peptide chain.

5.47 Glycine, valine, histidine, and serine could all form weak coordination compounds by using their amino nitrogens and carboxylate anions. More significant coordination would be expected with histidine and serine using the side chain nitrogens and oxygen, respectively.

5.49 Serotonin is derived from tryptophan; dopamine comes from tyrosine.

5.50 Fibrous proteins typically have high proportions of regular secondary structure such as α-helices or β-sheets. Typical amino acids that could be used are alanine, leucine, glycine, and serine.

5.52 The immobilized water of a protein molecule is locked in position by hydrogen bonding. The hydrophobic amino acid side chains are excluded from the water and tend to cluster together. This clustering holds portions of the polypeptides in a particular conformation.

Answers to "Practice Problems to Promote Perfection"

1. a. The pK_a value for each ionizable group is shown in bold.

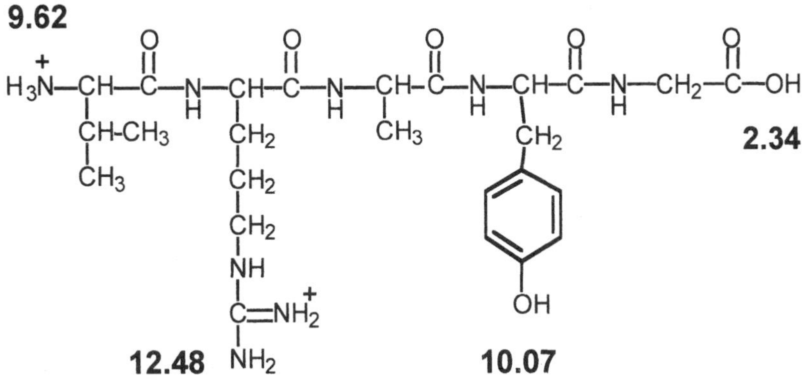

   b. Valylarginylalanyltyrosylglycine (Remove the "ine" and replace it with "yl.")

   c. We'd expect this peptide to have a high pI, since the side groups are either basic or neutral. To calculate the pI, determine the structure that has a net charge of zero, and average the pK_a values on either side of that structure. The best way to do this is to draw each form in order of pK_a, that is, as the peptide is titrated.

   The pI is the average of 9.62 and 10.07, or (9.62 + 10.07)/2 = 9.85. (Note that if this were a multiple-choice problem, one wrong choice would probably be 8.63, the average of *all* of the pK_a values.)

   d. To draw the titration curve, first label the *y*-axis "pH" and the *x*-axis "Equivalents OH⁻." Draw short plateaus at each pK_a value. Place a point at a pH value that is midway between each two pK_a values. Draw a smooth curve from the plateaus through each inflection point.

   e. At pH 7, the peptide has a +1 charge, and would move towards the cathode.

   f. Val–Arg–Ala–Tyr–Gly

   Since we have a fragment and not a single amino acid, we can rule out the N-terminus or the C-terminus determination methods. We can also rule out chymotrypsin, since the Tyr–Gly peptide bond remains intact, and trypsin, since the Arg–Ala bond remains intact. That leaves cyanogen bromide. The amino acid that is linked to the N-terminal side of Val must be Met, the amino acid whose C-side peptide bond is hydrolyzed by cyanogen bromide. Since we do not have a method that hydrolyzes at the C-terminal side of Gly, this fragment must be located at the C-terminal end of the protein.

2. a. The pK_a value for each ionizable group is shown in bold. Pro–His–Met–Ser–Phe:

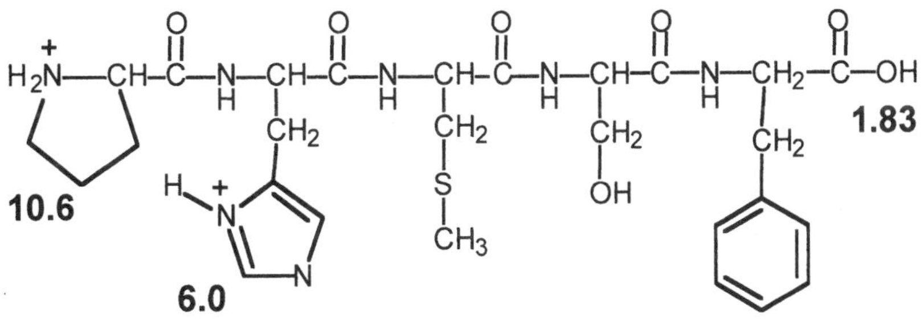

b. The pI is 8.3 (the average of 6.0 and 10.6).

c. At pH 12, the peptide will have a –1 charge, and will move towards the anode.

d. This peptide fragment could have been produced by treatment with chymotrypsin, since the C-side of Phe was hydrolyzed. If either Lys or Arg linked to the N-side of Pro, and this fragment was the C-terminal fragment, then it could have been produced by treatment with trypsin.

3.   Trp–Gly–Leu–Cys–Glu–Asn

   Solution to the Bonus Question included with Problem 3:
   The structure as drawn is impossible in aqueous solutions. At a pH low enough for both carboxylic acids to be protonated, the N-terminal amine would be protonated as well.

4.   Ser–Tyr–Ser–Lys–Glu–His–Phe–Arg–Val–Trp–Gly–Lys–Pro–Val

5.   Ser–Tyr–Ser–Lys–Phe–Glu–His–Lys–Phe–Gly–Arg–Trp–Gly–Lys–Pro–Val

Use this space for notes. Suggestion: practice drawing peptides and/or their titration curves.

6 Enzymes

Brief Outline of Key Terms and Concepts

6.1 PROPERTIES OF ENZYMES

ENZYMES are catalysts; most are proteins.
CATALYSTS modify the rate of a reaction because they provide an alternative reaction pathway that requires less ACTIVATION ENERGY than the uncatalyzed reaction.

ACTIVE SITE	TRANSITION STATE
SUBSTRATE	ACTIVITY COEFFICIENT
COFACTOR	APOENZYME
COENZYME	HOLOENZYME

6.2 CLASSIFICATION OF ENZYMES

OXIDOREDUCTASE	LYASE
TRANSFERASE	ISOMERASE
HYDROLASE	LIGASE

6.3 ENZYME KINETICS is the quantitative study of enzyme catalysis. Kinetic studies measure reaction rates (VELOCITY) and enzyme affinity for substrates and inhibitors.

Kinetics also provides insight into REACTION MECHANISMS: FIRST-ORDER, SECOND-ORDER, PSEUDO-FIRST-ORDER, ZERO-ORDER KINETICS; MOLECULARITY: UNIMOLECULAR, BIMOLECULAR

MICHAELIS-MENTEN KINETICS
MICHAELIS-MENTEN EQUATION

K_m (MICHAELIS CONSTANT)

TURNOVER NUMBER
SPECIFICITY CONSTANT, k_{cat}/K_m

DIFFUSION CONTROL LIMIT; CATALYTIC PERFECTION

INTERNATIONAL UNITS (IU), SPECIFIC ACTIVITY, KATAL

LINEWEAVER-BURK PLOTS
Lineweaver-Burk double-reciprocal plot: $1/v_0$ vs. $1/[S]$ to determine K_m and V_{max}

MULTISUBSTRATE REACTIONS
SEQUENTIAL REACTIONS
DOUBLE-DISPLACEMENT REACTIONS

ENZYME INHIBITION
REVERSIBLE *vs.* IRREVERSIBLE INHIBITION

Reversible Inhibitors
COMPETITIVE INHIBITORS compete with substrate for the active site.
NONCOMPETITIVE INHIBITORS can bind to either the enzyme or the enzyme-substrate complex.

UNCOMPETITIVE INHIBITORS bind only to the enzyme-substrate complex.
KINETIC ANALYSIS OF ENZYME INHIBITION
ALLOSTERIC ENZYMES

ENZYME KINETICS, METABOLISM, AND MACROMOLECULAR CROWDING
METABOLONS; METABOLIC FLUX

6.4 CATALYSIS

ORGANIC REACTIONS AND THE TRANSITION STATE
REACTION MECHANISMS; INTERMEDIATE;
REACTIVE INTERMEDIATES:
CARBOCATION, CARBANION, FREE RADICAL

CATALYTIC MECHANISMS
PROXIMITY AND STRAIN EFFECTS
ELECTROSTATIC EFFECTS
ACID-BASE CATALYSIS
COVALENT CATALYSIS

QUANTUM TUNNELING AND ENZYME CATALYSIS

THE ROLES OF AMINO ACIDS IN ENZYME CATALYSIS
The amino acid side chains in the active site of enzymes catalyze proton transfers and nucleophilic substitutions.
DYADS, TRIADS

THE ROLE OF COFACTORS IN ENZYME CATALYSIS
METAL IONS; COENZYMES; VITAMINS

EFFECTS OF TEMPERATURE AND pH ON ENZYME-CATALYZED REACTIONS
OPTIMUM TEMPERATURE; pH OPTIMUM

DETAILED MECHANISMS OF ENZYME CATALYSIS
SUBSTRATE SPECIFICITY, REACTION MECHANISM

COMPARTMENTATION

BIOCHEMISTRY IN PERSPECTIVE: ENZYMES AND CLINICAL MEDICINE
DIAGNOSTIC ENZYMES; ISOZYMES
THERAPEUTIC ENZYMES

ENZYMES: OVERVIEW

Enzymes are:
- proteins with specific, globular shapes (except for RNA molecules that are also catalytic; Chapter 18.)
- molecular machines
- biological catalysts that increase the rates of reactions that are essential to life. Without enzymes, these essential reactions would take place too slowly. Enzymes make life possible!

Enzyme function depends upon:
- having a shape that is complementary to reactant molecules
- having catalytic interactions between reactant molecules and the more or less flexible binding sites on the enzyme
- internal protein motions that extend throughout the molecule (for certain enzymes)

Enzymes function in crowded, gel-like conditions.

6.1 PROPERTIES OF ENZYMES

Enzymes are biological catalysts. Like all catalysts, enzymes:
- increase reaction rates
- stabilize the transition state, which lowers its energy, thus lowering the activation energy (E_a, the energy a substrate must have before it can be transformed into product, also called the free energy of activation, $\Delta G^{\ddagger}$).
- do not change the equilibrium constant or the ΔG (the thermodynamics) of a reaction. Enzymes – all catalysts – only affect the *kinetics*, or the rate, of a reaction. If a chemical reaction isn't *thermodynamically* favorable, i.e., if it's nonspontaneous with a positive (+) ΔG, an enzyme can't change this.

  ENZYMES – ALL CATALYSTS – HELP SPONTANEOUS REACTIONS ACHIEVE EQUILIBRIUM *FASTER*.

- regain their original form by the end of a reaction, and are ready to catalyze another reaction. Enzymes function at very low concentrations (nanomolar or picomolar) under physiological conditions.

How enzymes are unique and amazing:
- Enzymes catalyze reactions to phenomenally high rates (10^7-10^{19} times greater)
- Side products are rarely formed.
- Enzymes are *highly specific* to the reactions they catalyze and the substrates that it can bind, and this specificity is due to its unique **ACTIVE SITE.** Enzyme-substrate binding at the active site is via non-covalent interactions such as hydrogen bonding, electrostatic, and hydrophobic forces between its amino acid side chains and the substrate. Contact points at the active site have a particular orientation and a distinct shape, like a right-handed glove. For example, a particular enzyme might catalyze a reaction with L-alanine but not D-alanine or any other amino acid. The charge distribution within the active site is due to the location of amino acid side chains, which also participate in the catalytic mechanism.
- Enzymes stabilize the transition state via conformational changes made possible by the active site's unique shape and charge distribution that fit a specific substrate in a specific orientation. This fit forces the substrate to undergo conformation changes in a way that favors, or resembles, the transition state, so that the enzyme-substrate complex can proceed to product with a much lower activation energy.
- Enzymes can be regulated (because of their relatively large and complex structures) to conserve energy and raw materials.
- Enzymes work at moderate temperatures.

74

ENZYMES WORK IN NON-IDEAL CONDITIONS. Equilibrium constants for such reactions use **ACTIVITY***(a)*, or **EFFECTIVE CONCENTRATION**, instead of concentration. Activity includes the effect of intermolecular interactions, and the activity coefficient (γ) is a correction factor for concentration: $a = \gamma c$ [c = concentration (mol/L)]

> γ depends on: the size and charge of the species, and the ionic strength of the reaction solution

Enzyme function can be very different in a dilute buffer solution versus within a crowded cell. The equilibrium constant for a reaction in nonideal conditions includes a **NONIDEALITY FACTOR** (the ratio of the activity coefficients of the products and reactants)

6.2 CLASSIFICATION OF ENZYMES ACCORDING TO TYPE OF REACTION CATALYZED

(For further examples, see Table 6.1, page 189 of your text.)

| OXIDOREDUCTASES | HYDROLASES | ISOMERASES |
| TRANSFERASES | LYASES | LIGASES |

OXIDOREDUCTASE

Reaction Type: Oxidation-reduction

Names Include: dehydrogenase, oxidase, oxygenase, reductase, peroxidase, hydroxylase

Example: Succinate dehydrogenase catalyzes the transfer of electrons from succinate to FAD, producing fumarate and $FADH_2$ in the sixth step of the citric acid cycle. Electrons are removed from the carbons to form the alkene.

$$^-O_2C\text{-}CH_2\text{-}CH_2\text{-}CO_2^- + FAD \rightarrow {}^-O_2C\text{-}CH=CH\text{-}CO_2^- + FADH_2$$
Succinate   Fumarate

TRANSFERASE

Reaction Type: Transfer of functional groups

Names Include: transferase, kinase (phosphoryl group transfer), transaminase [*trans*+(group transferred)+*ase*], transmethylase, transcarboxylase

Example: Catechol N-methyltransferase catalyzes the transfer of a methyl group from S-adenosyl-methionine to norepinephrine. Used in the synthesis of epinephrine, a neurotransmitter.

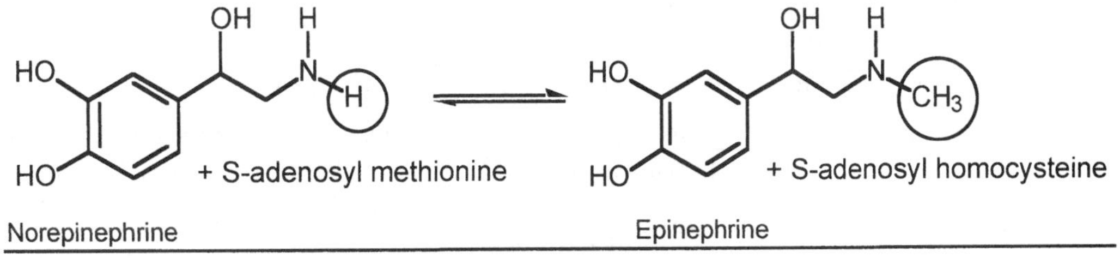

Norepinephrine   Epinephrine

HYDROLASE

Reaction Type: Hydrolysis (*hydro*-add water; *lysis*-cleave a bond)

Names Include: esterase, phosphatase, peptidase, protease

Serine proteases use the –CH_2OH side chain of serine to hydrolyze a peptide. Examples of serine proteases: trypsin, chymotrypsin, thrombin

Example: Thrombin hydrolyzes the peptide bond after an arginine. Converts fibrinogen into fibrin; fibrin then polymerizes to form a blood clot.

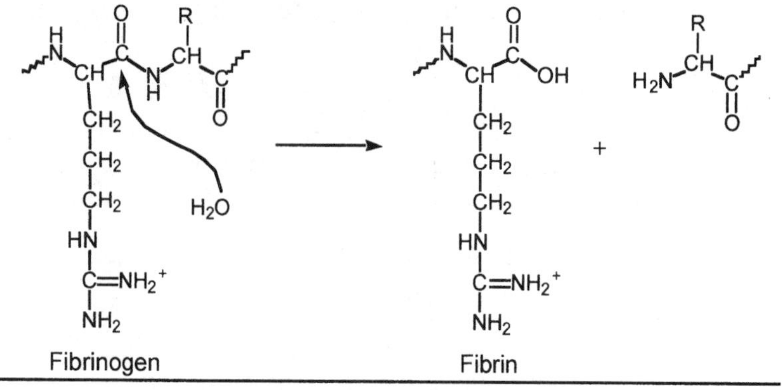

Fibrinogen → Fibrin

LYASE

Reaction Type: Removes a group and forms a double bond – or –

Adds a group to a double bond (the reverse reaction)

Names Include: decarboxylase, hydratase, dehydratase, deaminase, synthase, (name of molecule + lyase)

Example: Argininosuccinate lyase converts argininosuccinate to arginine and fumarate in urea synthesis. Removes a four-carbon dicarboxylic acid group, forming a double bond between the two central carbons.

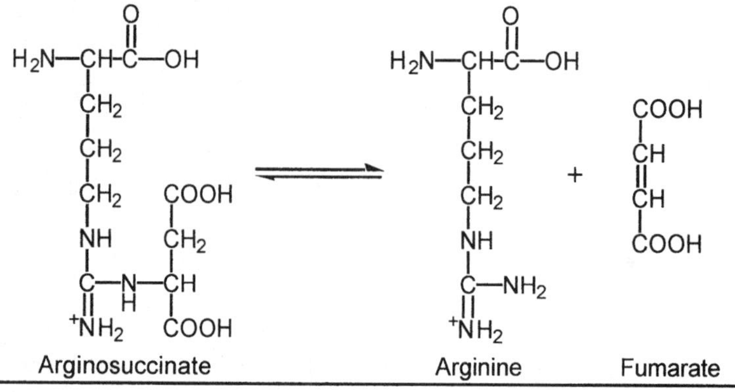

Arginosuccinate ⇌ Arginine + Fumarate

ISOMERASE

Reaction Type: Isomerization

Names Include: mutase, racemase, epimerase, (name of molecule + isomerase)

Example: Triose phosphate isomerase interconverts glyceraldehyde-3-phosphate and dihydroxyacetone phosphate, moving the carbonyl group from C-1 in glyceraldehyde to C-2 in dihydroxyacetone.

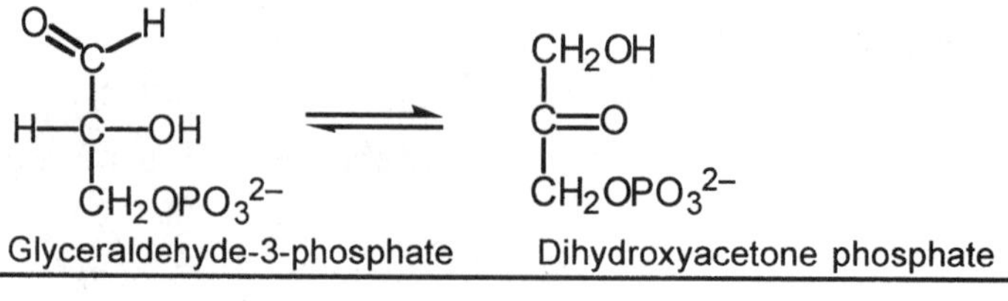

Glyceraldehyde-3-phosphate ⇌ Dihydroxyacetone phosphate

LIGASE

Reaction Type:	Bond formation (often includes removal of water; the reverse of hydrolysis); requires an energy source (ATP hydrolysis)
Names Include:	synthetase, carboxylase (name of molecule formed + ligase)
Example:	Glutamine synthetase catalyzes the formation of glutamine from glutamate and ammonia using ATP hydrolysis to make the reaction thermodynamically favorable.

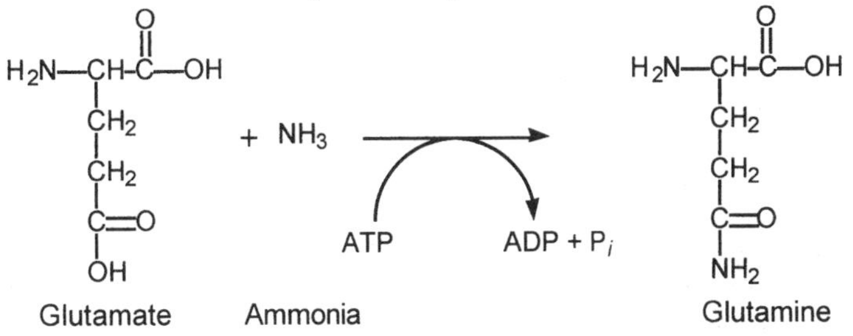

Glutamate   Ammonia   Glutamine

6.3 ENZYME KINETICS

The velocity (v), or rate, of any chemical reaction is the change in concentration of substrate (or product) as a function of time. The rate may also be expressed as:

$$v_0 = k[S]^n$$

where v_0 = initial velocity

n = the order of the reaction
k = the rate constant

The order of a reaction depends on the reaction mechanism, specifically, how many reactants (or substrates) are involved in the slowest step of the mechanism (since the rate can't go faster than the slowest step).

FIRST-ORDER KINETICS: For S → P, a simple first-order reaction, $n=1$, and $v_0 = k[S]$. If [S] is doubled, for example, v_0 should also double.

Consider the reaction: A + B → P. The general rate equation is:

$$v_0 = k[A]^m[B]^n$$ where m = the order with respect to A, n is the order with respect to B, and m+n is the order overall.

If the rate depends only upon the concentration of A, then $v_0 = k[A]^1[B]^0 = k[A]$, and the reaction is first-order in A, zero-order in B, and first-order overall.

SECOND-ORDER KINETICS: If the rate depends on the concentrations of both A and B, then $v_0 = k[A][B]$, and the reaction is first-order in A, first-order in B, and second-order overall.

Remember that **THE ORDER OF A REACTION MUST BE DETERMINED EXPERIMENTALLY.** There's no way to know the order of a reaction by just looking at the chemical equation.

PSEUDO-FIRST-ORDER KINETICS is a second-order reaction that behaves as if it's first order, typically because one reactant (such as water) is present in excess. (Doubling the excess reactant results in no change in the rate. Double excess is still excess...)

ZERO-ORDER KINETICS: the reaction rate doesn't depend on [S]. Adding more substrate won't increase the rate. Why not? If all of the active sites are filled (saturated) with substrate (and there's plenty of substrate around to hit any active site that becomes available), then adding more substrate won't increase the rate. The enzyme can't work any faster – it's at its maximum velocity (V_{max}).

MOLECULARITY is the number of colliding molecules in a single-step reaction: **UNIMOLECULAR**: 1 molecule (A→B) vs. **BIMOLECULAR**: 2 molecules (A+B→C+D)

MICHAELIS-MENTEN KINETICS

MICHAELIS-MENTEN EQUATION:	MICHAELIS-MENTEN PLOT:
$$v_0 = \frac{V_{max}[S]}{[S] + K_m}$$	v_0 vs. [S]

V_{max} = maximum velocity

K_m = Michaelis constant (experimentally determined) = [S] at half of V_{max}

When [S] = K_m, then the Michaelis-Menten equation becomes $v_0 = \frac{V_{max}[S]}{2[S]}$, or

$v_0 = \frac{V_{max}}{2}$. In other words,

K_m is equal to the substrate concentration when v_0 equals half of V_{max}.

So, K_m can be determined from the Michaelis-Menten plot of v_0 vs. [S].

Keep in mind that the Michaelis-Menten plot is actually a *series* of experiments that measure the *initial rate* of a reaction at a number of different substrate concentrations, with the same enzyme concentration in each experiment. Take a second look at Figure 6.3 on page 191. Figure (a) shows one experiment that measures the change in [P] over time. Figure (b) is a compilation of data from many experiments like (a), each data point representing one experiment that began with a different [S].

ASSUMPTIONS NEEDED TO DERIVE THE MICHAELIS-MENTEN EQUATION:

1. Rate of the reverse reaction, E+P→ES is ignored. This is valid as initial velocities are measured, when [P] is very low. (Recall that rate equations include the concentrations of reactants, not products. This assumption simplifies the derivation by removing [P].)

2. k_1 (the rate constant for E+S→ES) >>> k_{-1} (rate constant for ES→E+S)

3. Steady-state assumption: (the rate to form ES) = (the rate to degrade ES); The enzyme-substrate complex (ES) is in steady state, so [ES] remains constant as a function of time.

k_{cat} = TURNOVER NUMBER = V_{max}/(total enzyme concentration) = # substrate molecules that an enzyme molecule converts to product per unit time when enzyme is saturated

SPECIFICITY CONSTANT = k_{cat}/K_m = an indicator of substrate binding efficiency at low [S] ([S] << K_m), when the enzyme is saturated, and the reaction is second-order.

LOWER K_m = GREATER AFFINITY OF THE ENZYME FOR THE SUBSTRATE. Why is this true? Consider two enzyme-catalyzed reactions with the same V_{max} but different K_m values. For the reaction in which the enzyme has a greater affinity for substrate, it takes less substrate to get to the same rate (half of V_{max}). Conversely, an enzyme with a lower affinity for substrate needs a higher substrate concentration to work at the same velocity.

CATALYTIC PERFECTION – enzymes working at their **DIFFUSION CONTROL LIMIT** – their reaction rate is limited only by how fast the substrate can get to the active site.

INTERNATIONAL UNIT (I.U.) of enzyme activity = amount of enzyme to produce 1μmol of product per minute; **SPECIFIC ACTIVITY** = # I.U. per mg of protein; a measure of enzyme purification

KATAL (kat) = 1 mole of substrate converted to product per second

LINEWEAVER-BURK PLOTS: DETERMINATION OF K_m AND V_{max}
SIGNIFICANCE:

- Accurate values of V_{max} and K_m are much easier to determine using a straight-line Lineweaver-Burk plot as opposed to a Michaelis-Menten plot, in which the velocity approaches (but never reaches) a maximum value.

- Comparing Lineweaver-Burk plots of inhibited vs. uninhibited reactions can indicate the mechanism of enzyme inhibition.

Lineweaver-Burk turned the Michaelis-Menten plot into a straight line by rearranging the *reciprocal* of the Michaelis-Menten equation to fit the equation for a straight line, $y = mx + b$.

Michaelis-Menten equation:
$$v_0 = \frac{V_{max}[S]}{[S] + K_m}$$

Its Lineweaver-Burk reciprocal:
$$\frac{1}{v_0} = \frac{[S] + K_m}{V_{max}[S]}$$

$$\frac{1}{v_0} = \frac{[S]}{V_{max}[S]} + \frac{K_m}{V_{max}[S]} = \frac{1}{V_{max}} + \frac{K_m}{V_{max}[S]}$$

After rearranging slightly:
$$\frac{1}{v_0} = \frac{K_m}{V_{max}} \cdot \frac{1}{[S]} + \frac{1}{V_{max}}$$

Equation for a straight line:
$$y = m \quad x \quad + \quad b$$

A Lineweaver-Burk plot of: $\dfrac{1}{v_0}$ vs. $\dfrac{1}{[S]}$ is a straight line, with

$$m = \text{slope} = \frac{K_m}{V_{max}} \qquad b = y\text{-intercept} = \frac{1}{V_{max}}$$

Further algebra reveals: $\quad x\text{-intercept} = -\dfrac{1}{K_m}$

So, to calculate K_m and V_{max}, all that's needed are the x and y intercepts from the Lineweaver-Burk plot.[1] A Lineweaver-Burk plot is also called a *double-reciprocal* plot, since it's $\dfrac{1}{v_0}$ vs. $\dfrac{1}{[S]}$ instead of the Michaelis-Menten plot of v_0 vs. [S]. [Hint: To keep the intercepts straight, remember that the x-axis is a measure of substrate concentration, from which we get K_m, and that the y-axis is a measure of velocity.]

---

[1] Note that the x-intercept is just a mathematical trick to calculate K_m, since a negative substrate concentration has no real physical meaning. The calculation of V_{max} is also a mathematical trick, since a reaction rate at [S]=0 also has no physical meaning.

MULTISUBSTRATE REACTIONS

SEQUENTIAL:
All substrates must bind (**ORDERED** or **RANDOM**) before the reaction can occur.

DOUBLE-DISPLACEMENT REACTIONS (PING-PONG MECHANISM)
The first substrate binds and the first product is released *before* the second substrate binds.

ENZYME INHIBITION AND INHIBITORS

- Inhibitors bind to an enzyme and interfere with its activity. Inhibitors help to regulate enzyme activity. Inhibiting a key enzyme can result in inhibition of an entire pathway.

- Reversible inhibition: competitive, noncompetitive (pure and mixed), uncompetitive

- To determine the type of inhibitor, use a Lineweaver-Burk plot that includes both inhibited and uninhibited reaction data.

- Irreversible inhibitors bind to an enzyme covalently and permanently stops enzyme activity. Removing the inhibitor or increasing [S] will not restore enzyme activity.

MECHANISMS FOR REVERSIBLE ENZYME INHIBITION

COMPETITIVE: $\qquad$ E + I $\rightleftharpoons$ EI $\rightarrow$ *(no reaction)* $\qquad$ E + S $\rightleftharpoons$ ES $\rightarrow$ P + E
Competitive inhibitors bind to the active site of the free enzyme; they "compete" or prevent the substrate from binding to the enzyme, resulting in a change in K_m (since it affects substrate binding) but not V_{max} (since the effect of the inhibitor can be overcome by increasing [S]). In other words, it takes much more substrate to reach V_{max}, so K_m will increase.

NONCOMPETITIVE: $\quad$ E + I $\rightleftharpoons$ EI $\qquad\qquad\qquad$ E + S $\rightleftharpoons$ ES $\rightarrow$ P + E
ES + I $\rightleftharpoons$ EIS $\qquad\qquad$ EI + S $\rightleftharpoons$ EIS $\rightarrow$ *(no reaction)*
Noncompetitive inhibitors bind to a site other than the substrate binding site; they can remove both free enzyme (E) and ES. Increasing [S] can *sometimes partially* – but not completely – reverse noncompetitive inhibition.
PURE noncompetitive inhibitors change the V_{max} but not the K_m (the simplest case, but rare). **MIXED** noncompetitive inhibitors change both K_m and V_{max}.

UNCOMPETITIVE: $\qquad$ ES + I $\rightleftharpoons$ EIS $\qquad\qquad\qquad$ E + S $\rightleftharpoons$ ES $\rightarrow$ P + E
Uncompetitive inhibitors bind only to ES, and not to E. Since some of the enzyme is always tied up as ESI, V_{max} can't be attained, and is lower. K_m also *decreases*.
How can an inhibitor *increase* the affinity of an enzyme for its substrate, and still inhibit the reaction? Think on this: If the inhibitor binds to ES and removes ES, then LeChatelier's principle says that the equilibrium will shift to compensate for the change in concentration. The reaction (E + S $\rightleftharpoons$ ES) is shifted to the right because the inhibitor binds to and removes ES. So, the enzyme binds more substrate than expected (decreasing K_m), but the ES complex is waylaid by inhibitor before it can be turned into product (decreasing V_{max}). Adding more [S] can't overcome this – even though more ES would form, more ESI would also form. Uncompetitive inhibition is considered a type of noncompetitive inhibition.

SUMMARY OF REVERSIBLE ENZYME INHIBITION

TYPE:	BINDS TO:	AFFECTS:	L-B PLOT:[2]	NOTES:
COMPETITIVE	active site of E, not ES	higher K_m same V_{max}	lines intersect on y-axis	I is often similar to substrate's structure
NONCOMPETITIVE PURE	both E and ES (at a site other than the active site; need not resemble the substrate)	same K_m lower V_{max}	lines intersect on x-axis	rare; the rate to form EI equals the rate to form ESI
NONCOMPETITIVE MIXED		changes K_m lower V_{max}	x- and y- intercepts and slopes differ	the rate to form EI doesn't equal the rate to form ESI
UNCOMPETITIVE	ES, not E	lower K_m lower V_{max}	slopes are the same, x- and y- intercepts differ	typically for reactions with more than one S

ALLOSTERIC ENZYMES (Michaelis-Menten plot is sigmoidal rather than hyperbolic; Michaelis-Menten model does not explain the kinetic properties of allosteric enzymes.

ENZYME KINETICS, METABOLISM, AND MACROMOLECULAR CROWDING

Cell interiors are very crowded and highly heterogeneous, such numerous and varied intermolecular interactions result in lower rates of diffusion, higher effective concentrations, and other effects that are nonlinear and difficult to predict.

MACROMOLECULAR CROWDING: many macromolecules such as proteins, membranes, cytoskeletal components etc. impede molecular movement within a cell

Example of the complexity this introduces: Consider a slow substrate diffusion vs. greater efficiency via the formation of **MICROCOMPARTMENTS** (with high substrate concentrations) or METABOLONS (enzyme complexes that channel pathway intermediates from one enzyme to the next). Will the result be a greater or a lower efficiency? *In vivo* conditions are difficult to model *in vitro* and in silico.

METABOLIC FLUX — the rate of flow of metabolites (substrates, products, intermediates) along biochemical pathways

6.4 CATALYSIS

ORGANIC REACTIONS AND THE TRANSITION STATE

Reaction mechanisms and related terminology are the same as in organic chemistry:
TRANSITION STATE VS. INTERMEDIATE;
REACTIVE INTERMEDIATES: CARBOCATION, CARBANION, FREE RADICAL

The difference? The active site stabilizes the transition state, and has a greater affinity for the transition state than for substrates or products.

---

[2] Lineweaver-Burk plot of inhibited and uninhibited reactions: Increasing K_m makes the x-intercept ($-1/K_m$) less negative (-1/20 is less negative than $-1/10$). Lowering V_{max} makes the y-intercept($1/V_{max}$) larger (1/2 is larger than 1/4). The slope is K_m/V_{max}.

CATALYTIC MECHANISMS

PROXIMITY AND STRAIN EFFECTS; ELECTROSTATIC EFFECTS
The more tightly and efficiently that an active site can bind substrate while it's in its transition state, the faster the reaction rate will be.

ACID-BASE CATALYSIS
Amino acids with side chains that can act as either proton donors or acceptors depending upon their pK_a and its local environment: His, Asp, Glu, Tyr, Cys, Lys

COVALENT CATALYSIS
Nucleophilic amino acid side chains (like Ser) form weak covalent bonds to facilitate the reaction (example: serine proteases like chymotrypsin)

QUANTUM TUNNELING AND ENZYME CATALYSIS
Quantum tunneling is traveling through an energy barrier, rather than over it. This theory rises from quantum mechanics, and helps to explain the phenomenally high rates that enzymes can achieve.

THE ROLE OF AMINO ACIDS IN ENZYME CATALYSIS
Catalytic amino acids have polar or charged side chains:

Ser, Thr, Tyr, Cys, Gln, Asn, Glu, Asp, Lys, Arg, His

DYADS or TRIADS – groups of 2 or 3 amino acids that are positioned precisely for catalytic effects, such as polarizing a specific atom or group, or changing the pK_a of a nearby functional group.

Noncatalytic amino acids function in substrate orientation and transition state stabilization. Example: creation of a hydrophobic pocket of a specific size and shape

THE ROLE OF COFACTORS IN ENZYME CATALYSIS
Cofactors are non-protein components that add chemical functionality to enzymes, and thus increase the kinds of reactions that enzymes can catalyze. (Otherwise, the enzymes would be limited to the amino acid residues' R groups.)

APOENZYME VS. HOLOENZYME
In *A*poenzymes, the necessary cofactor or coenzyme is *A*bsent. (An apoenzyme is just the protein component.) *H*oloenzymes *H*ave the cofactor or coenzyme.

METALS: transition metals (Fe^{2+}, Cu^{2+})
alkali and alkaline earth metals (Na^+, K^+, Mg^{2+}, Ca^{2+})

Why are the transition metals good cofactors? Their high concentration of positive charge binds small molecules well. They can act as Lewis acids (i.e., accept electron pairs). Because they can interact with two or more ligands, they can form a substrate-metal ion complex in the active site, thus helping to orient the substrate properly, polarizing the substrate and promoting catalysis. Metals with two or more valence states can mediate redox reactions.

COENZYMES VS. VITAMINS
VITAMINS: water-soluble vs. lipid-soluble; many are precursors to coenzymes

The vitamin niacin is a precursor to the coenzyme NAD^+ (nicotinamide adenine dinucleotide), which functions as an intracellular electron carrier and as a hydride ($H{:}^-$) transfer agent.

Niacin   Oxidized NAD$^+$ Reduced NADH

The vitamin riboflavin is a precursor to the coenzymes FMN (flavin mononucleotide) and FAD (flavin adenine dinucleotide), which are components of flavoproteins (a sub-class of the oxidoreductases).

EFFECTS OF TEMPERATURE AND pH ON ENZYME-CATALYZED REACTIONS

Optimum temperature; pH optimum

Temperature and pH changes affect the tertiary structure of proteins, which must be very specific for enzymes to function at their maximum velocity. Optima vary according to enzyme function (consider enzymes that function at the different pH environments of the stomach at pH~3, or small intestine, at pH 5-8).

Higher temperatures generally increase rates, but denaturation of the enzyme can occur if the temperatures are too high.

DETAILED MECHANISMS OF ENZYME CATALYSIS: CHYMOTRYPSIN AND ALCOHOL DEHYDROGENASE

Note that these mechanisms resemble those that you wrote in organic chemistry, but the proximity of certain functional groups to each other and the geometry of the active site encourages reactions to happen that would not normally happen. It's like one of those romantic comedies: two people were meant to be together (their being together is thermodynamically favorable) but they'd never fall in love if they hadn't had something put them together, like being stuck in an elevator (or in an enzyme's active site). Again, enzymes can't make the impossible happen – they just make spontaneous reactions happen *faster*.

6.5 ENZYME REGULATION

Enzyme regulation: maintains an ordered state, conserves energy, and helps the cell to respond to environmental changes (i.e., modulates specific pathways in response to a cell's needs). Regulatory enzymes in a pathway are usually controlled by covalent modification or allosteric regulation.

GENETIC CONTROL

ENZYME INDUCTION: Enzymes are synthesized when needed.

REPRESSION: Enzyme synthesis is inhibited.

COVALENT MODIFICATION

- PHOSPHORYLATION (or DEPHOSPHORYLATION) to convert an enzyme between its active and inactive forms (Whether the phosphorylated enzyme is active or inactive depends on the specific enzyme.)

- ZYMOGENS (PROENZYMES) are converted into active enzymes by the irreversible cleavage of one or more peptide bonds.

ALLOSTERIC REGULATION

Effector molecules (ligands) bind to allosteric sites on the enzyme and trigger conformational changes.

Importance of the flexibility of proteins (binding of ligand to one subunit prompts a conformational change that's transmitted to other subunits)

HOMOTROPIC (ligand = substrate) vs. HETEROTROPIC allosteric effects

CONCERTED (symmetry) model vs. SEQUENTIAL model

> T vs. R forms: TAUT vs. RELAXED
> Inhibitors bind to and stabilize the Taut form;
> Activators bind to and stabilize the Relaxed form
>
> CONCERTED (SYMMETRY) MODEL: all T subunits change to R (or R→T) upon binding
> SEQUENTIAL MODEL: upon binding of an effector, subunits change sequentially

POSITIVE COOPERATIVITY – first ligand increases subsequent ligand binding
NEGATIVE COOPERATIVITY – first ligand reduces the enzyme's affinity for similar ligands

Many allosteric proteins are more complex than either of these models.
There are no simple rules that explain metabolic regulation.

COMPARTMENTATION

Enzymes, substrates and regulatory molecules are located in separate compartments (or areas) so that opposing pathways are physically separated, or located close together, so that a pathway can be more efficient.

Enzymes may also be attached to the cytoskeleton, or to a membrane.

Compartmentation solves problems:

1. Divide and control: separation of enzymes into different regions or compartments to use resources efficiently (e.g., to prevent a just-synthesized molecule from being degraded). This also gives the cell an additional level of regulation by controlling the transport of substrates, products, effectors, etc. across organelle membranes (example: mitochondria membranes).

2. Diffusion barriers: microenvironments that concentrate enzymes and/or substrates and so reduce diffusion barriers. Enzymes located close to each other can increase the efficiency of coupled reactions and/or of entire pathways (such as the electron transport chain); METABOLIC CHANNELING

3. Specialized reaction conditions (e.g., low pH within lysosomes)

4. Damage control: segregation of molecules that may be toxic to the cell

THE ORGANIC CONNECTION: NICOTINAMIDE ADENINE DINUCLEOTIDE (NAD⁺)

Take another look at the nicotinamide rings of NAD$^+$ and NADH. Which one would you expect to be more stable, that is, at a lower energy? Why?

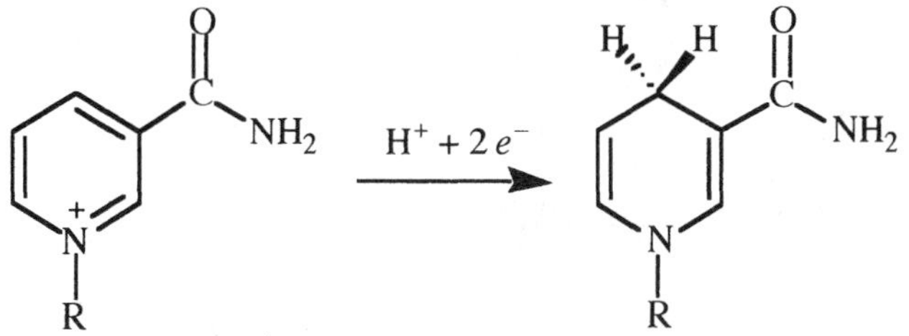

When the hydride (or, the H+ with two electrons) is added, it creates a tetrahedral carbon that interrupts the aromaticity of the ring. The nicotinamide ring of NAD$^+$ is aromatic and thus more stable and lower-energy than the non-aromatic ring of NADH. Since NADH is at a higher energy than NAD$^+$, this half-reaction would require an input of energy:

$$NAD^+ + H^+ + 2e^- \rightarrow NADH$$

Couple this with an oxidation half-reaction that releases energy, and we have a neat way of capturing that energy in the form of electrons. (Sound familiar? Refer to Chapter 4.)

AFTER STUDYING THIS CHAPTER, YOU SHOULD BE ABLE TO:

- Describe how enzymes work, general properties of enzymes, and factors that contribute to enzyme catalysis.

- Determine the class of an enzyme, given the reaction that the enzyme catalyzes.

- Given a biochemical reaction and reaction rate data at various substrate concentrations, determine the order for each substrate and the overall order of a reaction.

- Michaelis-Menten plots: Determine K_m.

- Lineweaver-Burk plots: Determine the type of enzyme inhibition, given data for both uninhibited and inhibited enzyme-catalyzed reactions. Determine V_{max} and K_m from the x- and y-intercepts.

- Compare enzyme efficiencies given k_{cat} and K_m data for various substrates.

- Describe methods of enzyme regulation.

Use this space to note any additional objectives provided by your instructor.

CHAPTER 6: SOLUTIONS TO REVIEW QUESTIONS

6.1   a.   activation energy – the minimum energy required to bring about a reaction

        b.   catalyst – a substance that alters the rate of a reaction but is not consumed by it

        c.   active site – the specific part of an enzyme that is directly responsible for catalysis

        d.   coenzyme – a small molecule needed to enable the enzyme to function

        e.   velocity of a chemical reaction – the change in concentration of a reactant with time

6.2   The important properties of enzymes are: high catalytic rates, a high degree of substrate specificity, negligible formation of side products, and capacity for regulation.

6.4   (a) oxidoreductase (b) transferase (c) lyase (d) isomerase (e) ligase

6.5   a.   reaction mechanism – a step-by-step description of the process that occurs during a reaction; typically illustrated stepwise with chemical structures of intermediates and/or transition states, and using curved arrows to show the flow of electrons

        b.   carbanion – nucleophilic carbon anions with 3 bonds and an unshared electron pair

        c.   free radical – highly reactive species with at least one unpaired electron

        d.   quantum tunneling – a quantum mechanical process in which a particle passes through an energy barrier instead of over it

        e.   reaction order – the sum of the exponents on the concentration terms in the rate expression; provides information relating to the reaction mechanism; must be determined experimentally

6.7   Three reasons why the regulation of biochemical processes are important are maintenance of an ordered state, conservation of energy, and responsiveness to environmental cues.

6.8   Negative feedback inhibition is a process in which the product of a pathway inhibits the activity of the pacemaker enzyme.

6.10   In the concerted model, the substrate and activators bind to the relaxed conformation. This binding shifts the equilibrium to the R conformation. In the sequential model, binding the activator molecule changes the conformation of the enzyme to a shape more favorable to binding substrate. When oxygen binds to hemoglobin, the first oxygen molecule binds slowly. It, however, introduces a conformational change that makes the sequential binding of the second, third, and fourth oxygen molecules much easier.

6.11   a.   vitamin – an organic molecule required by organisms in minute quantities; some vitamins are coenzymes required for the function of cellular enzymes

        b.   cofactor – the nonprotein component of an enzyme (either an inorganic ion or a coenzyme) required for catalysis

        c.   transition state – the unstable intermediate in catalysis in which the enzyme has altered the form of the substrate so that it now shares properties of both the substrate and the product

        d.   inhibitor – a molecule that reduces an enzyme's activity

        e.   allosteric enzyme – an enzyme whose activity is affected by the binding of effector molecules

6.13 Transition metal ions are useful as enzyme cofactors because they have concentrations of positive charge, can act as Lewis acids, and can bind to two or more ligands at the same time.

6.14 a. cooperativity – a property of allosteric enzymes in which binding of a ligand affects the enzyme's affinity for subsequent ligand binding

b. oxyanion – a negatively charged oxygen atom

c. apoenzyme – the protein portion of an enzyme that requires a cofactor to function in catalysis

d. holoenzyme – a complete enzyme consisting of the apoenzyme plus a cofactor, which is required for the protein to function

e. isozyme – one of two or more forms of an enzyme, all of which have the same enzyme activity but different amino acid sequences

6.16 Enzymes decrease the activation energy required for a chemical reaction because they provide an alternate reaction pathway that requires less energy than the uncatalyzed reaction. They do so principally because of the unique, intricately shaped active sites, which possess strategically placed amino acid side chains, cofactors, and coenzymes that actively participate in the catalytic process.

6.17 a. half-life – the time needed to consume half the reactant molecules

b. repression – the prevention of polypeptide synthesis

c. turnover number – the number of molecules of substrate converted to product each second per mole of enzyme

d. katal – measure of the rate of enzyme activity; 1 katal (kat) is equal to the conversion of one mole of substrate to product per second

e. noncompetitive inhibitor – an inhibitor molecule that binds to an enzyme, but not at the active site

6.19 The pK_a of the imidazole group of histidine is approximately 6. Therefore, the histidine side chain ionizes within the physiological pH range. The protonated form of histidine is a general acid, and the unprotonated form is a general base.

6.20 The amino acid residues that compose the active site are stereoisomers. Consequently, the active site is chiral and can bind only one form of an optically active compound.

6.22 The law of mass action assumes: linear reaction rates with respect to solute concentrations; a homogeneous reaction system; and random, independent interactions between molecules. However, macromolecular crowding increases the effective concentration, which impacts binding affinities, reaction rates, equilibrium constants, and diffusion rates. As a result, catalytic activities of enzymes *in vivo* are typically not predictable based upon *in vitro* studies of enzyme activity in dilute solutions.

6.23 a. oxygenase – an enzyme that catalyzes the addition of oxygen atoms into a substrate

b. epimerase – an enzyme that catalyzes the inversion of an asymmetric carbon atom of a sugar, converting it into one of its epimers (epimerization)

c. protease – an enzyme that catalyzes the cleavage of peptide bonds with the addition of water

d. hydroxylase – an enzyme that catalyzes the insertion of an oxygen atom into a C–H bond in order to form a new hydroxyl group

e. oxidase – an enzyme that catalyzes a reaction in which O_2 is reduced, but oxygen atoms do not appear in the product

6.25 The amino acid residues that most commonly participate in enzyme mechanisms are serine, threonine, tyrosine, cysteine, histidine, aspartate, glutamate, asparagine, glutamine, lysine, and arginine. Note that all of these amino acids have polar or charged side chains. See Figure 5.2 (p. 126) for the structures of these amino acids.

6.26 The roles played by amino acid side chains in enzymatic catalysis include: donating or accepting proton(s) in acid/base catalysis (His, Tyr, Cys, Glu, Asp, and Lys), acting as a nucleophile in covalent catalysis (Ser, Thr, Cys, Gln, Asn, Asp, and Glu), and having a polarizing effect on the substrate (Arg, Glu, Asp, Ser, His). Examples of enzymes that illustrate these roles are pepsin, the serine proteases, and adenylate kinase. Pepsin's active site contains two Asp carboxyl groups, one of which has a polarizing effect on the other, increasing its basicity. This initiates the acid-base hydrolysis of a peptide bond. The serine proteases also hydrolyze peptide bonds. This mechanism includes the deprotonation of serine's hydroxyl group and the use of the resulting oxyanion as a nucleophile. Adenylate kinase, which catalyzes the transfer of phosphoryl groups from ATP to other nucleotides, contains an Arg-Arg dyad whose polarizing effect converts phosphate into a good leaving group.

6.28 By physically separating enzymes that catalyze competing reactions, compartmentation prevents the inefficiency of futile cycles by allowing these enzymes to be regulated independently. Within microcompartments, metabolite channeling reduces the degree to which diffusion will limit reaction rates. Compartmentation can provide unusual environments for enzymes that require them, and can protect cellular components from damage due to toxic reaction products. Examples of compartmentation include: the separation of kinases and phosphatases, the attachment of multienzyme complexes to membranes or microfilaments, and the low pH within lysosomes.

6.29 The two major types of enzyme inhibitors are reversible and irreversible. Reversible inhibition does not destroy the enzyme. Adding substrate or removing the inhibitor can remove the inhibitory effect. Malonate, whose structure resembles that of succinate, is a reversible competitive inhibitor of succinate dehydrogenase. An irreversible inhibitor binds permanently (usually covalently) to the enzyme and destroys its catalytic ability. Removing irreversible inhibitors will not restore enzyme activity. Iodoacetate is an irreversible inhibitor that alkylates glyceraldehyde-3-phosphate dehydrogenase.

6.31 The three-dimensional shape of the active site is vital to enzyme function and its ability to stabilize the transition state. A very specific structure of an enzyme is required to ensure that the active site has the optimum shape, electrostatic environment, flexibility (to optimize proximity and strain effects), and precise locations of amino acid side chains that actively participate in the enzyme's catalytic mechanism. This complex precision of the active site structure is created and maintained by the intricate web of interactions between amino acid residues in the enzyme as a whole. The inefficiency of creating such a large molecule is tolerated because of the enormous advantage provided by the increase in reaction rates and enzyme efficiency that result.

CHAPTER 6: SOLUTIONS TO THOUGHT QUESTIONS

6.32 The data indicate that the reaction is first-order in pyruvate and ADP and second-order in P_i. The overall reaction is fourth-order.

6.34

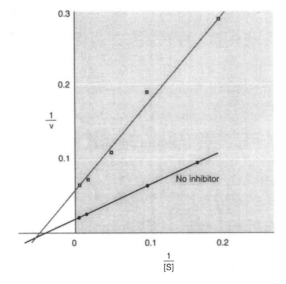

Intercept, horizontal axis $= -\dfrac{1}{K_m}$

Intercept, vertical axis $= \dfrac{1}{V_{max}}$

Slope $= \dfrac{K_m}{V_{max}}$

The type of inhibition being observed is noncompetitive.

6.35

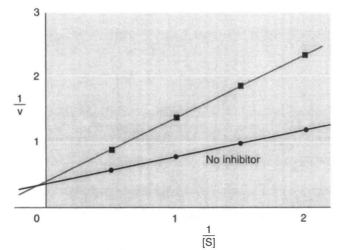

The inhibition is competitive.

6.37 The k_{cat}/K_m $(M^{-1}s^{-1})$ values for the substrates are as follows:

Substrate	k_{cat}/K_m
ethanol	0.5
1-butanol	1.02
1-hexanol	2.63
12-hydroxydodeconate	2.92
all-*trans*-retinol	3.9
benzyl alcohol	0.2
2-butanol	0.00114
cyclohexanol	0.0039

The larger the k_{cat}/K_m value, the faster the rate of the reaction. The substrate with the highest k_{cat}/K_m value is all-*trans*-retinol.

6.38 The inhibition is competitive.

6.40 As indicated on the graph (see answer to question 6.39), Inhibitor A is a competitive inhibitor and inhibitor B is a noncompetitive inhibitor. Competitive inhibitors have an affinity for E, and noncompetitive inhibitors have affinities for both E and ES.

6.41 Catalase has a higher affinity for the substrate and a higher turnover number than does the carbonic anhydrase.

6.43 This enzyme has a greater tendency to dehydrate malate to fumarate than the reverse reaction. In order for the citric acid cycle to operate with the enzyme production of malate, malate concentrations must be very low.

6.44 Crowding will tend to raise activity coefficients. If the substances are not allowed to diffuse from the reaction center the concentration in terms of moles per liter can be quite low but the local concentration can be considerably higher.

6.46 The rate is first order in each reactant. The overall rate expression is: Rate = k[A][B]. The overall reaction is second order.

6.47 NADH will react faster than NADD. The lighter the atom the easier it is to tunnel. Hydrogen is lighter than deuterium so NADH will react faster.

6.49 The magnesium ion has vacant d orbitals that can act as electron pair acceptors.

6.50 This serine residue is part of a serine triad. This makes the serine OH much more reactive. The diisopropylfluorophosphate fits into the active site next to the serine and reacts rapidly.

6.52 rate = $k[A]^2[B]$

[A]	[B]	Rate
0.1	0.01	1×10^6
0.1	0.02	2×10^6
0.2	0.01	4×10^6
0.2	0.02	8×10^6

6.53 V_{max} is the top of the curve. K_m is the substrate concentration at $V_{max}/2$. The line shown is curved. Lineweaver-Burk plots are straight lines and are more accurate.

6.55 The catalytic OH group would be a stronger nucleophile. In aqueous solution the OH would be solvated by water that must be lost before the OH can act as a nucleophile. This would require energy. In a water-free pocket of the enzyme, this desolvation is not necessary, so less energy is required and the OH is more reactive.

7 Carbohydrates

Brief Outline of Key Terms and Concepts

7.1 MONOSACCHARIDES
aldoses, ketoses; Fischer projections

MONOSACCHARIDE STEREOISOMERS
D- vs. L-sugars; enantiomers
diastereomer
epimer

CYCLIC STRUCTURE OF MONOSACCHARIDES
HEMIACETALS, HEMIKETALS
HAWORTH STRUCTURES
CONFORMATIONAL STRUCTURES (puckered
rings, chair structures)
MUTAROTATION
    anomer; anomeric carbon
    α- vs. β-carbon; relative stability of α- vs.
    β-anomers; equilibrium mixtures in water

REACTIONS OF MONOSACCHARIDES
OXIDATION
    aldonic acid, aldaric acid
    uronic acid
    lactone
    reducing sugars

REDUCTION; alditols
    ISOMERIZATION
        enediol
        epimerization
    ESTERIFICATION
        includes phosphoesters
    GLYCOSIDE FORMATION
        glycoside; glycosidic linkage
        acetal; ketal
        disaccharide; polysaccharide
    GLYCOSYLATION REACTIONS

IMPORTANT MONOSACCHARIDES
GLUCOSE   FRUCTOSE
GALACTOSE

MONOSACCHARIDE DERIVATIVES
URONIC ACIDS   AMINO SUGARS
DEOXY SUGARS

7.2 DISACCHARIDES
MALTOSE   CELLOBIOSE
LACTOSE   SUCROSE

7.3 POLYSACCHARIDES
and oligosaccharides

Functions: energy storage
                structural materials.

HOMOGLYCANS
STARCH: AMYLOSE, AMYLOPECTIN
GLYCOGEN
CELLULOSE
CHITIN

HETEROGLYCANS
N-GLYCANS, O-GLYCANS
GLYCOSAMINOGLYCANS (GAG)

7.4 GLYCOCONJUGATES

PROTEOGLYCANS

GLYCOPROTEINS; GLYCOPROTEIN FUNCTIONS

GLYCOCONJUGATES – carbohydrate linked covalently to proteins or lipids.

PROTEOGLYCANS are composed of relatively large amounts of carbohydrate (GAG units) covalently linked to small polypeptide components.

GLYCOPROTEINS – proteins linked to carbohydrate through *N*- or *O*-linkages.

7.5 THE SUGAR CODE

LECTINS: TRANSLATORS OF THE SUGAR CODE

THE GLYCOME
    glycoforms, glycomics
    microheterogeneity

FUNCTIONS OF CARBOHYDRATES: energy sources
structural elements
precursors in the synthesis of biomolecules
biological information storage

7.1 MONOSACCHARIDES

Name by type (aldose, ketose), number of carbons (pentose, hexose), or both (aldohexose, ketohexose).

Simplest sugars are trioses: glyceraldehyde, dihydroxyacetone.

FISCHER PROJECTIONS in organic chemistry are always shown as a cross, with the carbon understood to be at the center. When the carbon atom was labeled in the center, it was understood that that was not a Fischer projection but a line-bond structure that implied no stereochemistry whatsoever. In your text, however (as in many biochemistry texts), it's understood that any straight-chain carbohydrate drawn vertically (with the most oxidized carbon nearest the top), is a Fischer projection, even when the carbons are shown.

MONOSACCHARIDE STEREOISOMERS

HOW MANY STEREOISOMERS ARE POSSIBLE?

Because carbohydrates have many chiral centers, there are many stereoisomers. Identifying which carbons are chiral is the first step in determining the number of stereoisomers. Recall that for a carbon to be chiral, it must have *four different atoms or groups* attached to it.

The number of possible stereoisomers is related to the number of chiral carbons:
  number of stereoisomers = 2^n n = number of chiral carbons

Why is this true? *Each* chiral carbon can have two different arrangements of groups (that is, they can be mirror images of each other).[1] Glucose is an example of an aldohexose, which has four chiral carbons.

LET'S LOOK MORE CLOSELY AT GLUCOSE (AN ALDOHEXOSE):

Carbon 1 has only three substituents attached to it, and so it's *not* chiral.

Carbon 2 has four different groups attached to it, so it *is* chiral. Carbons 3, 4, and 5 are also chiral.

Carbon 6 has two hydrogens attached to it, and so it's *not* chiral.

Glucose, therefore, has four chiral carbons.

Number of isomers possible (including glucose) = 2^4 = 2 x 2 x 2 x 2 = **16**

Half of these isomers are enantiomers (mirror images) and are classified as D- or L-. So, of the 16 possible aldohexose isomers, eight are D-isomers and eight are L-isomers. For the structures of all eight D-isomers, see: Figure 7.3 in your text for Fischer projections, and the end of this Study Guide chapter for the Haworth structures.

---

[1] Organic chemists would differentiate between the two possible arrangements of groups by labeling each chiral carbon atom as *R* or *S*. (Remember them?)

ENANTIOMERS

D-Glucose   L-Glucose

The remaining isomers are diastereomers. Diastereomers are stereoisomers that are not mirror images of each other.

LET'S LOOK AT FRUCTOSE (A KETOSE) AS ANOTHER EXAMPLE:

The first carbon has two hydrogens attached to it and is therefore *not* chiral.

The second carbon has only three groups attached to it and is therefore *not* chiral.

The third, fourth, and fifth carbons *are* chiral.

The sixth carbon has two hydrogens attached to it and is therefore *not* chiral.

Thus, fructose has three chiral carbons, so it has $2^3 =$ 2 x 2 x 2 = 8 possible isomers (4 D- and 4 L-isomers)

D- VS. L-ENANTIOMERS are determined by comparing the Fischer projections to D- and L-glyceraldehyde. Compare the chiral carbons furthest from the carbonyl carbons. Most biologically active sugars are D-sugars.

ENANTIOMERS *vs.* DIASTEREOMERS *vs.* EPIMERS

Epimers are diastereomers that differ in configuration at only one chiral center.

Example: Which of the following monosaccharides are epimers of each other?

D-Glucose   D-Allose   D-Altrose

D-Glucose and D-allose are epimers of each other because they differ by only one chiral carbon (at carbon 3). D-Allose and D-altrose are also epimers of each other since they differ only at carbon 2. However, D-glucose and D-altrose are diastereomers and *not* epimers of each other because they're different at *both* carbons 2 and 3.

CYCLIC STRUCTURE OF MONOSACCHARIDES

Reversible reaction between an –OH and the aldehyde or ketone on the same molecule; formation of cyclic hemiacetals or hemiketals

Example: C-5 of glucose bonds with C-1, the aldehyde carbon. Approach can be from one side of the aldehyde or the other, so C-1 could either be "up" or "down". C-1 is the anomeric carbon atom.

MUTAROTATION – When α- and β-monosaccharides dissolve in water, they interconvert, forming an equilibrium mixture of open-chain, α– and β– pyranose and furanose forms. The more stable anomer is the most abundant in solution. The open-chain form can undergo redox reactions.

ANOMERS OF D-SUGARS*: α VS. β

β-D-Glucose

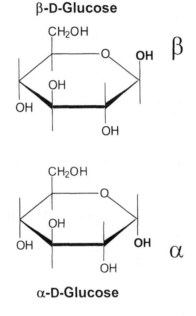

the beta boat sails atop "D" sea

the alpha fish swims beneath "D" sea

α-D-Glucose

*L-sugars are opposite: α is "up" and β is "down"

HAWORTH STRUCTURES OF PYRANOSE AND FURANOSE (CYCLIC) FORMS; DRAWING HAWORTH STRUCTURES FROM FISCHER PROJECTIONS

Haworth structures show the cyclic (pyranose or furanose) forms of a sugar. Note that for the sake of simplicity, biochemists typically don't include H labels in Haworth projections. This convention lets us focus on the location of the OH groups with less clutter.

1. Draw the 5- or 6-membered ring with the oxygen atom in its correct position.

2. Number the carbons in both the Fischer projection and the Haworth structure.

3. First, draw in the atoms attached to the anomeric carbon (that's the first carbon in aldoses and the second carbon in ketoses). For D-sugars, α is "down" and β is "up."*

4. The last carbon is always "up" in D-sugars. Draw that one next.

5. Imagine the Fischer projection "falling over" to the right. The OH groups that were on the left in the Fischer projection will be "up" in the Haworth structure. Remember that the OH on the carbon farthest from the C=O reacted to form the ring, so that carbon will not have an OH attached to it. In other words, the oxygen within the ring had been an OH, and the anomeric OH group had been the carbonyl oxygen.

6. Double-check your work by comparing the numbered carbons on both structures.

7. **PRACTICE.** Using the Fischer projections in Figure 7.3 on p. 230, practice drawing Haworth projections of the aldohexoses. The correct Haworth projections for all of the aldohexoses are at the end of this Study Guide chapter (just before the Solutions to Review Questions).

CONFORMATIONAL (CHAIR) STRUCTURES: THE ORGANIC CHEMISTRY CONNECTION

In organic, you learned that the most stable conformation of a six-membered ring is the chair conformation, and the equatorial positions are more stable than the axial positions. It should come as no surprise, then, that the monosaccharide with all of its substituents in the equatorial positions is β-D-glucose, the most common monosaccharide for both energy storage and as a building block for structural materials, namely, cellulose.

β-D-glucose

In this structure, the axial positions were omitted for clarity.

Remember that the axial positions were always "up" or "down" and the equatorial positions were labeled relative to the axial positions. In D–hexoses, carbon 6 is always drawn "up" in the Haworth projection. So, to draw the Haworth projection of β-D-glucose, first draw the 6-membered ring skeleton with the oxygen in the proper position. Then, add carbon 6 with its OH. The rest of the OH's can then be drawn in following the up-down-up-down-up pattern of the substituents in their equatorial positions. (Compare this chair with the Haworth projections on the previous page.)

REACTIONS OF MONOSACCHARIDES

OXIDATION OF ALDOSES TO:

    ALDONIC ACID: aldehyde to carboxylic acid

    URONIC ACID: terminal –CH_2OH to carboxylic acid

    ALDARIC ACID: BOTH aldehyde and terminal –CH_2OH to carboxylic acids
        (Note: To help remember this: "I'll dare Rick to oxidize BOTH ends!" I know, it's a stretch, but hey, it's something.)

    LACTONES can be formed from aldonic and uronic acids.

    REDUCING SUGAR – can be oxidized by Benedict's reagent (a weak oxidizing agent) only if it can exist in its open-chain (aldehyde) form.

REDUCTION of aldehyde and ketone groups to alcohols = **ALDITOLS**

ISOMERIZATION via an enediol intermediate
    Aldose-ketose interconversion
    Epimerization

ESTERIFICATION forms of phosphate and sulfate esters

Phosphate esters – important in metabolism; convert –OH, a terrible leaving group, into a good leaving group for nucleophilic substitution

Sulfate esters – in proteoglycan components of connective tissue; bind large amounts of water and small ions due to the many negative charges on sulfates; form sulfate bridges between carbohydrate chains

GLYCOSIDE FORMATION: cyclic hemiacetal to acetal; cyclic hemiketal to ketal

Glycosidic linkage, glycoside (fructoside vs. glucoside, furanoside vs. pyranoside)

The glycosidic linkage locks the ring in its cyclic form, so it can't oxidize or mutarotate. (So, β-methyl glucoside is not a reducing sugar.)

AGLYCONES – noncarbohydrate components of glycosides

Formation of **DISACCHARIDES, POLYSACCHARIDES**

IMPORTANT MONOSACCHARIDES

GLUCOSE – primary fuel for living cells, preferred energy source of brain cells and cells with few or no mitochondria, building block for cellulose, starch, glycogen

FRUCTOSE – a ketohexose isomer of glucose; fruit sugar; used as a sweetening agent (twice as sweet as sucrose)

GALACTOSE – an epimer of glucose; needed to synthesize lactose, glycolipids, certain phospholipids, proteoglycans, glycoproteins; can be made from glucose

In galactosemia, an enzyme needed to metabolize galactose is missing.

MONOSACCHARIDE DERIVATIVES

URONIC ACIDS (terminal –CH_2OH group is oxidized to a carboxylic acid)

D-glucuronic acid – combines with molecules to improve their water solubility (and help to remove waste products from the body)

L-iduronic acid = epimer of D-glucuronic acid; both are in components of connective tissues

AMINO SUGARS: AMINO GROUP REPLACES AN –OH GROUP (USUALLY ON CARBON 2)

Common in complex carbohydrate molecules that are attached to cellular proteins and lipids

Most common amino sugars (in animals): D-glucosamine, D-galactosamine

Acetylated amino sugars: *N*-acetylglucosamine, *N*-acetylneuraminic acid, sialic acids

DEOXYSUGARS: –H REPLACES AN –OH

Fucose – formed from D-mannose; found in glycoproteins (example: ABO blood group determinants on the surface of red blood cells)

2-Deoxy-D-ribose is the pentose sugar component in DNA.

7.2 DISACCHARIDES

DESIGNATION OF GLYCOSIDIC LINKAGES

Identify anomeric hydroxyl group (α or β) and the carbons that are linked. Example: α(1,4) means that carbon 1 in the α position of one monosaccharide is linked to carbon 4 of another monosaccharide.

DIGESTION

Digestion occurs in the small intestine. If enzymes are deficient, then in the large intestine, they draw in water (causing diarrhea) and/or are fermented by bacteria producing gas (causing bloating and cramps). Example: lactose intolerance caused by reduced lactase synthesis after childhood

REDUCING SUGARS

If one of the rings can convert to its open-chain form to regenerate the aldehyde, then the sugar is a reducing sugar. Lactose, maltose, and cellobiose are reducing sugars.

LACTOSE: galactose $\beta(1,4)$ glucose, found in milk[2]

MALTOSE: glucose $\alpha(1,4)$ glucose, degradation product of starch, malt sugar

CELLOBIOSE: glucose $\beta(1,4)$ glucose, degradation product of cellulose

SUCROSE: glucose $\alpha,\beta(1,2)$ fructose, energy source produced in plants. Since the glycosidic bond links both anomeric carbons, sucrose is a nonreducing sugar.

7.3 POLYSACCHARIDES: ENERGY STORAGE OR STRUCTURAL MATERIALS

Hundreds to thousands of sugar units; linear (unbranched) or branched

HOMOGLYCANS – MADE FROM ONLY ONE TYPE OF MONOSACCHARIDE
GLYCOGEN and **STARCH**: compact structures are ideal for energy storage.

GLYCOGEN: CARBOHYDRATE STORAGE IN VERTEBRATES

- more branches than amylopectin: as many as one every 4 at the core of the molecule, and one every 8-12 in the outer regions

- The ends of these many branches are non-reducing – since enzymes work on non-reducing ends, this structure provides great access to enzymes so that glucose can be rapidly mobilized from glycogen.

STARCH: CARBOHYDRATE STORAGE IN PLANTS

AMYLOSE forms long, tight left-handed helices, contains several thousand glucose residues, and contains only one reducing end.

AMYLOPECTIN contains a few thousand to 106 glucose residues, and its branches prevent helix formation

Iodine test: iodine inserts into helices of amylose and results in blue color

Digestion of starch (initial products are maltose, maltotriose, α-limit dextrins)
1. Mouth: salivary enzyme α-amylase
2. Small intestine: pancreatic α-amylase, other enzymes to convert to glucose
3. Glucose: from small intestine to bloodstream, to liver, then to rest of body

CELLULOSE: THE MOST ABUNDANT ORGANIC SUBSTANCE ON EARTH

Great strength comes from hydrogen bonding between extended linear cellulose chains.

---

[2] Lactose intolerance is rather normal among adults in the general population. Adults who drink a lot of milk (especially in the US) are the exception.

Microfibrils = parallel pairs of cellulose molecules (with up to 12,000 glucose residues) held together by hydrogen bonding; bundles of microfibrils contain about 40 pairs

Digestion: only by microorganisms that have cellulase; animals such as termites and cows have these microorganisms in their digestive tracts; if not digested, it's still important as dietary fiber

CHITIN: *N*-ACETYL GLUCOSAMINE

Strong hydrogen bonds between chitin chains; Chitin forms microfibrils

Types of chitin structures:

α-chitin: antiparallel bundles, most stable and common

β-chitin: parallel bundles

γ-chitin: mixture of parallel and antiparallel

β- and γ- are more flexible than α-chitin

SUMMARY OF HOMOGLYCANS

HOMOGLYCAN (OVERALL SHAPE)		SUGAR UNITS, GLYCOSIDIC BONDS	FUNCTION	FOUND IN
GLYCOGEN (helical, branched)		α-D-glucose α(1,4) with α(1,6) branches	energy storage	vertebrates (mostly in liver and muscle cells)
STARCH	**AMYLOSE** (left-handed helices)	α-D-glucose α(1,4), unbranched	energy storage	plants
	AMYLOPECTIN (branched every 20-25 residues)	α-D-glucose α(1,4) with α(1,6) branches		
CELLULOSE (extended linear chains with hydrogen bonding between chains)		β-D-glucose β(1,4), unbranched	structural, forms microfibrils and bundles	plants (both primary and secondary cell walls)
CHITIN		N-acetyl glucosamine β-(1,4), unbranched	structural	arthropod exoskeletons, fungi cell walls

HETEROGLYCANS – MADE FROM TWO OR MORE TYPES OF MONOSACCHARIDES

MAJOR CLASSES:

N-LINKED: attached to polypeptides by an N-glycosidic bond with the side chain amide group of Asn
3 types of Asn-linked oligosaccharides: high mannose, hybrid, complex

O-LINKED: attached to polypeptides by the –OH of Ser or Thr, or the –OH of membrane lipids

GLYCOSAMINOGLYCANS (GAGS) – PRINCIPAL COMPONENTS OF PROTEOGLYCANS

Linear; composed of disaccharide repeat units that contain a hexuronic acid (except keratan sulfate – contains galactose)

Classified according to: sugar residues, glycosidic linkages, presence and location of sulfate groups

FIVE CLASSES: hyaluronic acid   chondroitin sulfate

dermatan sulfate   heparin and heparan sulfate

keratan sulfate   (see table 7.1 on page 249)

GAG chains have *many* negative charges that repel each other and attract large volumes of water, making its final volume about 1000 times larger!

MUREIN (PEPTIDOGLYCAN)

Major structural component of bacterial cell walls, supplies strength and rigidity

Contains N-acetyl glucosamine (NAG), N-acetyl muramic acid (NAM), and several different amino acids

Three basic components of murein (peptidoglycan)

1. backbone: NAG-NAM disaccharide repeat units linked by β(1,4) glycosidic bonds
2. parallel tetrapeptide chains; each chain attached to N-acetyl muramic acid
3. peptide cross-bridges link the tetrapeptide chains of adjacent molecules

7.4 GLYCOCONJUGATES:

COVALENT LINKAGES OF CARBOHYDRATES TO PROTEINS AND LIPIDS

GLYCOLIPIDS = Oligosaccharide-containing lipids located on the outer surface of plasma membranes (See Chapter 11.)

PROTEOGLYCANS = GAG CHAINS LINKED TO CORE PROTEINS *(see Fig. 7.37 on p. 251)*

N- and *O*-glycosidic linkages, *very* high carbohydrate content

Located in extracellular matrix (intercellular material) of tissues

Large numbers of GAGs trap large volumes of water and cations

Contribute support and elasticity to tissues

Example: cartilage – strength, flexibility, resilience; part of meshwork (with matrix proteins such as collagen, fibronectin, laminin) that supports multicellular tissues

Genetic diseases associated with proteoglycan metabolism: mucopolysaccharidoses (example: Hurler's syndrome – dermatan sulfate accumulates)

GLYCOPROTEINS = PROTEINS COVALENTLY LINKED TO CARBOHYDRATE

N-glycosidic linkage to asparagine or *O*-glycosidic linkage to serine or threonine

Carbohydrate content varies (1% to more than 85% by weight)

ASPARAGINE-LINKED CARBOHYDRATE

N-glycosidic linkage between *N*-acetylglucosamine (GlcNAc) and asparagine

Core is constructed on a membrane-bound lipid molecule

HIGH-MANNOSE TYPE	GlcNAc and mannose
COMPLEX TYPE	GlcNAc and mannose (may contain fucose, galactose, sialic acid)
HYBRID TYPE	features of both high-mannose and complex types

MUCIN-TYPE CARBOHYDRATE

O-glycosidic linkage, most common is between N-acetylgalactosamine (GalNAc) and the hydroxyl group of Ser or Thr

Carbohydrate components vary in size and structure. (examples: Gal-β(1,3)-GalNAc, disaccharide found in antifreeze glycoprotein of Antarctic fish; complex oligosaccharides of blood groups such as the ABO system)

GLYCOPROTEIN FUNCTIONS AND THE ROLE OF THE CARBOHYDRATE COMPONENT

EXAMPLES OF GLYCOPROTEIN FUNCTIONS:
- Metal-transport proteins transferrin and ceruloplasmin
- Blood-clotting factors
- Complement (proteins involved in cell destruction during immune reactions)
- Hormones (example: follicle stimulating hormone (FSH) – stimulates development of eggs and sperm)
- Enzymes (example: ribonuclease (RNase) degrades ribonucleic acid)
- Integral membrane proteins {examples: Na^+-K^+-ATPase – ion pump in the plasma membrane of animal cells; major histocompatibility antigens (cell surface markers used to cross-match organ donors and recipients)}

GLYCAN COMPONENT STABILIZES PROTEIN MOLECULES:
Protects from denaturation (example: bovine RNase A vs. RNase B)

Increases resistance to proteolysis (carbohydrates on the protein's surface may shield the peptide chains from enzymes)

AFFECTS BIOLOGICAL FUNCTION
Saliva: high viscosity of salivary mucins is due to high content of sialic acid residues

Antifreeze glycoproteins in Antarctic fish retard the growth of ice crystals by hydrogen bonding with water molecules

COMPLEX RECOGNITION PHENOMENA
CELL-MOLECULE INTERACTIONS
Insulin receptor – binds to insulin and facilitates transport of glucose into cells, in part by recruiting glucose transporters to the plasma membrane

Glucose transporter – transports glucose into cells

CELL-VIRUS INTERACTIONS
gp120 (target cell binding glycoprotein of HIV) attaches to CD4 receptor on surface of several human cell types (removing oligosaccharides from gp120 reduces its binding to CD4 receptor)

CELL-CELL INTERACTIONS
Cell structure glycoproteins – components of glycocalyx (cell coat), important in cellular adhesion; CELL ADHESION MOLECULES (CAMs)

7.5 THE SUGAR CODE

CONTRIBUTIONS OF CARBOHYDRATES TO INFORMATION TRANSFER:

- posttranslational modification of proteins via glycosylation = after a protein is synthesized (translation), its structure is modified with the addition of carbohydrates

- oligosaccharides have a many more structural possibilities than peptides in that oligosaccharides may be branched, may be linked via an α- or β-glycosidic bond, or may be linked at different positions, e.g., a $(1 \rightarrow 3)$ vs. $(1 \rightarrow 4)$

- relatively inflexible chains allow for more specificity in ligand binding

LECTINS: TRANSLATORS OF THE SUGAR CODE

LECTINS are carbohydrate binding proteins (CBPs) that:

- are not antibodies and have no enzymatic activity
- consist of 2 or 4 subunits
- possess recognition domains that bind to specific carbohydrate groups via hydrogen bonds, van der Waals forces, and hydrophobic interactions

EXAMPLES OF BIOLOGICAL PROCESSES – CELL-CELL INTERACTIONS – THAT INVOLVE LECTIN BINDING:

INFECTIONS BY MICROORGANISMS
Bacterial lectins attach the bacteria cell to the host cell by binding to oligosaccharides on the cell's surface (example: gastritis and stomach ulcers)

MECHANISMS OF MANY TOXINS
Lectin-ligand binding initiates endocytosis into the host cell (example: cholera)

PHYSIOLOGICAL PROCESSES SUCH AS LEUKOCYTE ROLLING
SELECTINS are a family of lectins that act as cell adhesion molecules that slow the movement of neutrophils through blood vessels
INTEGRIN is a lectin that cause neutrophils to stop rolling (so they can migrate to the location where they're needed)

THE GLYCOME = THE TOTAL SET OF SUGARS AND GLYCANS THAT A CELL OR ORGANISM PRODUCES
Unlike proteins, there's no template for glycan synthesis – they're produced step-by-step on an assembly line in the ER and Golgi complex.

MICROHETEROGENICITY is the production of **GLYCOFORMS** – slightly different forms of glycan components of glycoproteins – possibly for cell- or tissue-specific purposes

Haworth Projections for α-D-Aldohexoses *(Pyranose Forms)*

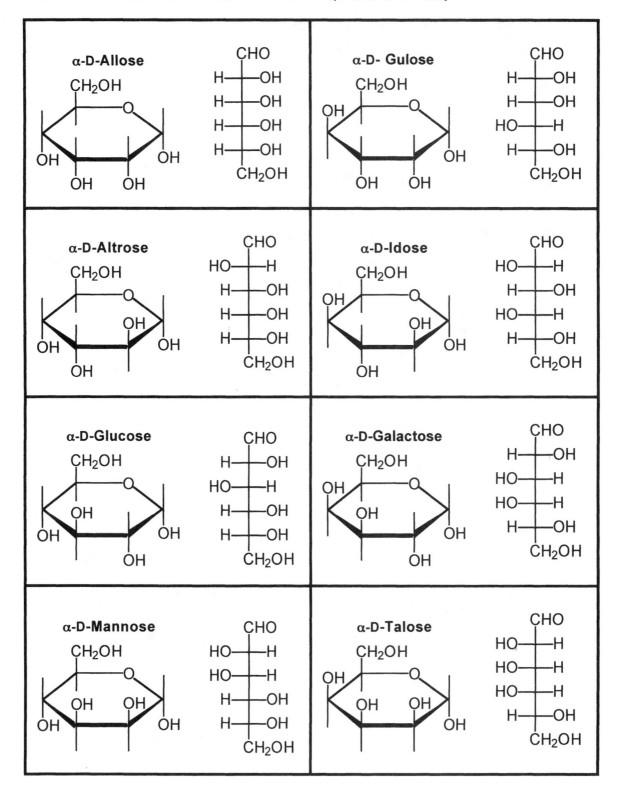

AFTER STUDYING THIS CHAPTER, YOU SHOULD BE ABLE TO:

- Convert a Fischer projection to a Haworth structure, or draw a Haworth structure given only the name of a compound.

- Identify pairs of compounds as epimers, anomers, enantiomers, diastereomers, or an aldose-ketose pair. Identify an anomer as α or β.

- Distinguish between reducing and nonreducing sugars.

- Determine the number of possible stereoisomers of a given sugar.

- Name a sugar given its structure. Include α or β and appropriate designations for any glycosidic linkages present.

- Draw the structure of a di-, tri-, or polysaccharide, given the names of the component mono- (or di-)saccharides and the type(s) of glycosidic linkages present.

- How do small differences in carbohydrate structure account for large differences in function? (For example, compare cellulose and glycogen.)

- Demonstrate an understanding of the role of glycans and glycosylation in information storage and transfer. Describe the advantages of sugars that make them well-suited for encoding information.

CHAPTER 7: SOLUTIONS TO REVIEW QUESTIONS

7.1 a. hemiacetal – one of the family of organic molecules with the general formula RR'C(OR')(OH) formed by the reaction of one molecule of alcohol with an aldehyde

   b. space-filling model – a molecular model or image that uses spheres to depict the space occupied by each atom (based upon van der Waals radii); bonds are not visible, but are implied between adjoining atoms (spheres)

   c. uronic acid – the product formed when the terminal CH_2OH group of a monosaccharide is oxidized to a carboxylic acid

   d. aldaric acid – the product formed when the aldehyde and the terminal CH_2OH group of a monosaccharide are oxidized to carboxylic acids

   e. lactone – a cyclic ester

7.2 a. Glucose and mannose are examples of epimers.

   b.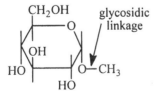

   c. Glucose is a reducing sugar.

   d. Ribose is a monosaccharide.

   e. α- and β-Glucose are anomers.

   f. Glucose and galactose are diastereomers.

7.4   a. Ribonuclease B is an example of a glycoprotein.

        b. Each proteoglycan contains glycosaminoglycans such as chondroitin sulfate and dermatan sulfate which are linked to a core protein via glycosidic linkages.

        c. Lactose is an example of a disaccharide.

        d. Heparin is a glycosaminoglycan.

7.5   a. alditol – a sugar alcohol; the product when the aldehyde or ketone group of a monosaccharide is reduced

        b. epimerization – the reversible interconversion of epimers

        c. enediol – a functional group that contains an OH group on each of the two alkenyl carbons; the intermediate formed during the isomerization reactions of monosaccharides

        d. acetal – the family of organic compounds with the general formula RCH(OR')2; formed from the reaction of a hemiacetal with an alcohol

        e. glycoside – the acetal of a sugar

7.7   a. Nonreducing, b. Nonreducing, c. Reducing, d. Nonreducing, e. Reducing

7.8   a. glycosylation – reactions catalyzed by glycosyl transferases that attach sugars to proteins or lipids

        b. glycation – nonenzymatic reactions in which reducing sugars react with nucleophilic nitrogens

        c. atherosclerosis – deposition of excess plasma cholesterol and other lipids and proteins on the inside walls of the arteries, decreasing the functional inside artery diameter

        d. AGE – advanced glycation end products – products that result upon glycation followed by further reactions, such as oxidations, rearrangements, and dehydrations

        e. deoxysugar – a sugar that lacks an OH group; for example, 2'-deoxyribose is a component of DNA

7.10

7.11 Ribulose has 2^2 or 4 possible stereoisomers. Carbons 3 and 4 in ribulose are chiral. Sedoheptulose has 2^4 or 16 isomers. Carbons 3, 4, 5, and 6 in sedoheptulose are chiral.

7.13 a. Raffinose consists of α-D-galactose with an α(1,6) linkage to glucose, which has an α,β(1,2) linkage to fructose. Its systematic name is: galactopyranosyl-α-(1,6)-glucopyranosyl-α,β(1,2)-fructofuranose.

        b. Raffinose is a nonreducing sugar.

        c. Raffinose is not capable of mutarotation.

7.14 a. Glycogen stores glucose.

        b. Glycosaminoglycans are components of proteoglycans.

   c. Glycoconjugates may serve as membrane receptors.

   d. Proteoglycans provide strength, support, and elasticity to tissue.

   e. Hormones such as FSH and enzymes such as RNAse are glycoproteins.

   f. Polysaccharides, or glycans, play important roles in the storage of carbohydrate (starch and glycogen) and the structure of plants (cellulose).

7.16   In glycoproteins carbohydrate moieties are most frequently linked to the amide nitrogen of asparagine and the hydroxyl oxygen of serine and threonine residues.

7.17   Chondroitin sulfate and proteoglycans are extensively negatively charged at physiological pH and as such are spread out, binding large amounts of water. The interwoven chains block the passage of large molecules. Smaller molecules can pass between the chains.

7.19   a. Hurler's syndrome – an autosomal recessive disorder in which an enzyme deficiency causes dermatan sulfate to accumulate, causing mental retardation, skeletal deformity, and death in early childhood

   b. heparin – a glycosaminoglycan with anticoagulant activity

   c. lectin – a carbohydrate-binding protein that is not an antibody and has no enzymatic activity

   d. *N*-glycan – a heteroglycan with a β-glycosidic bond between the core *N*-acetylglucosamine anomeric carbon and a side chain amide nitrogen of an asparagine residue in a protein

   e. O-glycan – a heteroglycan with a disaccharide core of galactosyl-β-(1,3)-N-acetylglucosamine linked to protein via an α-glycosidic bond to the OH oxygen of serine or threonine residues

7.20   Proteoglycans are extremely large molecules that contain a large number of glycosaminoglycan chains linked to a core protein. They are found primarily in extracellular fluids, where their high carbohydrate content allows them to bind large amounts of water. Glycoproteins are conjugated proteins in which the prosthetic groups are carbohydrate molecules. The carbohydrate groups stabilize the molecule through hydrogen bonding, protecting the molecule from denaturation, or shield the protein from hydrolysis. The carbohydrate groups on the glycoproteins on the surface of cells play an important role in a variety of recognition phenomena.

7.22   Amylose, glycogen, and amylopectin possess polymers of glucose residues that are linked by α-1,4-glycosidic linkages. Glycogen and amylopectin also possess branches that are connected to the α-1,4-linked chain by α-1,6-glycosidic linkages. Cellulose is an unbranched polymer of glucose residues linked by β-1,4-glycosidic linkages.

7.23   a. epimers (D-erythrose and D-threose)

   b. epimers (D-glucose and D-mannose)

   c. enantiomers (D-ribose and L-ribose)

   d. diastereomers (D-allose and D-galactose)

   e. aldose-ketose pair (D-glyceraldehyde and dihydroxyacetone)

CHAPTER 7: SOLUTIONS TO THOUGHT QUESTIONS

7.25

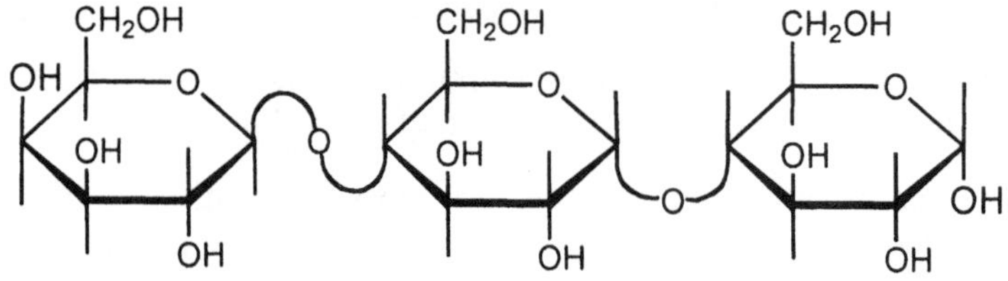

7.26 When steroids are conjugated with a uronic acid, the OH groups of the uronic acid form hydrogen bonds with the water. This structural feature increases the solubility of the conjugated steroid molecule.

7.28 Pathogenic organisms bind to the milk oligosaccharides instead of the oligosaccharides on the surface of the infant's intestinal cells, thus preventing infections.

7.29 Haworth furanose formulas for the four sugars are as follows:

a.

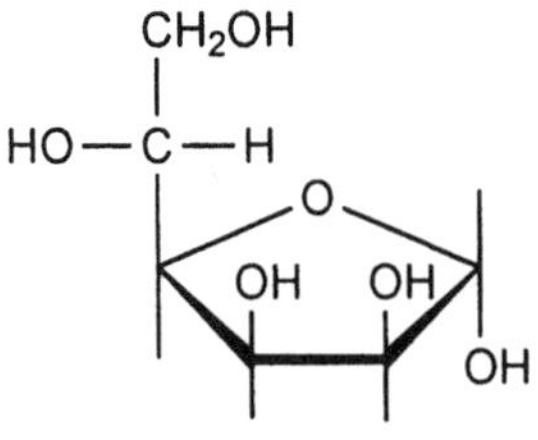

α-D-Mannofuranose 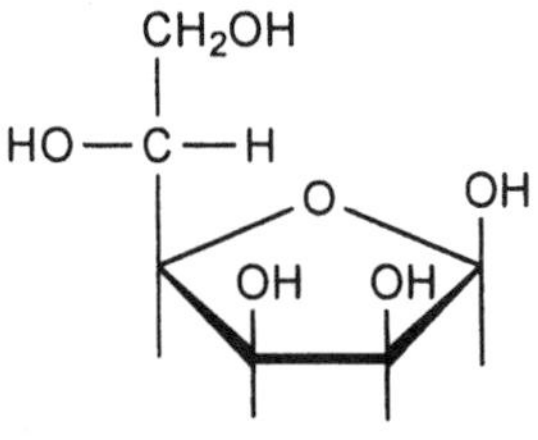   β-D-Mannofuranose

b.

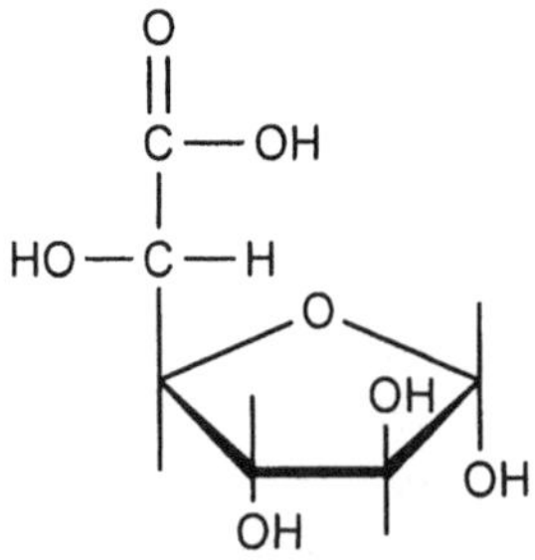

α-D-Furanoiduronic acid 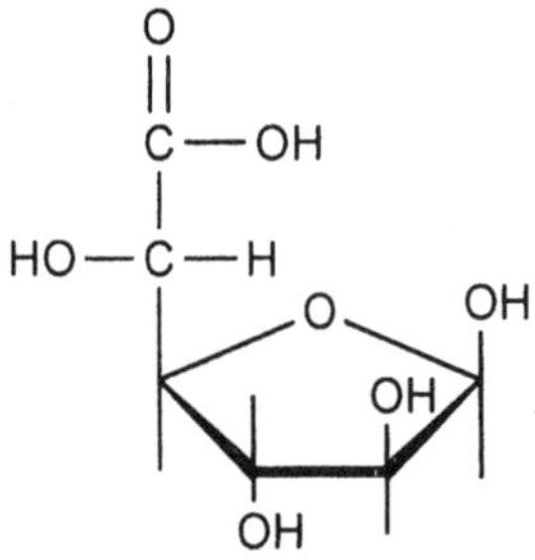   β-D-Furanoiduronic acid

c.

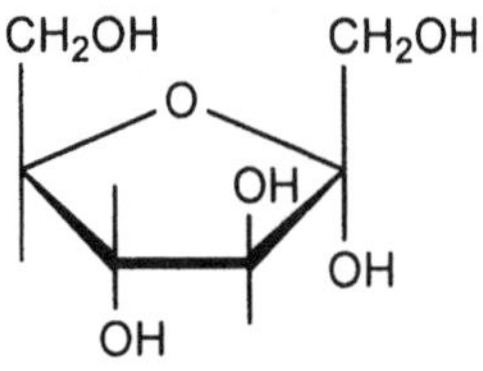

α-D-Fructofuranose 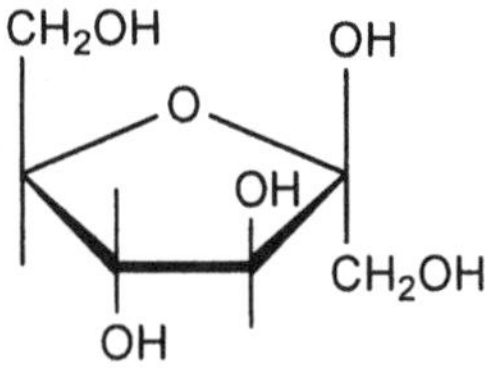   β-D-Fructofuranose

d.

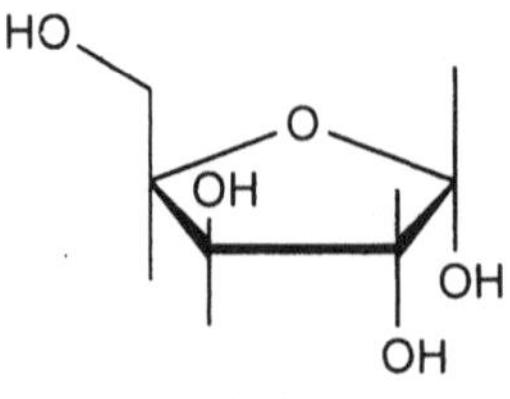

α-D-Xylofuranose 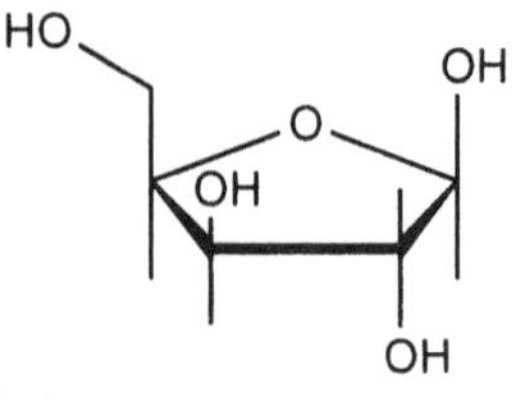   β-D-Xylofuranose

106

7.31 The names for the α-anomers in the answers to question 7.29 are as follows:

a) α-D-mannofuranose, b) α-D-furanoiduronic acid, c) α-D-fructofuranose, d) α-D-xylofuranose.

The names for the α anomers shown in the answers to question 7.30 are as follows: a) α-D-mannopyranose b) α-D-pyranoiduronic acid c) α-D-fructopyranose d) α-D-xylopyranose

7.32

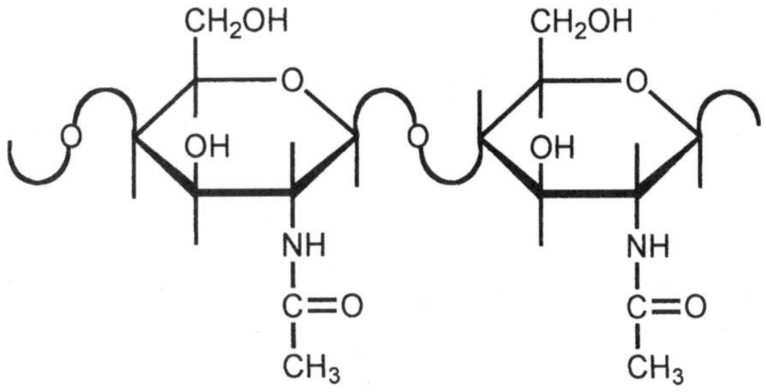

7.34 a.

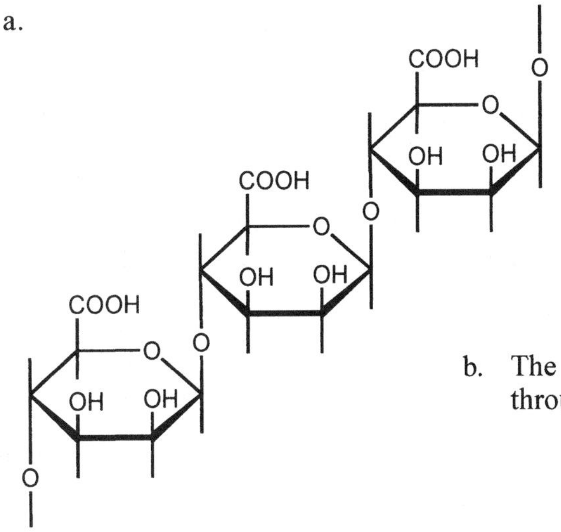

b. The polymer acts to immobilize water through extensive hydrogen bonding.

7.35 Mannuronic acid has four chiral centers, so it has a maximum number of isomers equal to 2^4 or 16.

7.37 In an individual with an AB blood type, one half of the antigens would be linked with galactose and the other half with N-acetyl-galactose amine.

7.38 Phosphate esters can form at positions 2-6 of an aldohexose because all these carbons bear alcoholic OH groups. In contrast, a phosphate at the anomeric carbon would be a mixed anhydride.

7.40 The sulfate group has several negatively charged oxygens that are capable of hydrogen bonding. Conjugation of sulfate with the hydrophobic molecule enhances solubility because of hydrogen bonding between the sulfate oxygens and water molecules in tissue fluids.

7.41 In order for the sugar to undergo mutarotation there must be a hemiacetal or hemiketal as part of the structure. Sucrose's anomeric carbons are linked in a full acetal.

7.43 Assuming a D family sugar, two structures are possible:

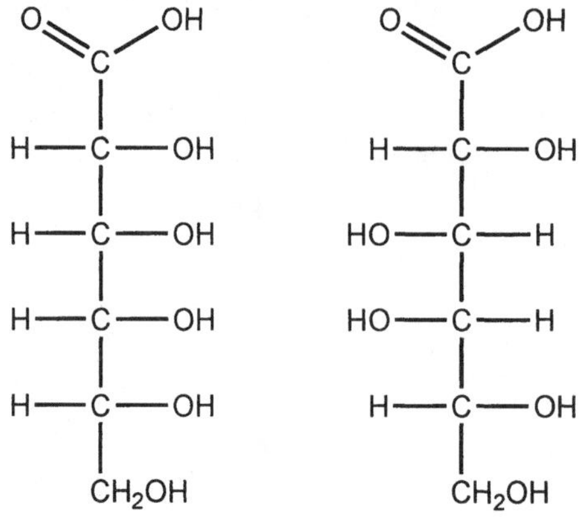

7.44 Three sugars are possible. Talose is the sugar produced by the epimerization of galactose at C2. The C4 epimer is glucose, and the C3 epimer is galactose. (Refer to Figure 7.3 to view the structures.)

7.46 Three α-anomeric ring forms of 3-ketoglucose are shown. The pyranose form is the most stable because the bond angles are less strained.

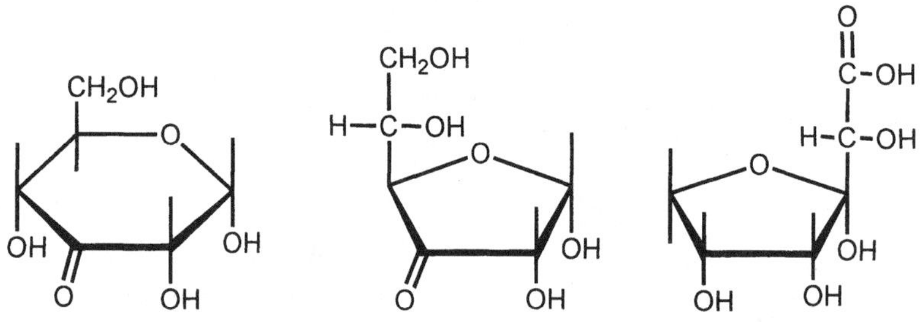

7.47 Moisture loss is prevented because of hydrogen bonding between the OH groups of sorbitol and water.

7.49 The structure of olestra is as follows:

7.50 The test for a reducing sugar requires an alkaline medium. Under these conditions, fructose converts to an aldose and gives a positive test.

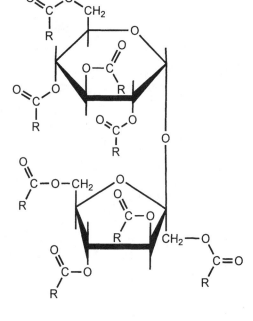

8 Carbohydrate Metabolism

Brief Outline of Key Terms and Concepts

8.1 GLYCOLYSIS
AEROBIC RESPIRATION
AEROBIC VS. ANAEROBIC ORGANISMS

THE REACTIONS OF THE GLYCOLYTIC PATHWAY
convert glucose to two pyruvate molecules. A
small amount of energy is captured in two ATP
and two NADH.
GLUCOSE SENSOR
ALDOL CLEAVAGE
PHOSPHOANHYDRIDE BOND
SUBSTRATE-LEVEL PHOSPHORYLATION
TAUTOMERS; TAUTOMERIZATION

THE FATES OF PYRUVATE
In the presence of oxygen, the cells of aerobic
organisms convert pyruvate into CO_2 and H_2O.
CITRIC ACID CYCLE
AMPHIBOLIC PATHWAY
ELECTRON TRANSPORT SYSTEM
DECARBOXYLATION
FERMENTATION
LACTIC ACID FERMENTATION

THE ENERGETICS OF GLYCOLYSIS
Enzymes of irreversible reactions:
HEXOKINASE,
PFK-1,
PYRUVATE KINASE

REGULATION OF GLYCOLYSIS
ALLOSTERIC EFFECTORS:
AMP, ATP, citrate, acetyl-CoA,
fructose-2,6-bisphosphate
fructose-1,6-bisphosphate
HORMONES: glucagon, insulin
cAMP – a second messenger

8.2 GLUCONEOGENESIS
– the synthesis of new glucose molecules from
noncarbohydrate precursors.
– occurs primarily in the liver.

GLUCONEOGENESIS REACTIONS are the reverse of
glycolysis except for three reactions that
bypass irreversible glycolysis reactions.
biotin   malate shuttle
substrate cycle   flux control

GLUCONEOGENESIS SUBSTRATES
Cori cycle   glucose-alanine cycle
glucogenic

GLUCONEOGENESIS REGULATION
– substrate availability
– allosteric effectors: ATP, AMP, fructose-2,6-
bisphosphate, acetyl-CoA
– hormones: insulin, glucagon, cortisol

8.3 THE PENTOSE PHOSPHATE PATHWAY
produces NADPH, ribose-5-phosphate, fructose-
6-phosphate and glyceraldehyde-3-phosphate.
HEXOSE MONOPHOSPHATE SHUNT
ANTIOXIDANT
TPP (THIAMINE PYROPHOSPHATE)

8.4 METABOLISM OF OTHER IMPORTANT SUGARS
FRUCTOSE METABOLISM bypasses two regulatory
enzymes: hexokinase and PFK-1.

8.5 GLYCOGEN METABOLISM

GLYCOGENESIS
glycogen synthase
glycogen branching enzyme
glycogenin

GLYCOGENOLYSIS
glycogen phosphorylase
debranching enzyme
LIMIT DEXTRIN

REGULATION OF GLYCOGEN METABOLISM
– several allosteric regulators
– hormones: glucagon, insulin, and epinephrine
– covalent modification:
forms of glycogen phosphorylase:
phosphorylase a – active form
phosphorylase b – inactive form

TERMS FROM IN-CHAPTER QUESTIONS:
glucose tolerance   Pasteur effect
malignant hyperthermia   hypoglycemia
von Gierke's disease   Cori's disease
hepatomegaly

STUDY HINTS/STRATEGY

Studying metabolism brings together what you've learned about structures of biomolecules and how energy flows in living systems.

An important tip in studying metabolism is to key in on the ultimate goal of the pathway and how each reaction brings you closer to the desired end product.

Begin by visualizing the structure of the starting material and follow the changes that occur until you get to the final product. This simplifies remembering the sequence or order of reactions. Focus on the structure of the intermediates. At each step, note any changes that occur to the molecule to understand the kind of reaction that's occurring. And, remember that each enzyme's name clues the type of reaction it catalyzes. For example, a kinase transfers phosphoryl groups.

Know yourself and your learning styles, and tap into your strengths. If you learn best by reading, create lists to study from. Visual learners will do well to create diagrams of the pathways. Those who learn best by hearing will want to find a study partner to talk through the pathways. Kinesthetic learners learn by doing, and practicing writing out the pathways might be the best bet. Most of us are a combination of the four. Try each style of learning - you might be surprised by which method clicks the best. Studying metabolism lends itself to all four learning styles, so it's a great subject to dive into. (It also gives meaning to everything we've learned so far!)[1]

Example: Glycolysis cleaves glucose to form two 3-carbon molecules. In this pathway, a kinase adds a *second* phosphoryl group to fructose-6-phosphate. *Why?* After cleavage, both halves will have a phosphoryl group that traps each half in the cell. Later, those phosphates will be transferred to ADP to form ATP.

PATHWAY	GENERAL FUNCTION	EACH PATHWAY IS ACTIVATED WHEN:
GLYCOLYSIS	glucose → 2 pyruvate makes 2 ATP, 2 NADH	…energy is needed: either anaerobic (pyruvate forms lactate) or aerobic (pyruvate enters the citric acid cycle)
GLUCONEOGENESIS	pyruvate → glucose requires 4 ATP, 2 GTP	…the liver needs to raise blood sugar levels and liver glycogen is depleted
GLYCOGENOLYSIS	glycogen → glucose	…muscles need glucose for energy or the liver needs to raise blood sugar levels
GLYCOGENESIS	glucose → glycogen requires 1 UTP/glucose	…glucose is in excess
PENTOSE PHOSPHATE PATHWAY	glucose-6-P → ribose-5-P + other sugars makes 2 NADPH	…NADPH is needed for lipid synthesis …ribose is needed for nucleotide synthesis

---

[1] Visual, Aural, Read/write, and Kinesthetic modes of learning are described on the VARK web page, http://vark-learn.com. VARK is a questionnaire that can be used to discover your learning preferences. If this link is outdated, try a general search on VARK at www.google.com. Copyright for VARK is held by Neil D. Fleming, Christchurch, New Zealand and Charles C. Bonwell, Green Mountain, Colorado, USA.

8.1 GLYCOLYSIS: THE OLDEST AND SIMPLEST WAY TO PRODUCE ENERGY

- Glycolysis provides energy (2 ATP), intermediates (2 pyruvate), and reducing power (2 NADH).
- It splits one glucose (with 6 carbons) into two pyruvates (with three carbons each).
- It doesn't need O_2 (it's anaerobic).
- It's **AMPHIBOLIC** - it functions in both anabolic and catabolic processes.
- It's activated when the cell needs energy, and inhibited when there's plenty of ATP.

THE OVERALL REACTION OF GLYCOLYSIS: ΔG = –103.8 KJ/MOL

$$D\text{-Glucose} + 2ADP + 2P_i + 2NAD^+ \rightarrow 2\text{Pyruvate} + 2ATP + 2NADH + 2H^+ + 2H_2O$$

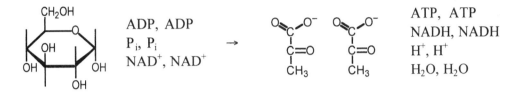

THE REACTIONS OF THE GLYCOLYTIC PATHWAY

STAGE ONE:
Glucose is phosphorylated and cleaved to form 2 glyceraldehyde-3-phosphate.

STAGE TWO:
Energy recapture stage (Oxidation-Reduction-Phosphorylation)
(SEE THE FOLLOWING TWO PAGES FOR REACTION DETAILS.)

THE FATES OF PYRUVATE

IF O_2 IS PRESENT:
pyruvate → acetyl-CoA → citric acid cycle ($2CO_2$, NADH, $FADH_2$)
NADH and $FADH_2$ → may transfer its electrons to H_2O via the electron transport system (ETS), and the energy released is used to make ATP

IF O_2 IS ABSENT:
The ETS and the citric acid cycle stop. The cell must rely on glycolysis to get the energy (ATP) it needs. To net 2 ATP, glycolysis needs a steady supply of NAD^+, which must be regenerated from NADH. Otherwise, the cell's NAD^+ supply would become depleted and glycolysis would stop.☹

IN MUSCLE CELLS AND SOME BACTERIA: HOMOLACTIC FERMENTATION

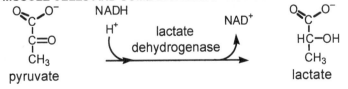

IN YEAST CELLS: ALCOHOLIC FERMENTATION

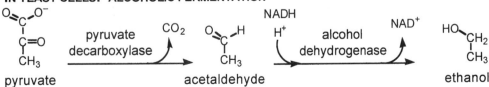

(Alcoholic fermentation is a decarboxylation followed by a reduction.)

111

GLYCOLYSIS

STAGE ONE: GLUCOSE IS PHOSPHORYLATED AND CLEAVED TO FORM TWO G-3-P

***Glycolysis occurs only in the cytoplasm.
The three irreversible reactions –
hexokinase, PFK-1, and pyruvate kinase
(in stage two) – are highly regulated.***

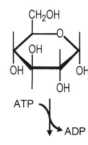

glucose

**1. HEXOKINASE, Mg^{2+} *or*
GLUCOKINASE(GK)(primarily in liver cells)**

Now this molecule is charged so that it
   cannot pass through the cell membrane
Investment of one ATP
IRREVERSIBLE!
Inhibited by ATP and glucose-6-phosphate
   (except for glucokinase)

ATP → ADP   **IRREVERSIBLE!**

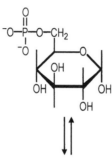

glucose-6-phosphate

---

2. PHOSPHOGLUCOSE ISOMERASE

Now C-1 can be phosphorylated
Enediol intermediate

fructose-6-phosphate

ATP → ADP   **IRREVERSIBLE!**

**3. PHOSPHOFRUCTOKINASE-1, Mg^{2+}
(PFK-1)**

Now both halves will have a phosphate
   after cleavage
Phosphates will make ATP later
IRREVERSIBLE!
Activated by fructose-2,6-bisphosphate
Inhibited by ATP and citrate

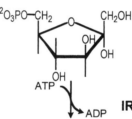

fructose-1,6-
bisphosphate

---

4. ALDOLASE

Aldol cleavage
Reverse aldol condensation

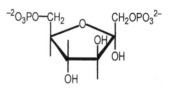

dihydroxyacetone
phosphate
(DHAP)

glyceraldehyde-3-
phosphate

---

5. TRIOSE PHOSPHATE ISOMERASE

Now the DHAP half won't be wasted

GLYCOLYSIS

STAGE TWO: ENERGY RECAPTURE STAGE (OXIDATION-REDUCTION-PHOSPHORYLATION)

*From this point on, there are **two** of each molecule below.*

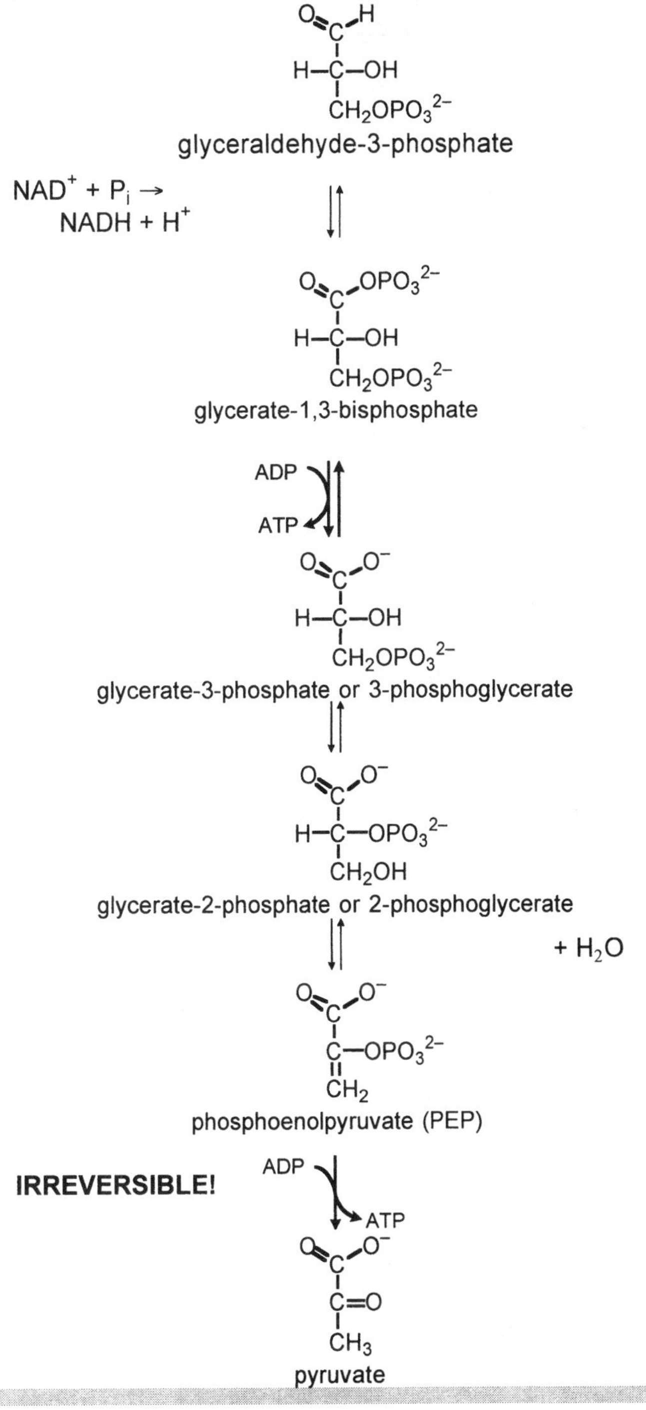

O=C–H
|
H–C–OH
|
$CH_2OPO_3^{2-}$

glyceraldehyde-3-phosphate

6. GLYCERALDEHYDE-3-PHOSPHATE DEHYDROGENASE

$NAD^+ + P_i \rightarrow$
$NADH + H^+$

Mechanism: Figure 8.5 (text)
Oxidation of aldehyde with reduction
  of NAD^+ to NADH
Phosphorylation without ATP
Product has a high phosphoryl group
  transfer potential

O=C–OPO_3^{2-}
|
H–C–OH
|
$CH_2OPO_3^{2-}$

glycerate-1,3-bisphosphate

7. PHOSPHOGLYCERATE KINASE, Mg^{2+}

ADP
ATP

Recovers earlier ATP investment
Substrate-level phosphorylation:
  transfer of phosphoryl group to
  ADP

O=C–O^-
|
H–C–OH
|
$CH_2OPO_3^{2-}$

glycerate-3-phosphate or 3-phosphoglycerate

8. PHOSPHOGLYCERATE MUTASE

Isomerization sets up PEP formation
Adds and removes phosphate

O=C–O^-
|
H–C–OPO_3^{2-}
|
CH_2OH

glycerate-2-phosphate or 2-phosphoglycerate

9. ENOLASE

High phosphoryl group transfer
  potential because without the
  phosphate, it's an enol
Dehydration reaction

+ H_2O

O=C–O^-
|
C–OPO_3^{2-}
‖
CH_2

phosphoenolpyruvate (PEP)

10. PYRUVATE KINASE, Mg^{2+}

IRREVERSIBLE!

ADP
ATP

Enol formed tautomerizes to the
  stable keto form
Net two ATP formed

IRREVERSIBLE!

O=C–O^-
|
C=O
|
CH_3

pyruvate

What happens next depends upon: whether or not oxygen is present,
the type of cell, and
the energy needs of the cell.

THE ENERGETICS OF GLYCOLYSIS

The three irreversible reactions have a significantly negative ΔG value. The other reactions have a ΔG value close to zero, and are reversible.

THE REGULATION OF GLYCOLYSIS

To regulate glycolysis, regulate the enzymes that catalyze the three irreversible reactions:

> hexokinase or glucokinase (GK)
> phosphofructokinase-1 (PFK-1)
> pyruvate kinase

For further details, see the comparison of glycolysis regulation and gluconeogenesis regulation later in this chapter.

8.2 GLUCONEOGENESIS: THE GENESIS OF NEW GLUCOSE

- is an *anabolic* pathway used to synthesize glucose from amino acids or lactate.
- requires energy: the hydrolysis of 4 ATP and 2 GTP.
- is the reverse of glycolysis except for reactions that bypass the three irreversible reactions of glycolysis
- is important because the brain and red blood cells need a steady supply of glucose (use glucose as their primary energy source).
- occurs primarily in the liver after glycogen is depleted

THE OVERALL REACTION OF GLUCONEOGENESIS: PYRUVIC ACID TO GLUCOSE

$$2\ C_3H_4O_3 + 4\ ATP + 2\ GTP + 2\ NADH + 2\ H^+ + 6\ H_2O \rightarrow$$
$$C_6H_{12}O_6 + 4\ ADP + 2\ GDP + 2\ NAD^+ + 6\ HPO_4^{2-} + 6\ H^+$$

GLUCONEOGENESIS REACTIONS (SEE THE FOLLOWING NEXT PAGE FOR REACTION DETAILS)

GLUCONEOGENESIS SUBSTRATES

GLUCOGENIC AMINO ACIDS

ALANINE CYCLE:

In exercising muscle, alanine transaminase transfers $-NH_3^+$ from glutamate to pyruvate:

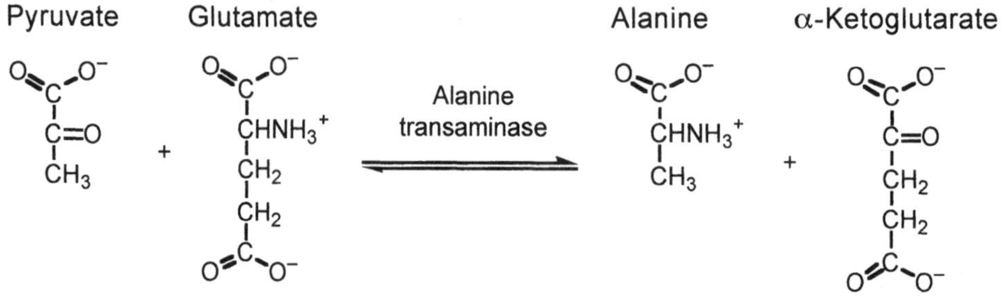

Alanine is transported to the liver, where it's converted back to pyruvate again via alanine transaminase in the reverse reaction.

(For lactate and glycerol, see the page that follows the gluconeogenesis reactions.)

REACTIONS OF GLUCONEOGENESIS THAT ARE NOT THE REVERSE OF GLYCOLYSIS (AND SO USE DIFFERENT ENZYMES)

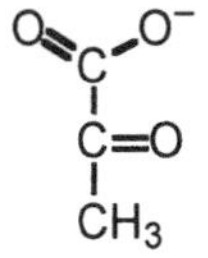

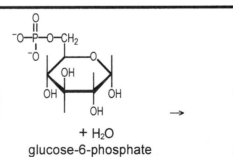

pyruvate

| Pyruvate carboxylase, biotin |

$ATP + H_2O \rightarrow ADP + P_i + H^+$

Mitochondria
Biotin is a coenzyme, CO_2 carrier
Certain amino acids can be used to make OAA.

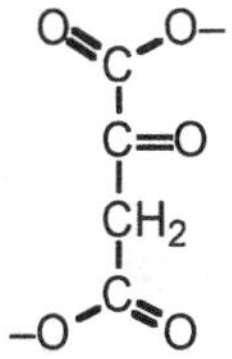

oxaloacetate (OAA)

| PEP carboxykinase |

$GTP \rightarrow GDP$

Mitochondria (some species), but OAA can't cross the membrane, so –
Needs malate shuttle:
OAA converts to malate, which is transported across the membrane, then converted back to OAA

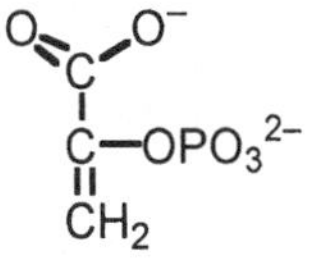

PEP

$+ CO_2$

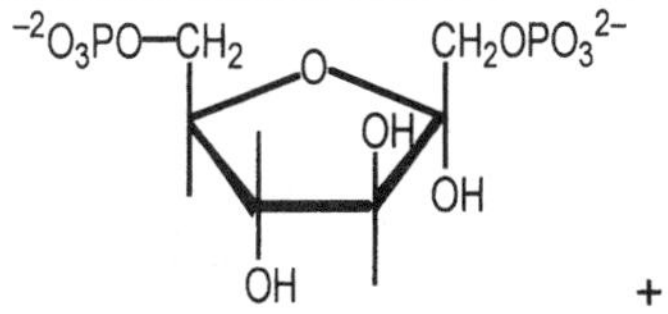

fructose-1,6-bisphosphate

$+ H_2O$

| Fructose-1,6-bisphosphatase |

Note that ATP isn't regenerated
Allosteric regulation:
    Activated by citrate
    Inhibited by fructose-2,6-bis-
      phosphate and AMP

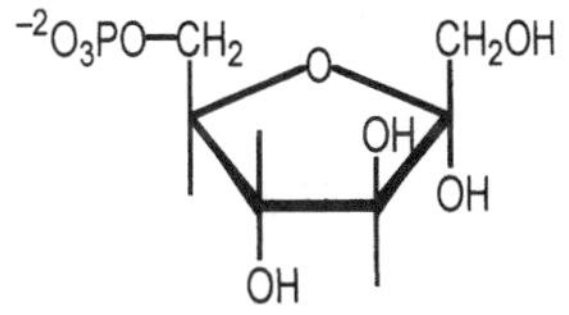

fructose-6-phosphate

$+ P_i$

| Glucose-6-phosphatase |

Only in the liver and the kidney; takes place in the endoplasmic reticulum
Glucose produced is released into the blood

$+ H_2O$
glucose-6-phosphate

$\rightarrow$

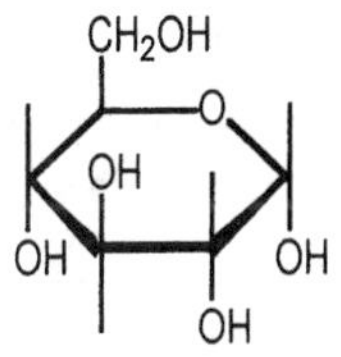

$+ P_i$
glucose

LACTATE: CORI CYCLE:

Exercising muscle: ....glucose to pyruvate to lactate (glycolysis/lactic acid fermentation); lactate enters bloodstream

Transport to liver: ....lactate to pyruvate to glucose (gluconeogenesis)

GLYCEROL → GLYCEROL-3-PHOSPHATE → DIHYDROXYACETONE PHOSPHATE (DHAP)

This occurs only in the liver, when cytoplasmic $[NAD^+]$ is high; uses an ATP but creates an NADH (which can produce ATP via the electron transport system). Glycerol is a product of fat catabolism.

SUMMARY OF REGULATION OF GLYCOLYSIS AND GLUCONEOGENESIS

COMPARTMENTATION:

Glycolysis occurs only in the cytoplasm, and several gluconeogenesis reactions occur in the mitochondria. Glycerol kinase is found only in the liver.

REGULATION BASED UPON THE NEEDS OF THE CELL:

Keep in mind the ultimate goal of each pathway, and their regulation will make sense. For example, glycolysis produces energy, reducing power, and intermediates. So, when the cell needs energy (when reserves are low), glycolysis is activated and gluconeogenesis is inhibited. When the cell has enough energy, glycolysis is inhibited and gluconeogenesis is activated. *Substrate cycle:* paired reactions that are coordinately regulated

THE LIVER MODULATES BLOOD SUGAR LEVELS

When blood sugar is low, the liver releases glucose into the bloodstream, first by degrading glycogen, then by gluconeogenesis (in response to glucagon).

When blood sugar is high, insulin inhibits gluconeogenesis and activates glycolysis in the liver. Also, hexokinase D in the liver isn't inhibited by glucose-6-phosphate, so it lets the liver remove glucose from the blood to store as glycogen.

REGULATION BY COVALENT MODIFICATION

PFK-2 and fructose-2,6-bisphosphatase is actually one enzyme that's bifunctional: without its phosphate, it's PFK-2, and with its phosphate, it's fructose-2,6-bisphosphatase.

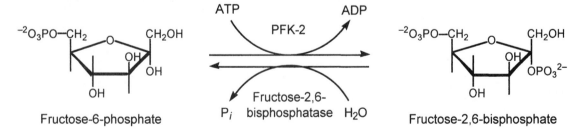

Fructose-6-phosphate   Fructose-2,6-bisphosphate

When blood glucose is high, insulin activates PFK-2, which increases fructose-2,6-bisphosphate levels, which:

*   activates PFK-1 (glycolysis) in the liver

*   inhibits fructose-2,6-bisphosphatase (inhibits gluconeogenesis)

REGULATION OF GLYCOLYSIS AND GLUCONEOGENESIS

WHEN ENERGY IS NEEDED ACTIVATE GLYCOLYSIS	TO STORE ENERGY; OR TO RAISE BLOOD GLUCOSE ACTIVATE GLUCONEOGENESIS
Glucose-6-phosphate 2 Pyruvate	2 Pyruvate (also: lactate, glycerol, some amino acids) ↓ Glucose-6-phosphate

PFK-1 activated by:
    AMP
    Fructose-2,6-bisphosphate

Pyruvate kinase activated by:
    AMP
    Fructose-1,6-bisphosphate
       (feed-forward control)

INSULIN stimulates the synthesis of:
    glucokinase
    PFK-1
    PFK-2

Pyruvate carboxylase activated by:
    Acetyl CoA (high during starvation;
       product of fatty acid catabolism)

Fructose 1,6-bisphosphatase activated by:
    ATP

GLUCAGON stimulates the synthesis of:
    PEP carboxykinase
    fructose-1,6-bisphosphatase
    glucose-6-phosphatase

INHIBIT GLUCONEOGENESIS

Pyruvate carboxylase inhibited by:
    Acetyl-CoA

Fructose-1,6-bisphosphatase
    inhibited by:
    AMP
    Fructose-2,6-bisphosphate

(Fructose-2,6-bisphosphate indicates
high levels of glucose.)

INHIBIT GLYCOLYSIS

Hexokinase inhibited by:
    ATP
    Glucose-6-phosphate
PFK-1 inhibited by:
    ATP
    Citrate
Pyruvate kinase inhibited by:
    ATP
    Acetyl-CoA

8.3 THE PENTOSE PHOSPHATE PATHWAY

ULTIMATE GOALS: 1. to make NADPH (reducing power) for syntheses and to help prevent oxidative damage (it's a great reducing agent)

2. to make sugar intermediates, especially ribose-5-phosphate, a component of nucleotides and nucleic acids

LOCATION: Cytoplasm; especially in cells which synthesize lipids and cells that are at high risk for oxidative damage
In plants, during the dark reactions of photosynthesis

OXIDATIVE PHASE: **ALL THREE REACTIONS ARE IRREVERSIBLE.**

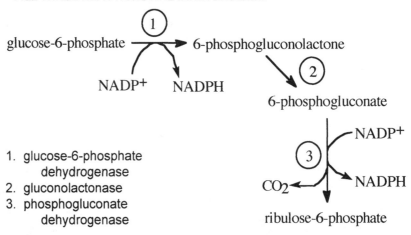

1. glucose-6-phosphate dehydrogenase
2. gluconolactonase
3. phosphogluconate dehydrogenase

REGULATION: Glucose-6-phosphate dehydrogenase (G-6-PD) is inhibited by NADPH and activated by glucose-6-phosphate and GSSG, which indicates oxidative damage and a need for NADPH.

A high carbohydrate diet triggers the synthesis of the enzymes G6PD and phosphogluconate dehydrogenase.

NONOXIDATIVE PHASE: ALL REACTIONS ARE REVERSIBLE.

What happens to ribulose-5-phosphate depends on the metabolic needs of the cells:

1. If the cell needs ribose-5-phosphate for nucleotide biosynthesis, ribose-5-phosphate isomerase will isomerize ribulose-5-phosphate.

2. If the cell only needs reducing power (NADPH), then the carbons of ribulose-5-phosphate will be recycled and converted into fructose-6-phosphate and glyceraldehyde-3-phosphate, intermediates of glycolysis. To completely recycle the ribulose-5-phosphate carbons via glycolysis, at least three molecules of ribulose-5-phosphate are needed initially. See the reactions outlined below.

3. If the cell needs ribose-5-phosphate but doesn't need reducing power (NADPH), the cell can also use the reverse reactions from the nonoxidative phase, starting with two fructose-6-phosphates and one glyceraldehyde-3-phosphate to make 3 ribose-5-phosphates.

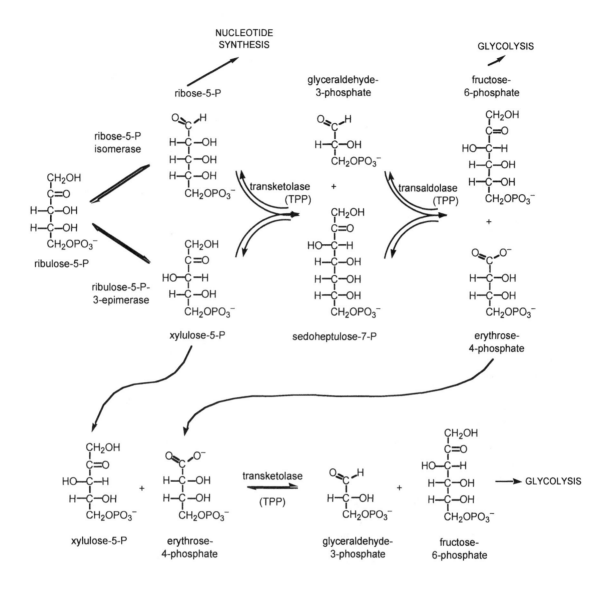

TRANSKETOLASE: TRANSFERS 2 CARBONS FROM A KETO SUGAR TO AN ALDO SUGAR

xylulose-5-P   ribose-5-P   glyceraldehyde-3-P   sedoheptulose-7-P

("P" = Phosphate)

119

TRANSALDOLASE: TRANSFERS 3 CARBONS FROM A KETO SUGAR TO AN ALDO SUGAR

$$
\begin{array}{ccccc}
& & \text{CH}_2\text{OH} & & \text{CH}_2\text{OH} \\
& & \text{C=O} & & \text{C=O} \\
\text{O=C–H} & & \text{HO–C–H} & & \text{HO–C–H} & & \text{O=C–H} \\
\text{H–C–OH} & + & \text{H–C–OH} & \rightleftharpoons & \text{H–C–OH} & + & \text{H–C–OH} \\
\text{CH}_2\text{OPO}_3^- & & \text{H–C–OH} & & \text{H–C–OH} & & \text{H–C–OH} \\
& & \text{H–C–OH} & & \text{CH}_2\text{OPO}_3^- & & \text{CH}_2\text{OPO}_3^- \\
& & \text{CH}_2\text{OPO}_3^- & & & &
\end{array}
$$

glyceraldehyde-3-P   sedoheptulose-7-P   fructose-6-P   erythrose-4-P

8.4 METABOLISM OF OTHER IMPORTANT SUGARS

FRUCTOSE METABOLISM

LIVER:	Four enzymes convert fructose to two G-3-P molecules, bypassing the regulatory enzymes hexokinase and PFK-1.
MUSCLE, ADIPOSE:	Hexokinase phosphorylates fructose (to fructose-6-phosphate).
GALACTOSE, MANNOSE:	Several reactions convert galactose to UDP-glucose. Hexokinase phosphorylates mannose to mannose-6-phosphate, which is isomerized to fructose-6-phosphate.

8.5 GLYCOGEN METABOLISM

Most glycogen is found in the muscle and liver ($\approx$10% liver mass, and $\approx$1% muscle mass). The actual amount of glycogen depends on the nutritional state of the organism.

Remember that glycogen is made of α(1,4)-linked glucose with α(1,6)-linked branches. Special enzymes are needed to form and degrade the branches. Glycogen degradation is not simply a reverse of glycogenesis.

The ends of all of those branches are nonreducing ends. Enzymes for both glycogenesis and glycogenolysis work only on the nonreducing ends. One huge glycogen molecule can be coated with working enzymes to release glucose very quickly when needed.

The ultimate goal of **GLYCOGENESIS** depends on where it takes place:

MUSCLE: stores excess glucose as glycogen: **ENERGY STORAGE.**

LIVER: removes glucose from the blood in response to insulin, which signals high blood glucose levels.

Glycogen synthase, and thus glycogenesis, is activated by glucose-6-phosphate, an indicator of excess glucose. High levels of ATP (as well as glucose-6-phosphate) inhibit the glycogenolysis by inhibiting glycogen phosphorylase.

The ultimate goals of **GLYCOGENOLYSIS** also depends on where it takes place:

MUSCLE: degrades glycogen to produce glucose-6-phosphate for **ENERGY** (➔ glycolysis).

LIVER: degrades glycogen to produce free glucose for **EXPORT** (to other cells), so glucose-6-phosphate is dephosphorylated and sent to the blood.

Glucagon signals **LOW BLOOD GLUCOSE,** and epinephrine signals an immediate need for energy. Glycogen phosphorylase, and thus glycogenolysis, is activated by AMP, an indicator that energy is needed.

GLYCOGENESIS: GENESIS OF GLYCOGEN
BEGINNING WITH GLUCOSE-6-PHOSPHATE

Other pathways to glycogen exist, namely glucose to C3 molecules to liver glycogen.

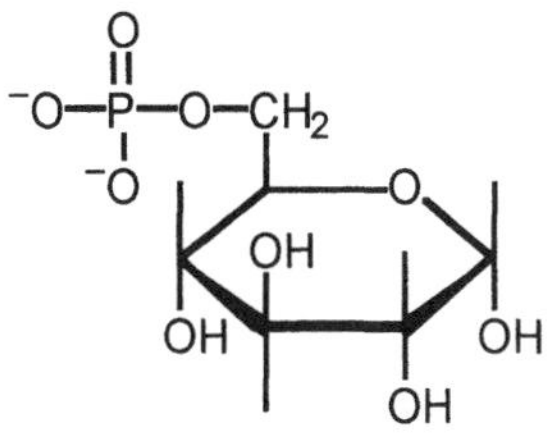

glucose-6-phosphate

PHOSPHOGLUCOMUTASE

Intermediate is glucose-1,6-bisphosphate

Mechanism involves a phosphoryl group attached to a Ser residue on the enzyme

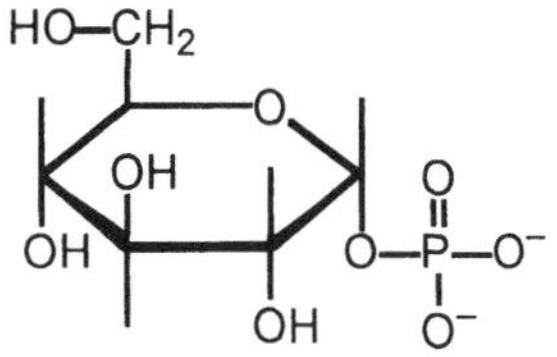

glucose-1-phosphate

UDP-GLUCOSE PYROPHOSPHORYLASE

$UTP \rightarrow PP_i$

UDP is a great leaving group

UDP-glucose is held more securely (than glucose alone) in the active site

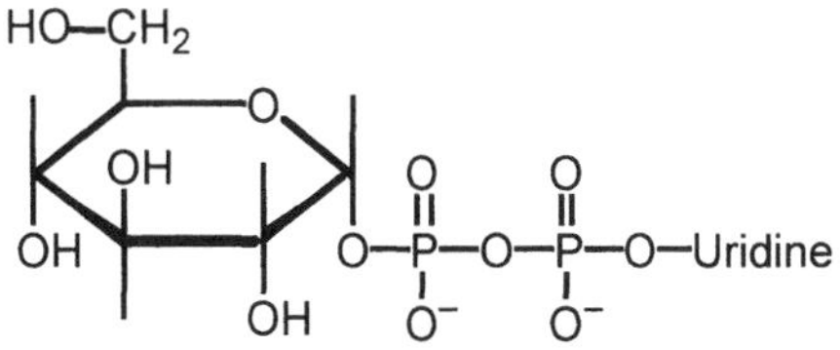

UDP-glucose

Irreversible PPi hydrolysis drives the previous reaction forward

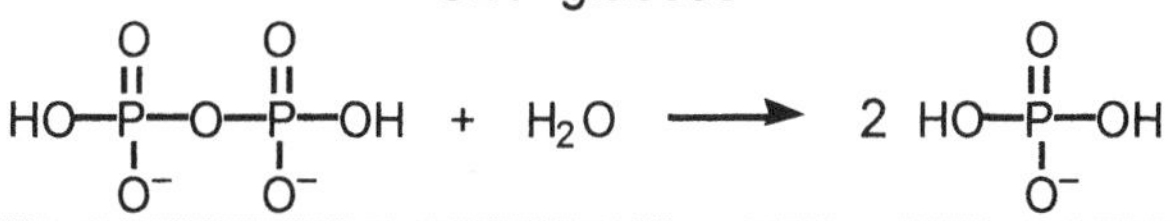

glycogen (n glucose residues) + UDP-glucose $\xrightarrow{\text{glycogen synthase}}$ glycogen (n+1 residues)

GLYCOGEN SYNTHASE

Creates $\alpha(1,4)$ glycosidic bonds

Needs glycogenin ("primer" protein) or an existing glycogen chain

Transfers glucose from UDP to glycogen chain

UDP + glycogen(n+1 residues)

BRANCHING ENZYME   creates $\alpha(1,6)$-linkages (branches)

GLYCOGENOLYSIS

glycogen (n residues) + P_i (HPO_4^{2-})

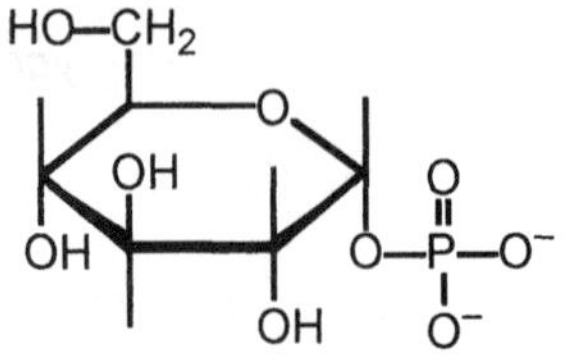

GLYCOGEN PHOSPHORYLASE

Cleaves $\alpha(1,4)$ linkages; stops 4
    glucose units before a branch

LIMIT DEXTRIN – glycogen
    molecule that's been
    degraded to its branch points

glycogen   +
(n-1 residues)   glucose-1-phosphate

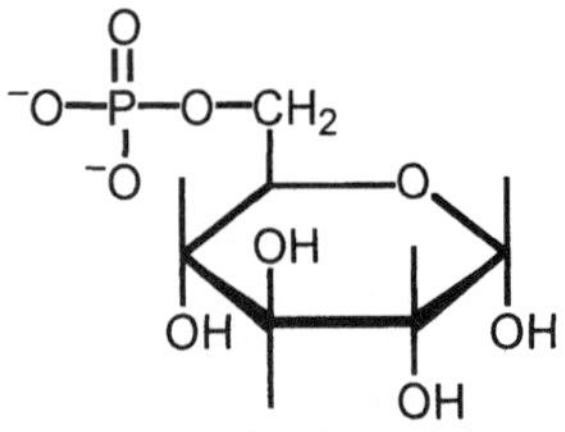

PHOSPHOGLUCOMUTASE

glucose-6-phosphate

DEBRANCHING ENZYME (AMYLO-α(1,6)-GLUCOSIDASE)

Transfers the outer three of the
    four glucose residues
    attached to a branch point to
    a nearby nonreducing end.

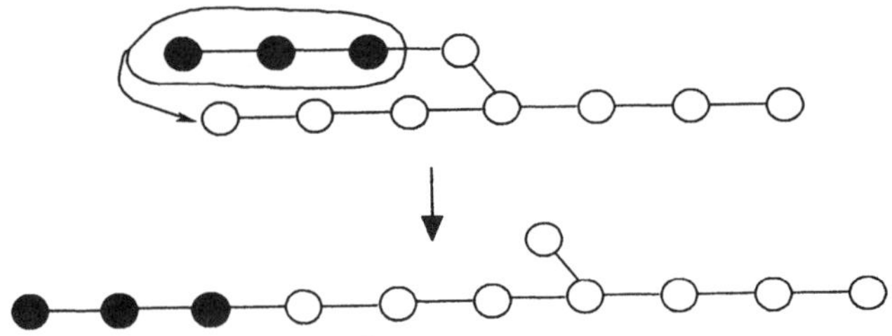

DEBRANCHING ENZYME (AMYLO-α(1,6)-GLUCOSIDASE)

Hydrolysis of $\alpha(1,6)$-linkages
Removes single glucose residue
    at each branch point
The unbranched polymer
    produced by 1,6-glucosidase
    is then degraded by glycogen
    phosphorylase.

+

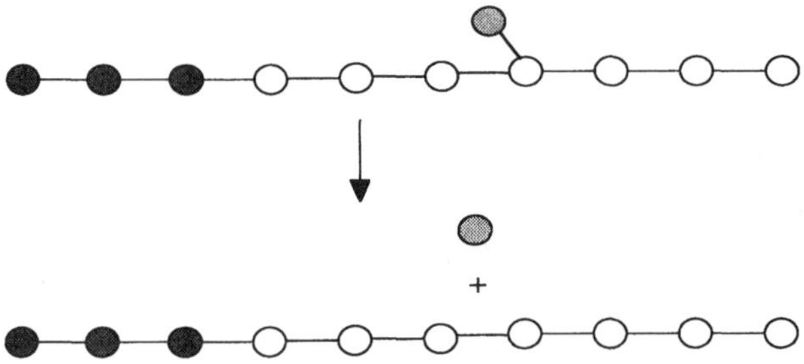

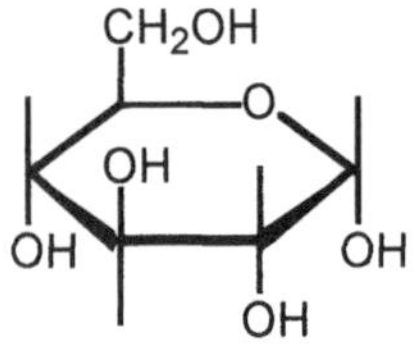

glucose

Regulation of Glycogen Metabolism

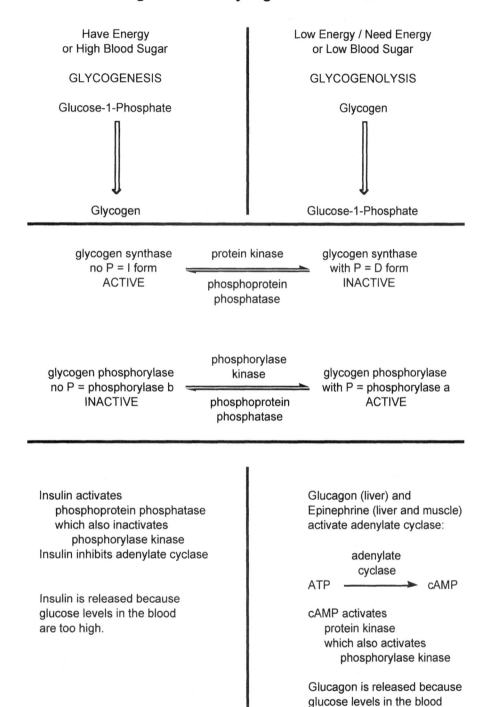

Have Energy or High Blood Sugar	Low Energy / Need Energy or Low Blood Sugar
GLYCOGENESIS	GLYCOGENOLYSIS
Glucose-1-Phosphate	Glycogen
⇓	⇓
Glycogen	Glucose-1-Phosphate

glycogen synthase
no P = I form
ACTIVE

protein kinase
⇌
phosphoprotein
phosphatase

glycogen synthase
with P = D form
INACTIVE

glycogen phosphorylase
no P = phosphorylase b
INACTIVE

phosphorylase
kinase
⇌
phosphoprotein
phosphatase

glycogen phosphorylase
with P = phosphorylase a
ACTIVE

Insulin activates
   phosphoprotein phosphatase
   which also inactivates
      phosphorylase kinase
Insulin inhibits adenylate cyclase

Insulin is released because
glucose levels in the blood
are too high.

Glucagon (liver) and
Epinephrine (liver and muscle)
activate adenylate cyclase:

$$ATP \xrightarrow{\text{adenylate cyclase}} cAMP$$

cAMP activates
   protein kinase
   which also activates
      phosphorylase kinase

Glucagon is released because
glucose levels in the blood
are too low.

Epinephrine is released in
response to stress: Fight or
flight! Need energy *FAST*!

Note: This page is a different version of Figure 8.20, p. 302 in your text.

CHAPTER 8: SOLUTIONS TO REVIEW QUESTIONS

8.1 Phosphorylation of glucose upon its entry into the cells prevents leakage of the molecule out of the cell and facilitates its binding to the active sites of enzymes.

8.2 a. Insulin – a hormone that stimulates glycogenesis and inhibits glycogenolysis.

b. Glucagon – a hormone that stimulates glycogenolysis and inhibits glycogenesis.

c. Fructose-2,6-bisphosphate – an effector molecule that activates PFK- 1 and stimulates glycolysis.

d. UDP-glucose – an activated form of glucose; a substrate for glycogen synthesis that is held more securely in the active site than glucose alone

e. cAMP – cyclic AMP – a second messenger molecule produced from ATP in response to glucagon or epinephrine

f. GSSG – the oxidized form of glutathione; an activator of glucose-6-phosphate dehydrogenase, which catalyzes a key regulatory step in the pentose phosphate pathway

g. NADPH – an important reducing agent; the reduced form of $NADP^+$, required for reductive processes (e.g. lipid biosynthesis) and antioxidant mechanisms (NADPH is a powerful antioxidant).

8.4 a. gluconeogenesis – cytoplasm and mitochondria

b. glycolysis - only in the cytoplasm

c. pentose phosphate pathway - cytoplasm

8.5 In glycolysis, the entry level substrates are sugars and the product is pyruvate. The main purposes of glycolysis are to provide the cell with energy and/or several metabolic intermediates. The substrates for gluconeogenesis are pyruvate, lactate, glycerol, and several amino or α-keto acids. Gluconeogenesis provides the body with glucose when blood glucose levels are low.

8.7 Such organisms utilize acetaldehyde as a hydrogen acceptor in order to regenerate NAD^+. Ethanol is the product of the reduction of acetaldehyde.

8.8 Under anaerobic conditions, pyruvate is reduced to lactate to regenerate NAD^+.

8.10 Glycolysis occurs in two stages. In stage I, glucose is phosphorylated and cleaved to two molecules of glyceraldehyde-3-phosphate. During this stage, two ATP molecules are consumed. In stage 2, each glyceraldehyde-3-phosphate is converted to pyruvate, a process in which four ATP and two NADH are produced.

8.11 Only (e), AMP, inhibits gluconeogenesis. Lactate, ATP, pyruvate, glycerol, and acetyl-CoA all stimulate gluconeogenesis.

8.13 Futile cycles are prevented by having the forward and reverse reactions catalyzed by different enzymes, both of which are independently regulated.

8.14 a. fermentation – the anaerobic metabolism or degradation of sugars

b. citric acid cycle – a biochemical pathway that degrades the acetyl group of acetyl-CoA to CO_2 and H_2O; in this process, three molecules of NAD^+ and one molecule of FAD are reduced

c. electron transport system – a series of electron carrier molecules that bind reversibly to electrons at different energy levels

    d.   tautomerization – a chemical reaction in which two tautomers are interconverted by the movement of a hydrogen atom and a double bond (e.g., keto-enol tautomerization)

    e.   aldol cleavage – the reverse of an aldol condensation; hydrolytic cleavage of a molecule in which the products are an aldehyde and a ketone

8.16   Glucose can be a substrate or a product of a number of key metabolic pathways depending upon the needs of the cell, and as such, glucose plays a central role in carbohydrate metabolism. When energy is needed, glucose undergoes glycolysis to form pyruvate, which can oxidize further either aerobically via the citric acid cycle and the electron transport chain to produce ATP, or anaerobically to form lactate (or ethanol in yeast cells and some bacteria). When energy is abundant, glucose is converted to glycogen via glycogenesis, or it may enter the pentose phosphate pathway to form pentose and other sugars. Also, pyruvate formed from glucose may react to form acetyl-CoA, a precursor for fatty acid synthesis as well as the citric acid cycle. Glucose may also be synthesized from pyruvate via gluconeogenesis, or from certain amino acids.

8.17

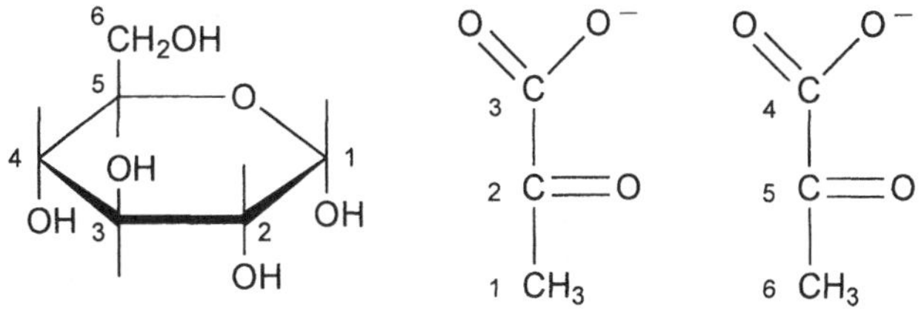

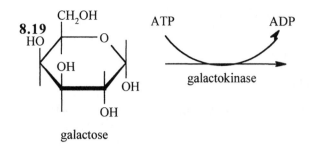

8.19

galactose

ATP → ADP
galactokinase

galactose-1-phosphate

UDP-glucose → glucose-1-phosphate
galactose-1-uridyltransferase

UDP-galactose

UDP-galactose-4-epimerase

UDP-glucose

PP_i → UTP
UDP-glucose
pyrophosphorylase

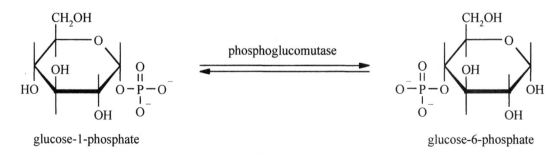

glucose-1-phosphate

phosphoglucomutase

glucose-6-phosphate

8.20   IN THE LIVER:

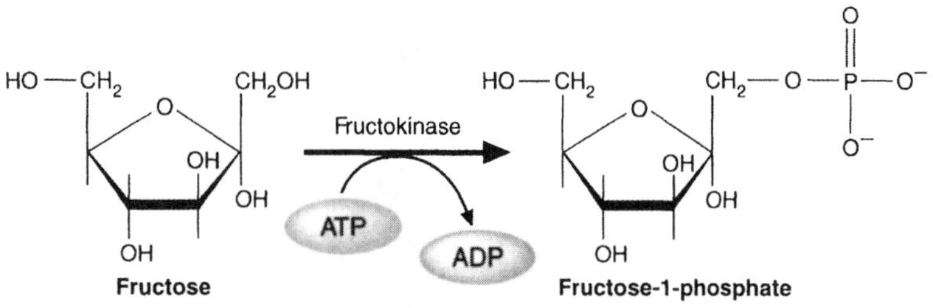

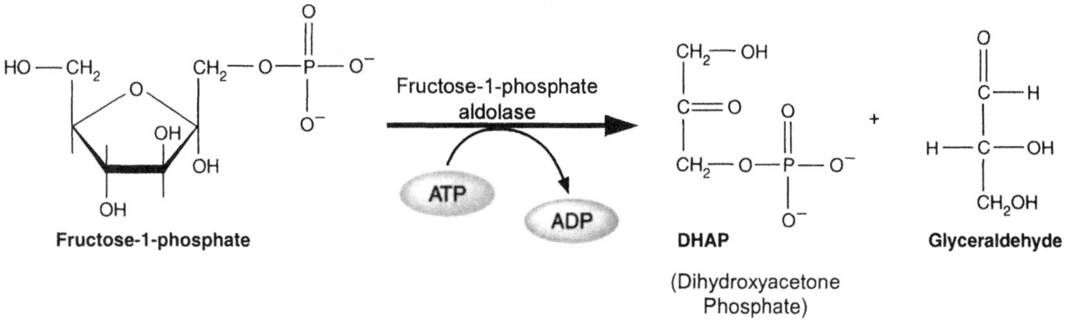

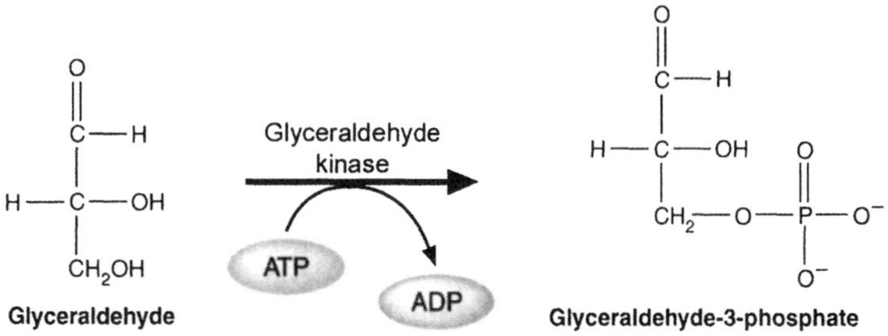

As in glycolysis, DHAP isomerizes to glyceraldehyde-3-phosphate in a reversible reaction catalyzed by triose phosphate isomerase.

8.22   The reaction in glycolysis that involves a redox reaction is the conversion of glyceraldehyde-3-phosphate to glycerate-1,3-bisphosphate, with the conversion of NAD^+ to $NADH + H^+$.

8.23   ATP hydrolysis at the beginning of glycolysis drives the pathway forward so that ATP is produced under varying conditions of substrate availability and product concentrations.

8.25   Sucrose is hydrolyzèd (with the addition of 1 H_2O) into 1 glucose and 1 fructose. Assuming that fructose is metabolized by the liver, the net result of the conversion of a molecule of sucrose to pyruvate would be: 4 pyruvate, 4 ATP, 4 NADH, $4H^+$, and $3H_2O$.

8.26   In the Cori cycle, lactate produced under anaerobic conditions in exercising muscle is passed to the liver, where it is converted back to pyruvate and then to glucose, which passes back to the muscle. In the muscle, glucose undergoes glycolysis to form pyruvate, which reacts with NADH to form lactate and NAD^+. The physiological function of the Cori cycle is to allow glycolysis to continue by regenerating NAD^+ so that under anaerobic conditions, exercising muscle can produce ATP for energy.

8.28 Insulin is released in response to high blood glucose levels. In the liver, insulin leads to the inhibition of glycogenolysis and the activation of glycogenesis, resulting in a decrease in blood glucose levels. Insulin also causes an increased cellular uptake of glucose. Glucagon is released in response to low blood glucose levels. In the liver, glucagon leads to the inhibition of glycogenesis and the activation of glycogenolysis, resulting in the release of glucose into the bloodstream.

8.29 Insulin and glucagon are produced by cells in the pancreas. Epinephrine and cortisol are produced by cells in the adrenal glands in response to stress.

8.31 The fate of pyruvate under anaerobic conditions is fermentation. In muscle cells, lactate is the only product, in yeast cells the products are ethanol and CO_2 and some microorganisms produce lactate and other acids or alcohols. All of these fermentation reactions result in the regeneration of NAD^+ so that glycolysis can continue. The fate of pyruvate under aerobic conditions is to be oxidized completely to form CO_2 and H_2O.

CHAPTER 8: SOLUTIONS TO THOUGHT QUESTIONS

8.32 In such an individual, following a carbohydrate meal blood glucose levels would be higher than normal. Recall that the kinetic properties of hexokinase D allow the liver to remove excess glucose from blood. Skeletal muscle would accumulate some additional glycogen, but most excess glucose would be used to synthesize triacylglycerol in adipocytes, a process that is promoted by insulin. A significant amount of liver glycogen is synthesized from glucose produced by gluconeogenesis.

8.34 In liver, fructose is metabolized more rapidly than glucose because its metabolism bypasses two regulatory steps in the glycolytic pathway: the conversion of glucose to glucose-6- phosphate and fructose-6-phosphate to fructose-1,6-bisphosphate. Recall that fructose-1-phosphate is split into glyceraldehyde and DHAP, both of which are subsequently converted to glyceraldehyde-3-phosphate.

8.35 Phosphoenolpyruvate has a high phosphoryl group transfer potential because the transfer of the enol phosphate to another molecule produces a vinyl alcohol. The vinyl alcohol tautomerizes rapidly to the keto-form making the transfer almost irreversible.

8.37 If gluconeogenesis and glycolysis were exactly the reverse of one another, futile cycles would be established and much energy would be wasted. In addition, it would be impossible for the body to store glycogen or release glucose into the blood as needed.

8.38 Ethanol is the most reduced molecule and acetate is the most oxidized. The degree of oxidation of an organic molecule can be correlated with its oxygen content, that is, acetate has more oxygen than does ethanol.

8.40 In high concentrations fructose can bypass most of the regulatory steps of the glycolytic cycle. Instead of being stored in glycogen molecules, the carbon skeletons of excess fructose molecules are converted through pyruvate and acetyl-CoA to fatty acids in triacylglycerol molecules.

8.41 The OH group itself is not very easily displaced. However conversion of an OH group to a phosphate ester is very easily accomplished. These esters are good leaving groups and when displaced have the same effect as if an OH had reacted.

8.43 Both glycogen and triacylglycerols are energy sources. One difference between the two molecules is in the speed with which energy can be mobilized. Glycogen can be converted to glucose and diverted to energy production very quickly (instant energy).

It takes longer to mobilize the fat reserves, but once activated they provide the energy for sustained effort.

8.44   Once muscle protein has been degraded to amino acids, a large percentage of them are converted to either oxaloacetate or pyruvate. Both of these molecules are substrates for the gluconeogenesis pathway.

8.46   The transamination reaction of glutamate that generates aspartate is as follows:

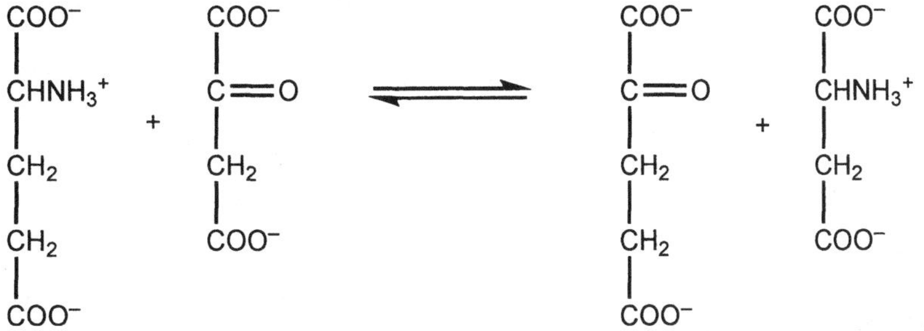

$$
\begin{array}{ccc}
\begin{array}{l}
COO^- \\
| \\
CHNH_3^+ \\
| \\
CH_2 \\
| \\
CH_2 \\
| \\
COO^-
\end{array}
&
+
\begin{array}{l}
COO^- \\
| \\
C=O \\
| \\
CH_2 \\
| \\
COO^-
\end{array}
&
\rightleftharpoons
\end{array}
\qquad
\begin{array}{l}
COO^- \\
| \\
C=O \\
| \\
CH_2 \\
| \\
CH_2 \\
| \\
COO^-
\end{array}
\quad
+
\begin{array}{l}
COO^- \\
| \\
CHNH_3^+ \\
| \\
CH_2 \\
| \\
COO^-
\end{array}
$$

Glutamate   Oxaloacetate   α-Ketoglutarate   Aspartate

8.47   Refer to Figures 8.13a and 8.13b, which illustrate the pentose phosphate pathway. Note that the ^{14}C label at C-2 of glucose-6-phosphate is C-1 of ribulose-5-phosphate as the result of a decarboxylation reaction. When ribulose-5-phosphate enters the nonoxidative phase of the pentose phosphate pathway the radioactive label will appear as C-1 of ribose-5-phosphate, xylulose-5-phosphate, sedoheptulose-7-phosphate and fructose-6-phosphate.

9 Aerobic Metabolism I: The Citric Acid Cycle

Brief Outline of Key Terms and Concepts

Overview
OBLIGATE ANAEROBE; AEROTOLERANT ANAEROBE
FACULTATIVE ANAEROBE
OBLIGATE AEROBE

9.1 OXIDATION-REDUCTION REACTIONS
CONJUGATE REDOX PAIR
STANDARD REDUCTION POTENTIAL ($E^{\circ\prime}$)

REDOX COENZYMES
NICOTINIC ACID (VITAMIN)
NAD(P) - NICOTINAMIDE ADENINE DINUCLEOTIDE (PHOSPHATE)
ELECTRON ACCEPTOR; HYDRIDE TRANSFER
RIBOFLAVIN (VITAMIN B_2)
FLAVIN MONONUCLEOTIDE
FLAVIN ADENINE DINUCLEOTIDE
FLAVOPROTEINS

AEROBIC METABOLISM
In living organisms, both energy-capturing and energy-releasing processes consist primarily of redox reactions.

In redox reactions, electrons move between an electron donor and an electron acceptor.

In many reactions, both electrons and protons are transferred.

In biological systems, most redox reactions involve hydride ion transfer (NADH/NAD$^+$) or hydrogen atom transfer (FADH$_2$/FAD).

9.2 CITRIC ACID CYCLE

CONVERSION OF PYRUVATE TO ACETYL-COA by the enzymes in the pyruvate dehydrogenase complex requires the coenzymes TPP, FAD, NAD$^+$, COENZYME A, AND LIPOIC ACID. TPP is THIAMINE PYROPHOSPHATE.

REACTIONS OF THE CITRIC ACID CYCLE
The CITRIC ACID CYCLE begins with the condensation of a molecule of acetyl-CoA with oxaloacetate to form citrate, which is eventually reconverted to oxaloacetate. In one turn of the cycle, 2 CO$_2$, 3 NADH, 1 FADH$_2$, and 1 GTP are produced.

FATE OF CARBON ATOMS IN THE CITRIC ACID CYCLE

THE AMPHIBOLIC CITRIC ACID CYCLE
An AMPHIBOLIC PATHWAY plays a role in both anabolism and catabolism. The citric acid cycle intermediates used in anabolic processes are replenished by several ANAPLEROTIC REACTIONS.

CITRIC ACID CYCLE REGULATION
The citric acid cycle is closely regulated, thus ensuring that the cell's energy and biosynthetic needs are met.

ALLOSTERIC EFFECTORS AND SUBSTRATE AVAILABILITY primarily regulate the enzymes citrate synthase, isocitrate dehydrogenase, α-ketoglutarate dehydrogenase, pyruvate dehydrogenase, and pyruvate carboxylase.

THE GLYOXYLATE CYCLE
Organisms in which the glyoxylate cycle occurs can use two-carbon molecules to sustain growth.

In plants, the glyoxylate cycle is in organelles called glyoxysomes.

As biomolecules are oxidized to CO$_2$, they transfer electrons to NAD$^+$ (and FAD), which is then reduced to NADH and FADH$_2$. In turn, they transfer electrons to the electron transport chain and ultimately to oxygen, which is reduced to H$_2$O. The energy released during this process pumps H$^+$ ions to the other side of the membrane. When they naturally flow back in, they are (typically) forced to pass through special channels that synthesize ATP. So, ATP captures energy that results from the transfer of electrons from fuel molecules to (ultimately) oxygen.

9.1 OXIDATION-REDUCTION REACTIONS

(Review Oxidation-Reduction Reactions in Chapter 1 (p. 21) of your text.)

$\Delta G^{\circ\prime} = -nF\Delta E^{\circ\prime}$ where $\Delta G^{\circ\prime} =$ the standard free energy
$n =$ number of electrons transferred
$F =$ Faraday's constant, 96.5 kJ/V·mol)
$\Delta E^{\circ\prime} =$ the difference in reduction potential between the electron donor and the electron acceptor under standard conditions

For redox reactions to be spontaneous (have a negative $\Delta G^{\circ\prime}$), they must have a positive $\Delta E^{\circ\prime}$. (Refer to the equation above; n and F are always positive, so for $\Delta G^{\circ\prime}$ to be negative, $\Delta E^{\circ\prime}$ must be positive.)

Electrons spontaneously transfer from the more negative $E^{\circ\prime}$ to the more positive $E^{\circ\prime}$. So, **the electron donor will have the more negative $E^{\circ\prime}$.**

When combining two half reactions, make sure that the overall equation is balanced in the number of electrons transferred.[1] When you reverse a half-reaction, also reverse the sign of its $E^{\circ\prime}$. To calculate $\Delta E^{\circ\prime}$, add the two values of $E^{\circ\prime}$ (including the sign-change of the reaction that you reversed).[2]

Let's use one of the redox coenzymes, NAD^+/$NADH$, and the reduction of acetaldehyde as an example. The two half reactions with their $\Delta E^{\circ\prime}$ values (as listed in Table 9.1, p. 311 of your text) are:

	$E^{\circ\prime}$
$NAD^+ + H^+ + 2e^- \rightarrow NADH$	-0.32 V
Acetaldehyde $+ 2H^+ + 2e^- \rightarrow$ ethanol	-0.20 V

For the $\Delta E^{\circ\prime}$ to be positive, and for the electrons to balance on both sides of the overall equation, we must reverse the first reaction:

	$E^{\circ\prime}$
$NADH \rightarrow NAD^+ + H^+ + 2e^-$	$+0.32$ V
Acetaldehyde $+ 2H^+ + 2e^- \rightarrow$ ethanol	-0.20 V
Acetaldehyde $+ H^+ + NADH \rightarrow$ ethanol $+ NAD^+$	$+0.12$ V[3]

TO IDENTIFY OXIDIZING AGENTS AND REDUCING AGENTS:

1. Locate each CONJUGATE REDOX PAIR. It often helps to draw the molecular structures.

    In the equation above, the two conjugate redox pairs are acetaldehyde/ethanol and NAD^+/$NADH$.

2.   In each conjugate redox pair, one member will be its oxidized form, and the other will be its reduced form. Label them. Remember that the oxidized form will have more oxygens (and/or less H's, and a higher oxidation state), and the reduced form will have less oxygens (and/or more H's, and a lower oxidation state).

---

[1] Remember: In calculating $\Delta E^{\circ\prime}$, don't multiply the value of $E0'$ by the number of electrons transferred or by the coefficient in the balanced equation. This is different from ΔG° and ΔH°calculations.

[2] This method avoids the confusion of which $E^{\circ\prime}$ to subtract from which. If you feel more comfortable with a different method, by all means, use it.

[3] We've already seen this reaction as the second step of homolactic fermentation, catalyzed by alcohol dehydrogenase. (This is also hiding out in Question 9.2, page 314 of your text.)

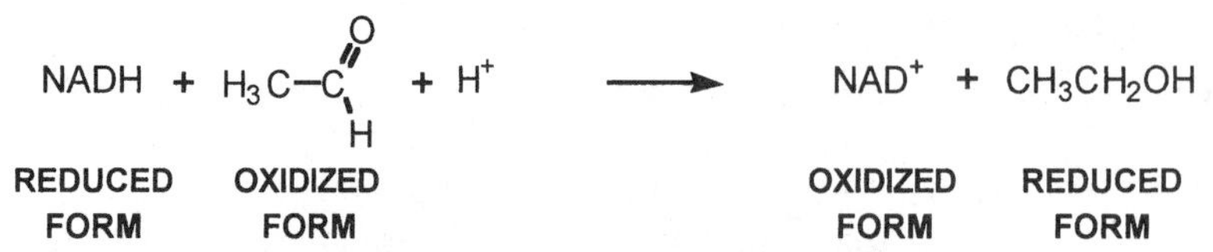

$$NADH + H_3C-\overset{\overset{\displaystyle O}{\|}}{C}-H + H^+ \longrightarrow NAD^+ + CH_3CH_2OH$$

| REDUCED FORM | OXIDIZED FORM | | OXIDIZED FORM | REDUCED FORM |

3. Check your labels using the balanced equation (above). Each side of the equation should have one reduced form and one oxidized form.

4. Of the two reactants, the reduced form is the reducing agent - it gives away its extra electrons, reducing the other guy (and becoming oxidized in the process). Conversely, the oxidized form is the oxidizing agent - it takes electrons from the other guy, oxidizing it (and becoming reduced in the process).

REDOX COENZYMES

NICOTINIC ACID

NAD(P), NICOTINAMIDE ADENINE DINUCLEOTIDE (PHOSPHATE), is an electron carrier; see "The Organic Connection" in Chapter 6, p. 85 of this Study Guide for more information on $NAD^+/NADH$.

RIBOFLAVIN (vitamin B$_2$)
RIBOFLAVIN is the vitamin precursor for **FLAVIN MONONUCLEOTIDE (FMN)** and FLAVIN ADENINE DINUCLEOTIDE (**FAD**), prosthetic groups for the **FLAVOPROTEINS**.

FMN has a three-ring isoalloxazine group that stabilizes a radical intermediate, and can transfer electrons one at a time. The structure of FAD is FMN with another phosphate, another sugar, and an adenine, and can also transfer one electron at a time.

FMN – FLAVIN MONONUCLEOTIDE

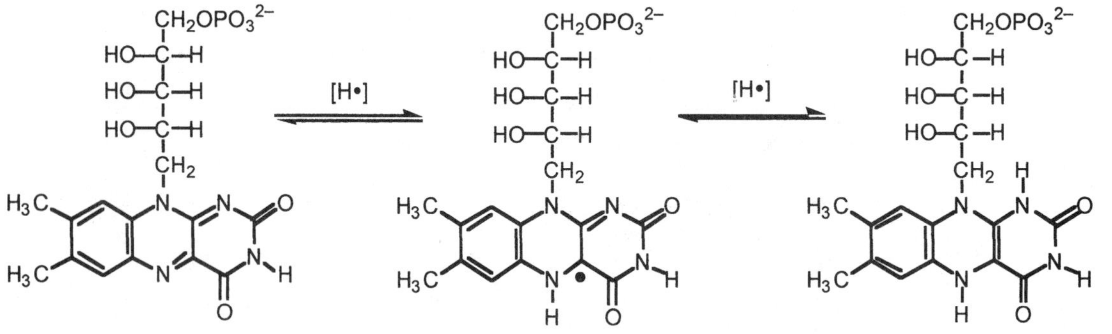

| FMN | FMNH• | FMNH$_2$ |
| (oxidized or quinone form) | (radical or semiquinone form) | (reduced or hydroquinone form) |

9.2 CITRIC ACID CYCLE

THE GOALS OF THE CITRIC ACID CYCLE ARE TO:

1. Convert fuel molecules to **ENERGY**: Produce reducing power (3 NADH and 1 FADH$_2$ per acetyl-CoA), which can enter the electron transport chain and provide the energy to synthesize ATP. One GTP (which can easily be converted to ATP) is produced directly, although this is a small energy bonus compared to the number of ATP that are ultimately produced from the electron transport chain.

2. Produce **INTERMEDIATES**: The carbon precursors for lipids, sugars, amino acids, and nucleic acids are all derived from citric acid cycle intermediates.

OVERALL REACTION

Acetyl-CoA + 3NAD$^+$ + FAD + GDP + P$_i$ + 2H$_2$O →

$$2CO_2 + 3NADH + FADH_2 + CoASH + GTP + 3H^+$$

In eukaryotes, the citric acid cycle takes place in the mitochondria. Remember that glycolysis took place in the cytoplasm.

COENZYMES

THIAMINE PYROPHOSPHATE (TPP) - decarboxylates and transfers aldehyde groups; used by pyruvate dehydrogenase (E$_1$ of the pyruvate dehydrogenase complex) and the α-ketoglutarate dehydrogenase complex

LIPOIC ACID - carries hydrogens or acetyl groups; used by dihydrolipoyl transacetylase (E$_2$ of the pyruvate dehydrogenase complex) and dihydrolipoyl transsuccinylase of the α-ketoglutarate dehydrogenase complex

FAD is used by dihydrolipoyl dehydrogenase (E$_3$ of the pyruvate dehydrogenase complex and of the α-ketoglutarate dehydrogenase complex), and succinate dehydrogenase.

NAD is used by dihydrolipoyl dehydrogenase (E3 of the pyruvate dehydrogenase complex and of the α-ketoglutarate dehydrogenase complex), isocitrate dehydrogenase, and malate dehydrogenase

COENZYME A (CoASH) - carries acetyl groups via a thioester bond (Acetyl-CoA); used by dihydrolipoyl transacetylase (E$_2$ of the pyruvate dehydrogenase complex) and dihydrolipoyl transsuccinylase of the α-ketoglutarate dehydrogenase complex

ACETYL-CoA

Acetyl-CoA is an acetyl group bound to coenzyme A by a thioester bond. When studying reaction mechanisms that involve acetyl-CoA, it's helpful to remember that the linkage is a thioester. Compare the structure of acetyl-CoA below with the structure of CoASH in Figure 9.9, p. 318 of your text.

{Note that coenzyme A is abbreviated CoASH. To be consistent, instead of "acetyl-CoA" we should probably say "acetyl-S-CoA" (or even "CoAS-acetyl"), but, acetyl-CoA is just easier, and is in common usage.}

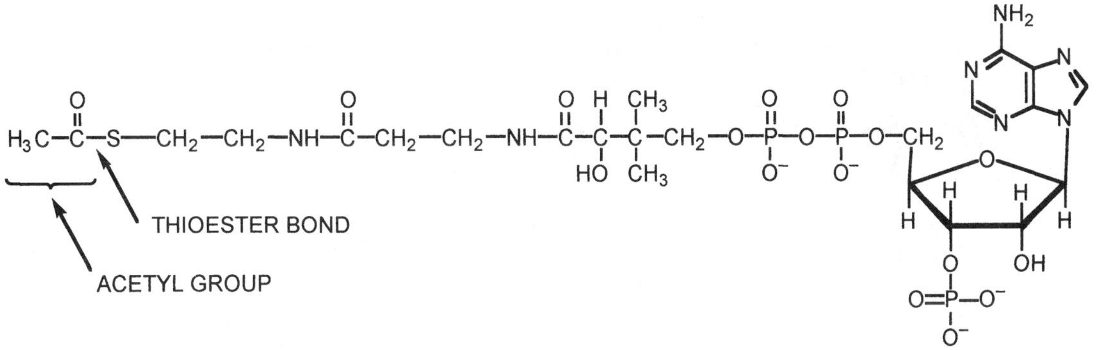

CONVERSION OF PYRUVATE TO ACETYL-CoA   ΔG°′ = –33.5 KJ/MOL

OVERALL REACTION: AN OXIDATIVE DECARBOXYLATION

pyruvate + NAD$^+$ + CoASH → acetyl-CoA + NADH + CO$_2$ + H$_2$O + H$^+$

PYRUVATE DEHYDROGENASE COMPLEX = MULTIPLE COPIES OF:

E_1: pyruvate dehydrogenase (or pyruvate decarboxylase); needs TPP

E_2: dihydrolipoyl transacetylase; needs lipoic acid and CoASH

E_3: dihydrolipoyl dehydrogenase; needs FAD and NAD^+

MECHANISM OF THE PYRUVATE DEHYDROGENASE COMPLEX

(This is simplified to emphasize the changes to the acetyl group. For details regarding the linkage to the TPP thiazole ring, see Figure 9.10 of your text.)

E_1: Decarboxylation of pyruvate to form hydroxyethyl-TPP (HETPP):

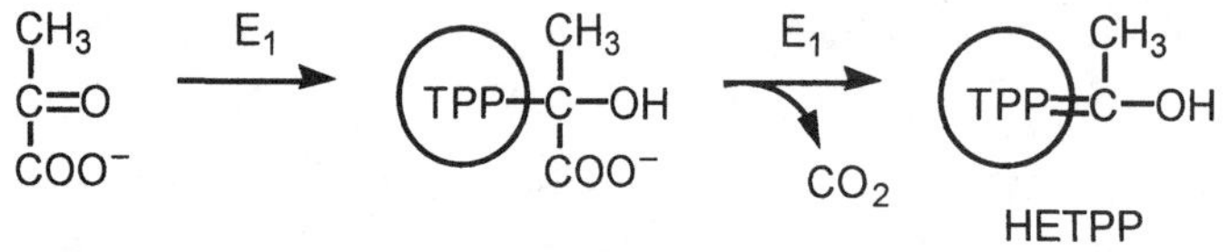

E_2 converts the hydroxyethyl group of HETPP to acetyl-CoA:

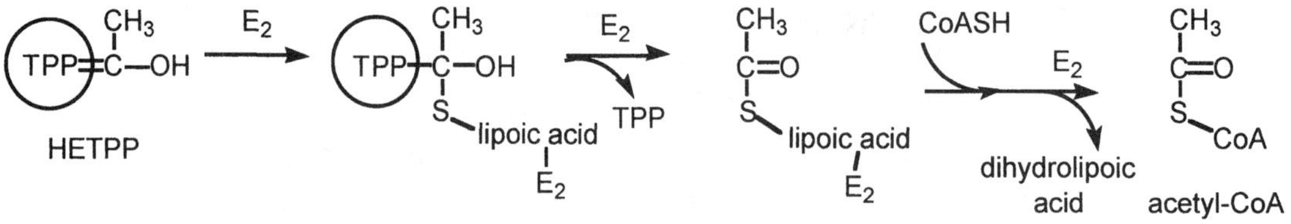

E_3 regenerates lipoic acid; FAD is also regenerated spontaneously.

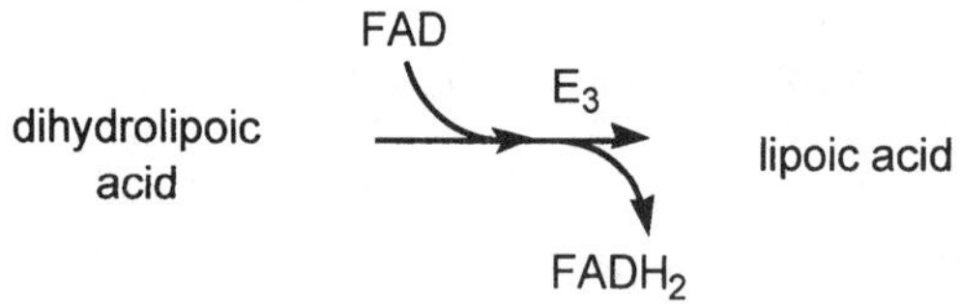

$$FADH_2 + NAD^+ \rightarrow FAD + NADH + H^+$$

REGULATION:

In general, inhibitors are products and indicators that the cell has plenty of energy (e.g., ATP), while activators are substrates and indicators that the cell needs energy (e.g., AMP)

PRODUCT INHIBITION: Acetyl-CoA and NADH are inhibitors.

ALLOSTERIC: ATP inhibits; AMP, NAD^+, CoASH are activators.

COVALENT MODIFICATION: The products acetyl-CoA and NADH activate a kinase, which phosphorylates and inactivates the pyruvate dehydrogenase complex. Substrates pyruvate, CoASH, and NAD^+ inhibit this kinase, removing an inhibitor (that's not quite the same as activating a reaction, but it helps). Low levels of ATP activate a phosphoprotein phosphatase, which dephosphorylates and activates the pyruvate dehydrogenase complex.

REACTIONS OF THE CITRIC ACID CYCLE TAKE PLACE IN THE MITOCHONDRIAL MATRIX
Acetyl-CoA brings in two carbons, and two carbons leave as CO_2.
Oxaloacetate is regenerated so the cycle can continue.

THE CITRIC ACID CYCLE

ACETYL-COA BRINGS IN TWO CARBONS, TWO CARBONS LEAVE AS CO_2

Acetyl-CoA enters the cycle.

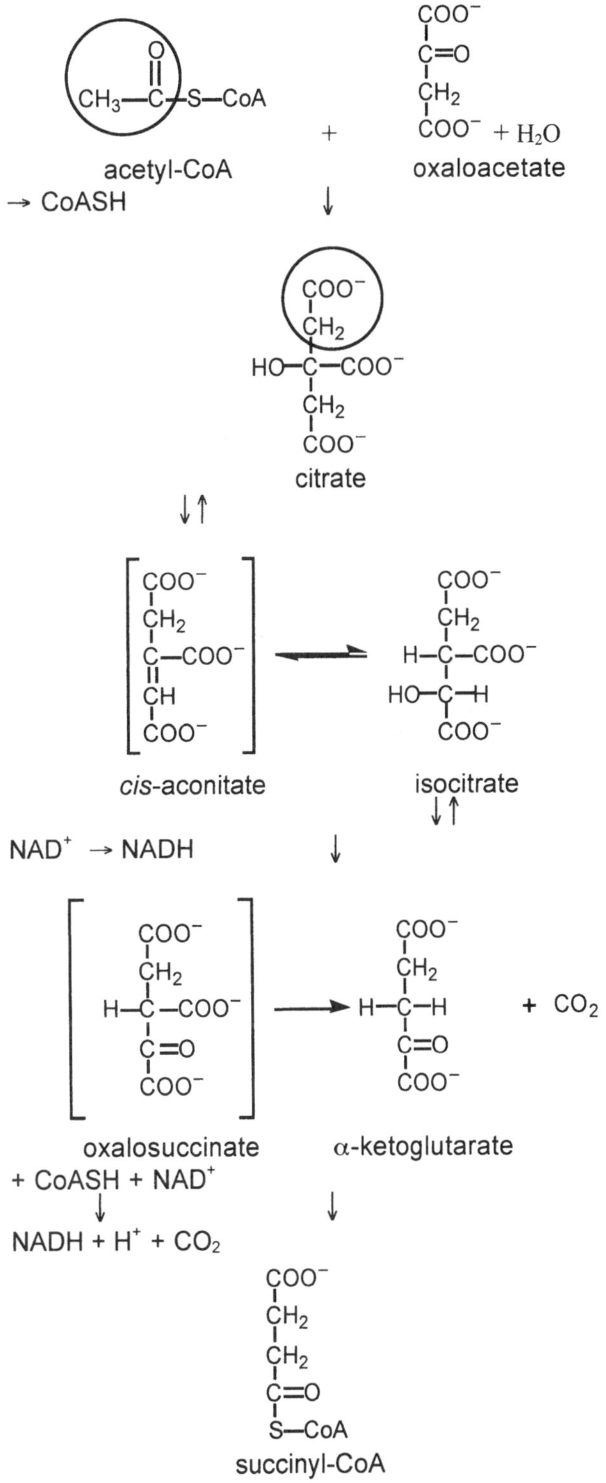

acetyl-CoA   +   oxaloacetate $+ H_2O$

1. Citrate synthase

Aldol condensation,
   goodbye CoASH
Large negative ΔG
Note that the tertiary alcohol can't
   be oxidized.

$\rightarrow$ CoASH

citrate

2. Aconitase

Now the alcohol is secondary, and
   can be oxidized.
Mechanism: dehydrate then
   rehydrate. Intermediate is *cis-*
   aconitate.
This reaction is stereospecific.
   Only the product with the
   stereochemistry shown is
   formed.

*cis-*aconitate   isocitrate

3. Isocitrate dehydrogenase

Oxidation of isocitrate to form an
   oxalosuccinate intermediate
Reduction of NAD^+ to NADH
Decarboxylation

$NAD^+ \rightarrow$ NADH

oxalosuccinate α-ketoglutarate $+ CO_2$

4. α-Ketoglutarate dehydrogenase complex

Oxidation of α-ketoglutarate
Reduction of NAD^+ to NADH
Decarboxylation
Large negative ΔG
Similar to pyruvate dehydrogenase
   complex, hello CoASH
Allosterically regulated

$+$ CoASH $+ NAD^+$
$\downarrow$
$NADH + H^+ + CO_2$

succinyl-CoA

CITRIC ACID CYCLE, CONTINUED...

REGENERATION OF OXALOACETATE TO CONTINUE THE CYCLE

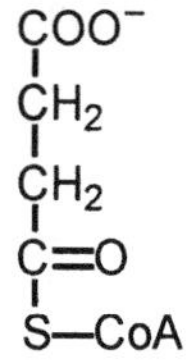

$$
\begin{array}{c}
COO^- \\
| \\
CH_2 \\
| \\
CH_2 \\
| \\
C=O \\
| \\
S—CoA
\end{array}
$$

succinyl-CoA

---

5. Succinate thiokinase

Cleavage of thioester bond,
 goodbye CoASH

Substrate-level phosphorylation of
 GDP in mammals (other
 organisms use ADP)

$GTP + ADP \rightleftharpoons GDP + ATP$

Succinate is symmetrical ! Carbons
 1 & 4 are now identical, as are
 carbons 2 & 3.

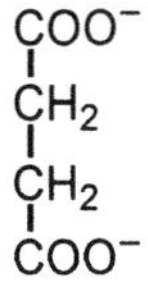

$GDP + P_i \rightarrow GTP$ ↓↑

$$
\begin{array}{c}
COO^- \\
| \\
CH_2 \\
| \\
CH_2 \\
| \\
COO^-
\end{array}
$$

succinate

---

6. Succinate dehydrogenase

Oxidation of succinate

Reduction of FAD to $FADH_2$

Bound tightly to inner mitochondrial
 membrane (This is the only citric
 acid cycle enzyme not in the
 mitochondrial matrix.)

Subsequent transfer of electrons from
 $FADH_2$ to Coenzyme Q drives this
 reaction forward.

$FAD \rightarrow FADH_2$ ↓↑

$$
\begin{array}{c}
COO^- \\
\diagdown \\
C \\
\| \\
C \\
\diagup \\
COO^-
\end{array}
$$

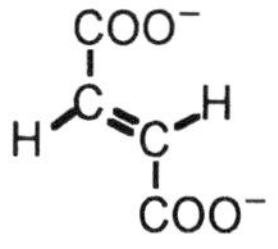

fumarate

| 7. Fumarase (also called |
| fumarate hydratase) |

Stereospecific hydration

This reaction sets up the formation
 of OAA in the next step.

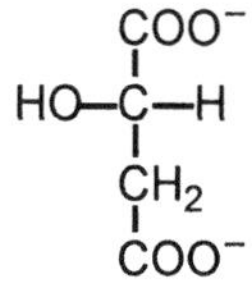

↓↑

$$
\begin{array}{c}
COO^- \\
| \\
HO—C—H \\
| \\
CH_2 \\
| \\
COO^-
\end{array}
$$

malate

Malate dehydrogenase

Compare the structure of OAA with
 that of succinate. It took three
 steps to change a CH_2 to a
 $C=O$.

$\Delta G^{\circ\prime} = + 29$ kJ/mol , BUT the next
 reaction with another molecule
 of acetyl-CoA pulls this reaction
 forward by removing OAA (so
 the *actual* ΔG will be negative).

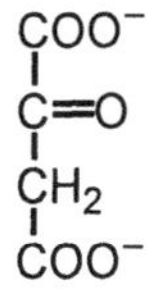

$NAD^+ \rightarrow NADH + H^+$ ↓↑

$$
\begin{array}{c}
COO^- \\
| \\
C=O \\
| \\
CH_2 \\
| \\
COO^-
\end{array}
$$

oxaloacetate (OAA)

FATE OF CARBON ATOMS IN THE CITRIC ACID CYCLE

To trace specific (labeled) carbon atoms through the citric acid cycle:

Draw out the structures of the citric acid cycle on one page, as in Figure 9.8 of your text. Circle the labeled carbon in each structure, following it around the citric acid cycle. When you get back to citrate, switch to drawing a square around the labeled carbon. For each turn of the cycle, use a different shape (different colored markers are even better). This is a neat way to see where specific carbon atoms end up with each turn of the cycle.

When you get to succinate, you have to choose between carbons 1 and 4 (or between carbons 2 and 3), since succinate is symmetrical, and those carbons are equivalent. Try repeating the exercise above, choosing an equivalent (but different) carbon. It's interesting to see how that will change the number of turns of the cycle it takes to lose the labeled carbon as CO_2.

Try using this method to answer Question 9.5 on page 326 of your text.

THE CITRIC ACID CYCLE IS AMPHIBOLIC (I.E., BOTH CATABOLIC AND ANABOLIC).

Catabolic is obvious: 2 carbons enter (acetyl-CoA) and they leave oxidized (CO_2).

To be anabolic, the citric acid cycle must produce intermediates. Wait! If 2 carbons enter and 2 CO_2 leave, how can the citric acid cycle produce intermediates? There are two possible solutions:

1. There must be a way to bypass the CO_2-generating steps of the cycle. This is the glyoxylate cycle, used by plants (and discussed later).
2. There must be additional ways for carbon to enter the cycle (other than via acetyl-CoA). Mammals use this option. **ANAPLEROTIC REACTIONS** are reactions that replenish citric acid cycle intermediates.

ANAPLEROTIC REACTIONS

- pyruvate → oxaloacetate → citric acid cycle (instead of pyruvate → acetyl-CoA)

  Pyruvate carboxylase is activated by acetyl-CoA (which indicates that there isn't enough OAA). Excess OAA can be used for gluconeogenesis. (Remember that the first two steps of gluconeogenesis are pyruvate → oxaloacetate → PEP.)

- certain fatty acids (also certain amino acids) → succinyl-CoA

- glutamate → α-ketoglutarate

- aspartate → oxaloacetate

CITRIC ACID CYCLE INTERMEDIATES ARE USED IN THESE BIOSYNTHESES:

Amino acids and proteins (from oxaloacetate and α-ketoglutarate); heme and chlorophyll (from succinyl-CoA)

Glucose (from oxaloacetate)

Fatty acids and cholesterol (from acetyl Co-A)

Pyrimidines (from oxaloacetate) and purines (from α-ketoglutarate)

CITRIC ACID CYCLE REGULATION

Remember the ultimate goals of the citric acid cycle, and its regulation will make sense. When the cell needs energy or intermediates, the cycle will be activated. The cycle will be inhibited when the cell has plenty of ATP, and its energy needs are being met. It will also be inhibited at low substrate concentrations – including NAD^+, FAD, and ADP.

low NADH/ NAD^+ ratio[4]   high NADH/ NAD^+ ratio

low ATP/ADP ratio   high ATP/ADP ratio

Metabolic branch points: Look at Figure 9.10 in your text (or at the list above). The four intermediates mentioned are oxaloacetate, acetyl-CoA, α-ketoglutarate and succinyl -CoA. So, it makes sense that their enzymes are closely regulated. It also makes sense that the two CO_2-generating reactions would be closely regulated.

CITRATE SYNTHASE   acetyl-CoA + oxaloacetate → citrate + CoASH

Activated by: Inhibited by:
acetyl-CoA, oxaloacetate, ADP, NAD^+ citrate, succinyl-CoA, ATP, NADH

ISOCITRATE DEHYDROGENASE   isocitrate + NAD^+ → α-ketoglutarate + NADH + CO_2

Activated by: high ADP, NAD^+ Inhibited by: ATP, NADH

Isocitrate dehydrogenase is also important for citrate metabolism. Only citrate can penetrate the mitochondrial membrane; acetyl-CoA can't. Why would citrate want to leave the mitochondrion? The cell's energy needs are being met, so citrate isn't needed for the citric acid cycle. Citrate can be used to carry acetyl-CoA from the mitochondria to the cytosol, where it can be used in fatty acid synthesis. Citrate activates the first reaction of fatty acid synthesis, and recall that citrate inhibits glycolysis (at PFK-1). Figure 9.12 in your text is a great summary of citrate metabolism.

α-KETOGLUTARATE DEHYDROGENASE

α-ketoglutarate + NAD^+ → succinyl-CoA + NADH + CO_2 + H^+

Inhibited by: NADH

Will the citric acid cycle make intermediates or will it make energy? To make intermediates, pyruvate enters the cycle as oxaloacetate (using pyruvate carboxylase). To make energy, pyruvate enters the cycle as acetyl-CoA (using pyruvate dehydrogenase). Obviously, these two enzymes will be closely regulated as well.

Acetyl-CoA activates pyruvate carboxylase and inhibits pyruvate decarboxylase. Why does this make sense? If intermediates are being made, oxaloacetate levels drop and acetyl-CoA will accumulate. Activating pyruvate carboxylase replenishes oxaloacetate; inhibiting pyruvate dehydrogenase prevents more acetyl-CoA from being made.

---

[4] Watch out for the quintessential trick question, which plays with these ratios. Make sure that this makes sense to you, so that you will immediately see that a *low* NADH/NAD^+ ratio and a **high** NAD^+/NADH ratio are the same, that they both indicate a need for energy, and that the citric acid cycle will be activated.

THE GLYOXYLATE CYCLE BYPASSES CO₂-GENERATING STEPS
TWO ACETYL-COA REACT TO FORM ONE MALATE

The glyoxylate cycle occurs in plants, some fungi, algae, protozoans, and bacteria.

Glyoxylate enzymes occur in the cytoplasm, except in plants, where they occur in glyoxysomes.

---

1. A glyoxysome-specific isozyme of citrate synthase

...same as the citric acid cycle...

---

2. A glyoxysome-specific isozyme of aconitase

...same as the citric acid cycle...

---

3. Isocitrate lyase

Aldol cleavage

---

4. Malate synthase

Also requires H2O

Net:

    Four carbons in,
    zero carbons out !!

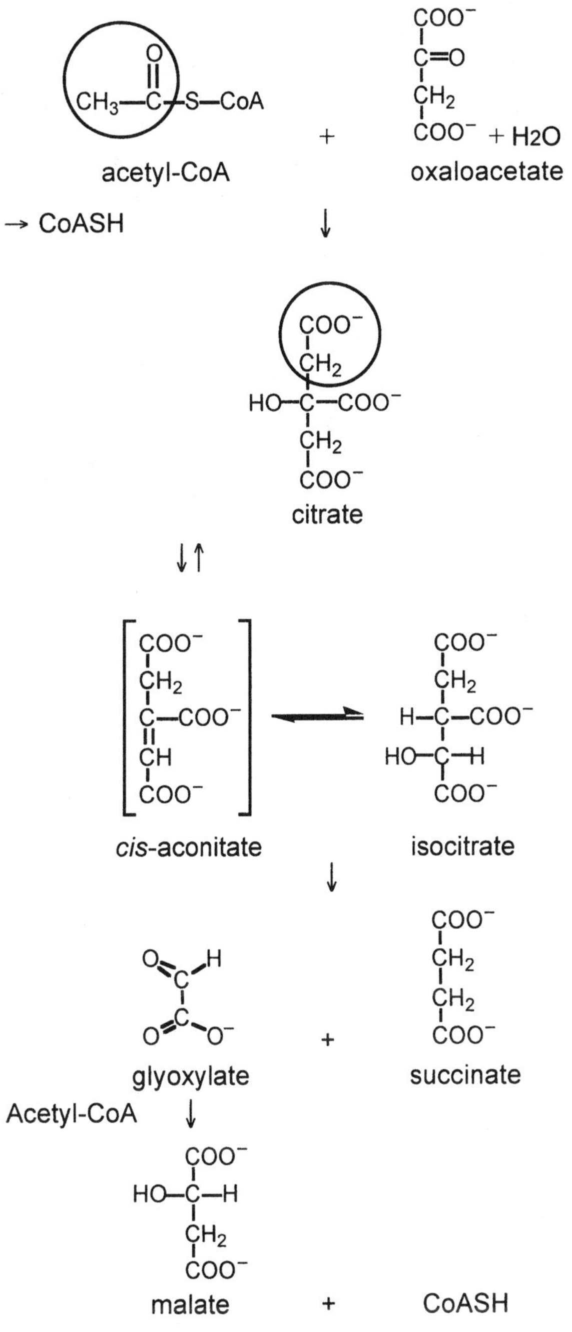

acetyl-CoA + oxaloacetate + H2O

→ CoASH

citrate

cis-aconitate ⇌ isocitrate

glyoxylate + succinate

Acetyl-CoA

malate + CoASH

Succinate and malate can each continue around the citric acid cycle to form the intermediate(s) needed

AFTER STUDYING THIS CHAPTER, YOU SHOULD BE ABLE TO:

- Calculate $\Delta E^{\circ\prime}$ for a reaction from half-cell potential ($\Delta E^{\circ\prime}$) data. Determine whether a given redox reaction will proceed in the direction written.

- Identify redox reactions. Identify oxidizing and reducing agents. Identify whether a molecule is oxidized or reduced in a given reaction.

- Calculate $\Delta G^{\circ\prime}$ using the equation: $\Delta G^{\circ\prime} = -nF\Delta E^{\circ\prime}$ (F = 96.5 kJ/V·mol)

- Identify oxidation states of atoms in molecules.

- Trace specific carbons through the pathways learned to date.

- Determine whether the citric acid cycle will be activated or inhibited, given one or more of the following data:

    high/low levels of specific effectors (such as high ATP levels)

    ratios of concentrations of effectors or indicators of energy levels (such as high NAD^+/NADH or low AMP/ATP)

    specific energy conditions (running vs. resting, for example)

Use this space to note any additional objectives provided by your instructor.

CHAPTER 9. SOLUTIONS TO REVIEW QUESTIONS

9.1   a.   Anaerobes are organisms that grow without utilizing oxygen in energy generation.

b.   Anaplerotic reactions replenish substrates used in biosynthetic processes.

c.   Glyoxysomes are plant organelles that possess glyoxylate cycle enzymes.

d.   Reduction potential is the tendency for a specific substance to gain electrons.

e.   An electron donor and its acceptor are a conjugate redox pair.

9.2   Ancient earth possessed an atmosphere that contained methane, ammonia and was devoid of oxygen. With the development of photosynthesis, oxygen was released into the atmosphere. Subsequently, this oxygen reacted with methane to form carbon dioxide and with ammonia to form molecular nitrogen. The continued release of oxygen produced an oxidizing atmosphere consisting of primarily oxygen, nitrogen, and carbon dioxide.

9.4   Contracting muscle converts large amounts of ATP to ADP in the muscle cells. The drop in ATP concentration stimulates two key regulatory enzymes of the citric acid cycle. (1) Citrate synthetase catalyzes the condensation of acetyl-CoA and oxaloacetate to give citrate. ATP is an allosteric inhibitor of this enzyme. As concentrations of ATP drop, this enzyme becomes more active. (2) Isocitrate dehydrogenase, which converts isocitrate to α-ketoglutarate, is inhibited by high concentrations of ATP and activated by high concentrations of ADP. A reduced ATP concentration also stimulates the conversion of pyruvate to acetyl-CoA catalyzed by pyruvate dehydrogenase.

9.5   The citric acid cycle is an important component of aerobic respiration. The NADH and $FADH_2$ produced during oxidation-reduction reactions of the cycle donate electrons to the mitochondrial ETC. Citric acid cycle intermediates are also used as biosynthetic precursors.

9.7   The glyoxylate cycle is a modified version of the citric acid cycle that allows certain organisms (e.g., plants or some microorganisms) to grow by utilizing two-carbon molecules such as acetyl-CoA, acetate, or ethanol. The glyoxylate cycle allows for the net synthesis of larger molecules from two-carbon molecules because the two decarboxylation reactions of the citric acid cycle are bypassed.

9.8   Of all the conditions listed only (d) low citrate concentration indicates a low cell energy status and would increase flux through the citric acid cycle:. The remaining conditions indicate that the cell's energy requirements are currently being met, and would reduce flux through the citric acid cycle.

9.10   The components of the molecules shown are the oxidized and reduced versions of riboflavin, both of which are linked to a phosphate group. The riboflavin contains a reduced form of ribose attached to an isoalloxazine ring system at C1; the phosphate group is attached at C5. To convert these molecules into the redox coenzyme FMN, the nitrogenous base adenine and an additional phosphate group must be added.

9.11   The balanced equations for each of the reactions of the citric acid cycle are:

1. $C_2H_3O\text{-SCoA} + C_4H_2O_5^{2-} + H_2O \rightarrow CoASH + C_6H_5O_7^{3-} + H^+$
       ACETYL-CoA   OXALOACETATE   CITRATE

2. $C_6H_5O_7^{3-} \rightarrow C_6H_5O_7^{3-}$
       CITRATE   ISOCITRATE

3. $C_6H_5O_7^{3-} + NAD^+ \rightarrow C_5H_4O_5^{2-} + NADH + CO_2$
       ISOCITRATE   α-KETOGLUTARATE

4. $C_5H_4O_5^{2-} + NAD^+ + CoASH \rightarrow C_4H_4O_3^{3-}-SCoA + NADH + CO_2$
   α-KETOGLUTARATE   SUCCINYL-CoA

5. $C_4H_4O_3^{3-}-SCoA + GDP + HPO_4^{2-} \rightarrow C_4H_4O_4^{2-} + GTP \text{ (or ATP)} + CoASH$
   SUCCINYL-CoA   (OR ADP) P_i SUCCINATE

6. $C_4H_4O_4^{2-} + FAD \rightarrow C_4H_2O_4^{2-} + FADH_2$
   SUCCINATE   FUMARATE

7. $C_4H_2O_4^{2-} + H_2O \rightarrow C_4H_4O_5^{2-}$
   FUMARATE   L-MALATE

8. $C_4H_4O_5^{2-} + NAD^+ \rightarrow C_4H_2O_5^{2-} + NADH + H^+$
   L-MALATE   L-OXALOACETATE

9.13 The three mechanisms of control of the irreversible reactions of the citric acid cycle are: substrate availability, product inhibition, and competitive feedback inhibition. Citrate synthase catalyzes reaction 1, and is sensitive to availability of substrates oxaloacetate and acetyl-CoA. Also, succinyl-CoA, a product of reaction 4, is a competitive feedback inhibitor for citrate synthase. Citrate, NADH, and ATP are all allosteric inhibitors as well. Isocitrate dehydrogenase, which catalyzes the next irreversible reaction (3), is activated by high ADP and NAD^+ and inhibited by ATP and NADH. The third irreversible reaction (4) is catalyzed by α-ketoglutarate dehydrogenase, which is activated by AMP, and inhibited by products succinyl-CoA and NADH.

9.14 Oxygen is required for the electron transport chain (ETC):

$$\tfrac{1}{2}O_2 + NADH + H^+ \rightarrow H_2O + NAD^+$$

Without oxygen, the ETC shuts down, and NADH accumulates at the expense of NAD^+. Both high NADH and low NAD^+ levels inhibit the citric acid cycle. Low NAD^+ levels would also impact the two key regulatory enzymes that utilize NAD^+ as a substrate.

9.16   a. flavoprotein – a class of enzymes that catalyze redox reactions and require a flavin coenzyme such as FMN or FAD

        b. FAD – flavin adenine dinucleotide – a coenzyme for the flavoproteins; an electron carrier that can accept two H• atoms to form $FADH_2$

        c. FMN – flavin mononucleotide – a coenzyme for the flavoproteins; an electron carrier that is capable of transferring one H• atom at a time

        d. lipoic acid – a coenzyme that functions as a carrier of hydrogens or acetyl groups; required by the pyruvate dehydrogenase complex and the α-ketoglutarate dehydrogenase complex

        e. TPP – thiamine pyrophosphate – the coenzyme form of thiamine (vitamin B1); functions in decarboxylation and/or aldehyde group transfer reactions

9.17 Pyruvate dehydrogenase complex is a large multienzyme complex that contains three enzyme activities, each of which is present in multiple copies. E_1 is pyruvate dehydrogenase (or pyruvate decarboxylase) with TPP, E_2 is dihydrolipoyl transacetylase with lipoic acid and CoASH, and E_3 is dihydrolipoyl dehydrogenase, with NAD^+ or FAD. The pyruvate dehydrogenase complex in mammalian cells contains 60 copies of E_2 and 20-30 copies each of E_1 and E_3.

9.19 In the citric acid cycle, stage I, reactions 1-4, two carbon atoms enter the cycle as acetyl-CoA, and two carbon atoms leave the cycle as two molecules of CO_2. The products of stage I are: succinyl-CoA, 2NADH, $2CO_2$, and H^+. In stage II, reactions

5-8, oxaloacetate is regenerated from succinyl-CoA. The products of stage II are: L-oxaloacetate, CoASH, $FADH_2$, NADH, H^+, and either ATP or GTP.

9.20 Excess citrate is transported to the cytoplasm and cleaved to regenerate oxaloacetate and acetyl-CoA. Acetyl-CoA may then be used to synthesize fatty acids.

9.22 Unlike organisms that are capable of the glyoxylate pathway, animals cannot use two-carbon molecules as precursors in gluconeogenesis. One pathway to gluconeogenesis begins with pyruvate reacting to form oxaloacetate, then phosphoenolpyruvate. Since the decarboxylation of pyruvate to form acetyl-CoA is irreversible, and there is no biosynthetic pathway from acetyl-CoA to pyruvate, it cannot be formed from two-carbon molecules. Gluconeogenesis may use oxaloacetate as a precursor. However, oxaloacetate cannot be synthesized solely from acetyl-CoA. When acetyl-CoA enters the citric acid cycle by reacting with oxaloacetate, it adds two carbons to the cycle. However, two carbons are removed as CO_2 in subsequent reactions. Thus, there is no net addition of carbon atoms to the cycle. If oxaloacetate is used for gluconeogenesis, it must be replenished via a citric acid cycle intermediate larger than acetyl-CoA, or by its synthesis from certain amino acids. The glyoxylate cycle, not present in animals, does allow gluconeogenesis from 2-carbon molecules because it bypasses the citric acid cycle reactions that liberate CO_2.

9.23 The net equation for the citric acid cycle is:

$$Acetyl\text{-}CoA + 3NAD^+ + FAD + GDP(or\ ADP) + P_i + 2H_2O \rightarrow$$
$$2CO_2 + 3NADH + FADH_2 + CoASH + GTP(or\ ATP) + 2H^+$$

9.25 Biosynthetic pathways that utilize citric acid cycle intermediates as precursors include the syntheses of: glucose from oxaloacetate; pyrimidines from oxaloacetate, purines from α-ketoglutarate, fatty acids and cholesterol from citrate; porphyrin, heme, or chlorophyll from succinyl-CoA; the amino acids Asp, Lys, Thr, Ile, and Met from oxaloacetate; and the amino acids Glu, Gln, Pro, and Arg from α-ketoglutarate.

CHAPTER 9. SOLUTIONS TO THOUGHT QUESTIONS

9.26 In substrate-level phosphorylation, ADP is converted to ATP by the direct transfer of a phosphoryl group from a high energy compound. The only reaction in the citric acid cycle that involves this type of reaction is the cleavage of succinyl-CoA to form succinate, CoASH, and GTP. Another example of a substrate level phosphorylation is the glycolytic reaction that converts phosphoenolpyruvate and ADP to pyruvate and ATP.

9.28 The biosynthesis of glutamate from pyruvate is shown below:

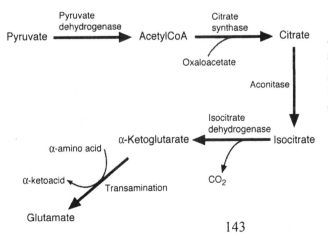

Also, the oxaloacetate (in the second reaction) may be generated from pyruvate in a reaction catalyzed by pyruvate carboxylase.

9.29 Glutamate is converted to α-ketoglutarate in a transamination reaction. The α-ketoglutarate enters the citric acid cycle and is eventually converted to CO_2 and H_2O.

9.31 The presence of oxygen makes possible a much higher metabolic rate. Oxygen is the ultimate electron acceptor. In its absence metabolic intermediates accumulate and disable the energy generation systems of the body. The organism then dies.

9.32 Three enzymes that require thiamine are pyruvate dehydrogenase, a-ketoglutarate dehydrogenase and transketolase. Thiamine is involved in decarboxylation and acyl group transfer reactions. Absence of decarboxylation reactions would prevent pyruvate from being decarboxylated to acetyl-CoA. The body would then lack two carbon units for synthesis and energy production. Pyruvate accumulates as lactate. The overall results of thiamine deficiency are lack of energy, muscle wasting and acidosis.

9.34 Malonate is a competitive inhibitor of succinate dehydrogenase. It competes reversibly with succinate for the active site in the enzyme. Increasing the concentration of succinate should reverse the effect of the malonate.

9.35 Carboxylation of pyruvate produces oxaloacetate, a citric acid cycle intermediate. Increasing the concentration of one of the intermediates stimulates the cycle and more energy is produced.

9.37 When oxygen levels are reduced products of the citric acid cycle accumulate. Energy generation is primarily by glycolysis. The end product of glycolysis is pyruvate. In order to regenerate the NAD^+ required for the reaction pyruvate is reduced to lactate.

9.38 Inhibition of pyruvate dehydrogenase kinase, an enzyme that contributes to the regulation of pyruvate dehydrogenase, has the effect of increasing the conversion of pyruvate molecules to acetyl-CoA, thereby decreasing lactate levels.

9.40 The acetyl-CoA is used to produce energy, so its carbons are not available for synthesis of oxaloacetate. (Also, acetyl-CoA brings two carbons into the citric acid cycle, and two CO_2 molecules are produced, leaving no net addition of carbons to make oxaloacetate.) The required additional oxaloacetate molecules are synthesized by pyruvate carboxylase from pyruvate and carbon dioxide.

9.41 $\Delta G° = -nF\Delta E°$

The half cells for the two reactions are: $\Delta E°$

$$S + 2H^+ + 2e^- \rightarrow H_2S \qquad\qquad -0.23V$$

$$\tfrac{1}{2}O_2 + 2H^+ + 2e^- \rightarrow H_2O \qquad\qquad +0.82V$$

$$NADH \rightarrow NAD^+ + H^+ + 2e^- \qquad +0.32V$$

For the formation of hydrogen sulfide the reaction would be:

$$S + H^+ + NADH \rightarrow H_2S + NAD^+ \quad = +0.09V$$

$\Delta E° = -0.23$ **V** $+ 0.32$ **V** $= +0.09$ **V**

$\Delta G° = $ **2(96485 J/ mol V)(0.09)** $=$ **17,367 J**

For the formation of water the reaction would be

$$\tfrac{1}{2}O_2 + H^+ + NADH \rightarrow H_2O + NAD^+ = +1.14 V$$

$\Delta E° = +0.82V + 0.32V = 1.14$ **V**

$\Delta G° = $ **2(96485 J/ mol V)(1.14 V)** $=$ **219985 J**

The difference in energy yield is 219985 - 17367 = 202.618 kJ or 200 kJ

10 Aerobic Metabolism II: Electron Transport and Oxidative Phosphorylation

Brief Outline of Key Terms and Concepts

10.1 ELECTRON TRANSPORT
AEROBIC RESPIRATION

ELECTRON TRANSPORT AND ITS COMPONENTS
The **ELECTRON TRANSPORT CHAIN (ETC)** is located in the inner mitochondrial membrane of eukaryotic cells. Electrons are transferred from NADH and $FADH_2$ through the following complexes, ultimately reducing O_2 to H_2O. Free energy released in the ETC pumps H^+ from the matrix into the intermembrane space, creating a proton gradient across the membrane. Electron carriers include: IRON-SULFUR CENTERS; NONHEME IRON PROTEINS; COENZYME Q (UQ, UBIQUINONE) and CYTOCHROMES.
COMPLEX I (NADH DEHYDROGENASE COMPLEX)
COMPLEX II (SUCCINATE DEHYDROGENASE COMPLEX)
COMPLEX III (CYTOCHROME B_1 COMPLEX; Q CYCLE)
COMPLEX IV (CYTOCHROME OXIDASE)

ELECTRON TRANSPORT INHIBITORS
Antimycin A inhibits cyt b (complex III). Rotenone and amytal inhibit NADH dehydrogenase (complex I). CO, N^{3-}, CN^- inhibit cytochrome oxidase (complex IV).

10.2 OXIDATIVE PHOSPHORYLATION

THE CHEMIOSMOTIC COUPLING THEORY
The proton (H^+) gradient (ΔpH) created by the ETC is also an electrical potential, Δp, the PROTONMOTIVE FORCE. Protons may only cross the membrane through specific channels (ATP synthase), resulting in the synthesis of ATP. UNCOUPLER; IONOPHORE

ATP SYNTHESIS
ATP SYNTHASE converts energy from the protonmotive force into rotational energy, which causes changes in conformation: L(loose), T(tight), O(open). ADP and P_i bind, and ATP is released.

CONTROL OF OXIDATIVE PHOSPHORYLATION
ATP synthase is inhibited by high [ATP] and activated by high [ADP]; RESPIRATORY CONTROL is control by [ADP]. The ATP mass action ratio is controlled in part by transport proteins in the inner membrane:

ADP-ATP TRANSLOCATOR
PHOPHATE TRANSLOCASE ($H_2PO_4^-/H^+$ SYMPORTER)
The P/O RATIO reflects the coupling between electron transport and ATP synthesis. The measured maximum P/O ratios for NADH and $FADH_2$ are 2.5 and 1.5, respectively.

THE COMPLETE OXIDATION OF GLUCOSE
GLYCEROL PHOSPHATE SHUTTLE
MALATE-ASPARTATE SHUTTLE
Aerobic oxidation of glucose yields 29.5 to 31 ATP molecules.

UNCOUPLED ELECTRON TRANSPORT
UNCOUPLING PROTEINS (UCP1 or THERMOGENIN; also UCP2, UCP3)
Dissipation of the proton gradient releases energy as heat: NONSHIVERING THERMOGENESIS

10.3 OXYGEN, CELL FUNCTION, AND OXIDATIVE STRESS
RESPIRATORY BURST

REACTIVE OXYGEN SPECIES (ROS)
ROS, unstable RADICALS that are derivatives of O_2, ROS include SUPEROXIDE RADICAL, HYDROXYL RADICAL, SINGLET OXYGEN. ROS form as a normal by-product of metabolism, as O_2, a DIRADICAL, accepts electrons one at a time during reduction. Other causes include radiation exposure. REACTIVE NITROGEN SPECIES (RNS) are often classified as ROS because the synthesis of these species is often linked. Examples of RNS: NO (nitric oxide), nitrogen dioxide, and peroxynitrite.

ANTIOXIDANT ENZYME SYSTEMS
The major enzymatic defenses against oxidative stress are SUPEROXIDE DISMUTASE (SOD), GLUTATHIONE PEROXIDASE [part of the GLUTATHIONE-CENTERED SYSTEM (GSH/GSSG)] PEROXIREDOXIN (PRX), and CATALASE. The TRX-centered system includes TRX (thioredoxin), TR (thioredoxin reductase) and PRX.

ANTIOXIDANT MOLECULES protect cell components from oxidative damage. Prominent antioxidants include GSH and the dietary components α-tocopherol, β-carotene, and ascorbic acid.

10.1 ELECTRON TRANSPORT AND ITS COMPONENTS

WHY BOTHER?

Energy. AEROBIC RESPIRATION (using oxygen to generate energy from food) yields much more energy from nutrients than does fermentation. This energy is used to do work such as synthesizing ATP, pumping Ca^{2+} into the mitochondrial matrix, and generating heat in brown adipose tissue. ATP hydrolysis fuels endergonic reactions, runs molecular machines, and in general, acts as the energy "currency" for the cell.

WHY OXYGEN?

O_2 PROPERTIES THAT ARE RELEVANT TO AEROBIC METABOLISM:

O_2 is everywhere (almost), and diffuses easily across cell membranes.

O_2 is highly reactive so it accepts electrons readily. (That could also be bad; O_2 can form ROS.)

ELECTRON TRANSPORT AND ITS COMPONENTS

ELECTRON DONORS

NADH and $FADH_2$ (from glycolysis, citric acid cycle, and fatty acid oxidation)

INTERMEDIARIES THAT CAN TRANSFER 1 ELECTRON AT A TIME: FMN, FAD, UBIQUINONE (UQ), IRON-SULFUR CENTERS and CYTOCHROMES (contain Fe and/or Cu).

IRON-SULFUR CENTERS are held in place by Cys residues in nonheme iron proteins, and occur as 2Fe-2S or 4Fe-4S (not counting the cysteinyl sulfurs).

FMN, **FAD**, and **UQ** have resonance-stabilized radicals. (FMN and FAD: see Redox Coenzymes, pp. 312-314 of your text.)

UBIQUINONE (UQ) (COENZYME Q)

This UQ has six isoprenoid units (as some bacteria). Mammalian UQ, with ten isoprenoid units, is Q_{10}.

The long hydrocarbon tail allows UQ to diffuse within the inner membrane, so it's mobile – it travels to carry electrons between the donors at complex I or II and the acceptor at complex III. The semiquinone intermediate is a resonance-stabilized radical.

| Ubiquinone (UQ) (oxidized or quinone form) | Ubisemiquinone (UQH•) (radical or semiquinone form) | Ubiquinone (UQH$_2$) (reduced or hydroquinone form) |

CYTOCHROMES: a series of electron transport proteins that carry an Fe-containing heme prosthetic group (similar to the heme in hemoglobin).

Example: cyt c (Fe^{3+}) + 1e$^-$ $\rightarrow$ cyt c (Fe^{2+})
               oxidized form   reduced form

ELECTRON TRANSPORT LOCATION: Eukaryotes: in the inner membrane of the mitochondria
                                   Aerobic Prokaryotes: in the plasma membrane

ELECTRON TRANSPORT COMPLEX	PATH OF ELECTRONS		RESULT
	FROM e^- DONOR:	TO e^- ACCEPTOR:	
COMPLEX I **NADH DEHYDROGENASE COMPLEX:** FMN, ~7 iron-sulfur centers	from NADH	→FMN →FeS →FeS →UQ→ $\boxed{UQH_2}$	$4H^+$ transported from the matrix to the intermembrane space new e^- carrier: UQH_2
COMPLEX II **SUCCINATE DEHYDROGENASE COMPLEX:** Succinate dehydrogenase, FAD (bound), 2 Fe-S proteins	from succinate	→FAD →FeS centers (several) →UQ → $\boxed{UQH_2}$	$2e^-$ transferred from succinate (forms fumarate) new e^- carrier: UQH_2
In some cell types: Glycerol-3-phosphate dehydrogenase (on the intermembrane-space-side surface of the inner membrane)	from cytoplasmic NADH	→DHAP →FAD →FeS centers (several) →UQ→ $\boxed{UQH_2}$	$2e^-$ transferred from cytoplasmic NADH new e^- carrier: UQH_2
Acyl-CoA dehydrogenase (matrix side of inner membrane) (first step of fatty acid oxidation)	from fatty acyl-CoA	→FAD →FeS centers (several) →UQ→ $\boxed{UQH_2}$	2 e^- transferred from fatty acyl-CoA new e^- carrier: UQH_2
COMPLEX III (Q CYCLE) **CYTOCHROME bc_1 COMPLEX** 1 Fe-S center cyt b_L cyt b_H cyt c_1 cyt c (loosely attached to intermembrane-space side)	from UQH_2 (space-side)	$1e^-$ →Fe-S →cyt c_1→ $\boxed{cyt\ c}$	$UQH_2 + 2cyt\ c_{ox}(Fe^{3+})$ $+2H^+$ matrix → UQ + $2cyt\ c_{red}(Fe^{2+}) + 4H^+$ cytosol
	$UQH•$ (space-side, from UQH_2 above)	$1e^-$→cyt b_L →cyt b_H →UQ→ $\boxed{UQ•}$ matrix-side	$1e^-$ transferred from UQH_2 new e^- carrier: cyt c
	from UQH_2 (space-side)	$1e^-$→cyt b_L →cyt b_H →UQ• matrix-side→ $\boxed{UQH_2}$	$1e^-$ transferred from UQH• (space-side) new e^- carrier: UQH• (matrix-side) $1e^-$ transferred from UQH_2 new e^- carrier: UQH_2 Transport of $4H^+$ to intermembrane space
COMPLEX IV **CYTOCHROME OXIDASE** cyt a, cyt a_3 2Cu: Cu_A near cyt a; Cu_B near cyt a_3	from cyt c $1e^-$ at a time	→Cu_A →cyt a →cyt a_3 →Cu_B → O_2→ $\boxed{H_2O}$	$O_2+ 4H^+ + 4e^- →2H_2O$ $2H^+$ (per cyt c) transported to the intermembrane space

ELECTRON TRANSPORT INHIBITORS

Inhibitors have been used to elucidate the order of the ETC components.

ANTIMYCIN A inhibits cyt b (Complex III).

ROTENONE and AMYTAL inhibit NADH dehydrogenase (Complex I).

CARBON MONOXIDE, AZIDE, and CYANIDE (CO, N^{3-}, CN^-) inhibit cytochrome oxidase (Complex IV).

10.2 OXIDATIVE PHOSPHORYLATION = USING ETC ENERGY TO MAKE ATP

THE CHEMIOSMOTIC THEORY

H^+ ions are transported from the matrix to the intermembrane space during ETC electron transfer. This creates a proton gradient (ΔpH) that is also an electrical potential, the PROTONMOTIVE FORCE (Δp), across the inner membrane.

The H^+ ions in the intermembrane space can only flow back to the matrix (down the proton gradient) through special channels. As they pass through the channels, ATP synthesis occurs. Yay! ☺

Evidence for the chemiosmotic theory:

1. Working mitochondria take in O_2 and expel H^+.
2. If the mitochondrial membrane is disrupted, the ETC continues but ATP synthesis stops.
3. UNCOUPLERS and IONOPHORES allow H^+ to leak across the membrane (kind of like a free route around a toll booth). Uncouplers carry H^+ across, and ionophores form channels through the membrane. Examples: Dinitrophenol is an uncoupler; gramicidin A is an ionophore. (See Uncoupled Electron Transport, next page.)

ATP SYNTHESIS

ATP SYNTHASE is a molecular motor that resembles a lollipop. Keep in mind:

ATP SYNTHASE MAY *LOOK* LIKE A LOLLIPOP, BUT IT *ACTS* LIKE A MOTOR.

Think of ATP synthase primarily as a motor. Knowing that it looks like a lollipop can be confusing when studying its function. Twirl a lollipop, and all of it spins. Turn the shaft, or rotor (the lollipop stick), in ATP synthase and the stout, roundish stator stays still. THE "LOLLIPOP" CANDY PART IS HELD IN PLACE AND DOES NOT ROTATE.

As a motor, it converts the PROTONMOTIVE FORCE (the energy from the H+ gradient across the inner membrane) into a rotational force. This rotational force causes conformational changes that result in ATP synthesis.

STUDY HINT: Spend some time comparing Figures 14 and 15 on pp. 350-351 of your text.[1] Think of ATP synthase as a MOTOR and refer to these figures to connect its structure (Fig. 14) with its production of ATP (Fig. 15). Again, *the α,β-hexamer* (located on the matrix side of the inner membrane) *is stationary – it does not rotate.*

---

[1] Figure 10.14, *ATP Synthase from* Escherichia coli, shows the location of each subunit, with an arrow that identifies the 12-part rotor. Figure 10.15, *ATP Synthesis Model,* shows only the catalytic sites on the β subunits of the stator. Fig. 14 labels each subunit, with the α,β-hexamer on the matrix side. Fig. 15, a close-up of the catalytic action at the three β subunits, looking down at ATP synthase from the matrix side. The α-subunits and the central shaft (the g and e subunits) have been omitted for clarity.

For example, in Fig. 15 (p. 351 of your text), the section that faces the top of the page changes its conformation from L to T in the second step – it did not turn with the shaft.

Draw your own version of these figures, with the subunits labeled. Use colors to identify the rotor vs. the stator subunits.

ATP Synthase Components and Their Functions

F_0 unit   transmembrane H^+ channel
contains subunits a, b, and c in the ratio $a:b_2:c_{12}$
c ring = 12 c subunits

F_1 unit   active ATP synthase with three nucleotide-binding sites
located on the matrix side of the inner mitochondrial membrane
contains 5 subunits: 3 $\alpha\beta$ dimers, γ, δ, and ϵ

Note that F_1 subunits are named with Greek letters, and the F_0 subunits are not. Also, the units F_0 and F_1 identify the structure of ATP synthase, but not the function. Both the rotor and the stator include subunits of both F_0 and F_1.

ROTOR ϵ, γ, c_{12} The c ring and the shaft (ϵ, γ) rotate 120° for each H^+ that passes through.

STATOR a, b_2, δ, α_3, β_3 a contains the entry and exit channels for H^+
b and δ prevent the three $\alpha\beta$-hexamers from rotating
β subunit – site of catalysis; has three possible conformations: L, T, and O

Rotation changes the protein conformations, which changes the binding affinity.
Conformation L: Loose: binds ADP and P_i weakly
Conformation T: Tight: brings substrates closer, facilitates ATP formation
Conformation O: Open: nonbinding, expels ATP ONLY AFTER another ADP and P_i bind to the adjacent β subunit

MECHANISM:

1. ADP and P_i bind to the β subunit in the L conformation.

2. H^+ enters through the channel in the a subunit, and is passed from an Arg to an Asp residue in c, causing a conformational change and motion that results in the entire c ring to rotate by 120°. The torque causes the shaft to rotate, which causes a conformational change in β from L to T, and ATP synthesis occurs.

3. The shaft continues to rotate, changing the *adjacent* β from T to O, and an ATP is released. Meanwhile, the ATP made in step #2 is still in its β- in the T conformation. After another ADP and P_i bind, this ATP will be released.

After three H^+ ions pass through, a full turn is made and an ATP is released.

ADP's list of things to do: *Meet little P_i in the $\beta(L)$.*
When it changes to $\beta(T)$, link up with P_i to become ATP. (Remember that H_2O will fall out but it's o.k..)
Wait for the change to O to go.

CONTROL OF OXIDATIVE PHOSPHORYLATION

P/O RATIO: moles P_i used (to make ATP) per O_2 reduced to H_2O. This ratio correlates the ETC with ATP synthesis. The max P/O ratio measured for NADH is 2.5 and for $FADH_2$ is 1.5.

RESPIRATORY CONTROL: control of aerobic respiration by ADP (that is, there has to be enough ADP and P_i available to be able to make ATP)

ATP SYNTHASE: inhibited by ATP, activated by ADP and P_i.

Ratio of ATP and ADP controlled by:

ADP-ATP TRANSLOCATOR: ships ATP out (of the matrix) and imports ADP (into the matrix from the intermembrane space). Remember that the intermembrane space is more positive (because of all of those H^+ ions) and the matrix is more negative. So, shipping out ATP (with its extra negative charge) is favored.

PHOSPHATE TRANSLOCASE: imports $H_2PO_4^-$ (into the matrix) with an H^+ from the intermembrane space. (This is driven by the proton gradient.)

For each ATP synthesized, FOUR H^+ total must pass through the membrane: three to turn the ATP synthase rotor plus one to provide the phosphate.

FOR THE COMPLETE OXIDATION OF GLUCOSE:

The electrons captured by NADH (from glycolysis in the cytoplasm) has to be transported into the mitochondrion, since NADH can't cross the inner mitochondrial membrane.

GLYCEROL PHOSPHATE SHUTTLE

Converts cytoplasmic NADH (from glycolysis) to inner membrane $FADH_2$

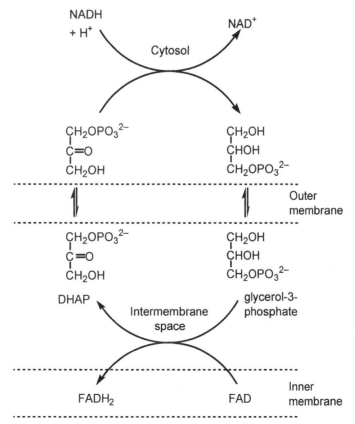

MALATE-ASPARTATE SHUTTLE

Converts NADH and an H^+ from the intermembrane space into a matrix NADH and H^+, so it can enter the ETC. See Figure 10.17b, page 355 of your text. Because of that H^+ import, the number of ATP produced per NADH is a bit less.

ONE MORE SHUTTLE

Transports two citric-acid-cycle-produced ATPs from the matrix to the cytoplasm (so they can be used) cost $2H^+$ (total), so that also reduces the amount of ATP that can be produced from glucose oxidation.

The Complete Oxidation of Glucose:

$$C_6H_{12}O_6 + 6O_2 + 30ADP + 30P_i \rightarrow 6CO_2 + 6H_2O + 30ATP$$

UNCOUPLED ELECTRON TRANSPORT AND HEAT GENERATION

Sometimes a little leak is a good thing. When uncouplers and ionophores allow H^+ ions to cross the membrane, the energy that would have been used to make ATP is released as heat. That's **NONSHIVERING THERMOGENESIS**: "generating heat without shivering." Brown fat cells have many mitochondria that literally are little furnaces. In their inner membranes, they contain **THERMOGENIN (UNCOUPLING PROTEIN 1** or **UCP1)**.

Norepinephrine is a neurotransmitter (released from neurons that end in brown fat) that ultimately causes fat molecules to hydrolyze. The fatty acids produced activate UCP1. Other uncoupling proteins are UCP2 (in most mammals, linked to body weight regulation) and UCP3 (linked to thermogenic effect of thyroid hormone; also detected in muscle).

10.3 OXYGEN, CELL FUNCTION, AND OXIDATIVE STRESS

REACTIVE OXYGEN SPECIES (ROS) ARE FORMED FROM O_2

$O_2^{-\bullet}$	**SUPEROXIDE RADICAL**	$^1O_2$	**SINGLET OXYGEN** (an unpaired electron absorbs energy and shifts to a higher orbital, i.e., an excited state); formed from superoxide or peroxide
$\bullet OH$	**HYDROXYL RADICAL** – very reactive, can initiate a chain reaction; can be formed from peroxide	H_2O_2	**HYDROGEN PEROXIDE**, H-O–O-H (can cross membranes) $Fe^{2+} + H_2O_2 \rightarrow Fe^{3+} + \bullet OH + OH^-$

These radicals may be new to you (with the possible exception of peroxide). They're so reactive they can't be bought – to use them in a lab you'd have to make them just prior to reacting them. Hydrogen peroxide isn't a radical – yet – but its O–O single bond can react to form radicals. (By the way, that's why drugstore peroxide is in a brown bottle – light energy would degrade it.) ROS formation is linked to aging, radiation exposure, smoking, and certain diseases.

Oxygen, or dioxygen (O_2) is itself a diradical, and accepts electrons one at a time during reduction. So, ROS formation is also a normal byproduct of metabolism.

ROS react as soon as they're formed, oxidizing and damaging whatever they touch: inactivating enzymes, depolymerizing polysaccharides, breaking DNA, destroying membranes, etc. Macrophages and neutrophils seek out microorganisms and damaged cells and destroy them with a **RESPIRATORY BURST** – ROS created specifically for this purpose.

REACTIVE NITROGEN SPECIES (RNS) are often classified as ROS because the synthesis of these species is often linked. Nitric oxide, •NO, can damage proteins with sulfhydryl groups or iron-sulfur centers. Other examples of RNS are nitrogen dioxide, and peroxynitrite.

ANTIOXIDANT ENZYME SYSTEMS DESTROY ROS

The major enzymatic defenses against oxidative stress are **SUPEROXIDE DISMUTASE (SOD)**, **GLUTATHIONE PEROXIDASE** [part of the **GLUTATHIONE-CENTERED SYSTEM (GSH/GSSG)**] **PEROXIREDOXIN (PRX)**, and **CATALASE**.

SUPEROXIDE DISMUTASE (SOD): $2 O_2^- + 2H^+ \rightarrow H_2O_2 + O_2$

Two types in humans: Cu-Zn isozyme in the cytoplasm, Mn isozyme in the mitochondrial matrix. Lou Gehrig's disease is caused by a mutation in the gene that codes for cytosolic Cu-Zn isozyme of SOD.

GLUTATHIONE PEROXIDASE: $2 GS\text{–}H \rightarrow GS\text{–}SG$

Controls cellular peroxide levels; contains selenium

$2\ GSH + R\text{–}O\text{–}O\text{–}H \rightarrow G\text{–}S\text{–}S\text{–}G + R\text{–}OH + H_2O$
(If R = H, then ROOH is simply hydrogen peroxide so ROH is water.)

Glutathione reductase regenerates GSH

$$G\text{–}S\text{–}S\text{–}G + NADPH + H^+ \rightarrow 2\ GSH + NADP^+$$

NADPH is from the pentose phosphate pathway, isocitrate dehydrogenase, malic enzyme

PEROXIREDOXINS (PRX) DETOXIFY PEROXIDES

by reacting peroxides with a cysteine residue –SH. Peroxide is reduced, and the thiol is oxidized to sulfinate. Cys –SH is regenerated by another –SH-containing protein like thioredoxin (TRX). The TRX-centered system includes TRX (thioredoxin), TR (thioredoxin reductase) and PRX.

CATALASE (CONTAINS HEME) CATALYZES $2 H_2O_2 \rightarrow O_2 + 2 H_2O$

Step 1. $H_2O_2 + Fe(III)\text{–}enzyme \rightarrow H_2O + O{=}Fe(IV)\text{–}enzyme$
Step 2. $H_2O_2 + O{=}Fe(IV)\text{–}enzyme \rightarrow O_2 + H_2O + Fe(III)\text{–}enzyme$

ANTIOXIDANT MOLECULES

Antioxidant molecules can accept an electron from ROS to form a radical that is stabilized by resonance and won't be as reactive. Great examples of the stabilized radicals are FMNH• and UQH• (structures appear in this chapter, p. 147). **β-Carotene** is an important antioxidant in membranes and a precursor to vitamin A in animals. Other important antioxidants: vitamin C (ascorbic acid), GSH, and α-tocopherol (vitamin E).

α-Tocopherol
(Vitamin E)

ABBREVIATIONS AND ACRONYMS

ATP	adenosine triphosphate	NAD	nicotinamide adenine dinucleotide
cyt	cytochrome	RNS	reactive nitrogen species
DNP	dinitrophenol (an uncoupler)	ROS	reactive oxygen species
ETC	electron transport chain	SMP	submitochondrial particles
FAD	flavin adenine dinucleotide	SOD	superoxide dismutase
FMN	flavin mononucleotide	UQ	ubiquinone, or coenzyme Q

AFTER STUDYING THIS CHAPTER, YOU SHOULD BE ABLE TO:

- Explain how certain electron carriers such as UQ can transfer electrons one at a time.

- Describe the prominent features of each complex in the ETC, including the enzymes, cofactors, and coenzymes present.

- Trace electrons through the each complex of the ETC.

- Identify which metabolites would accumulate should a given ETC enzyme be inhibited.

- Describe how the proton gradient is formed and its function in metabolism.

- Describe the chemiosmotic coupling theory.

- Outline the mechanism of ATP synthesis, and correlate each step with specific structural features of ATP synthase.

- Describe the mode of action of uncoupling agents, and predict their effect on the metabolic pathways affected as well as on the organism.

- Describe ROS – how they are formed, their mode of action, and how they are destroyed.

Use this space to note any additional objectives provided by your instructor.

CHAPTER 10: SOLUTIONS TO REVIEW QUESTIONS

10.1   a. chemical coupling hypothesis – postulates that a high-energy intermediate generated by electron transport is used to drive ATP synthesis

        b. chemiosmotic coupling theory – postulates that a proton gradient created by the mitochondrial electron transport system drives ATP synthesis

        c. ionophore – a hydrophobic molecule that inserts into membrane and dissipates osmotic gradient

        d. respiratory control – the regulation of aerobic respiration by ADP

        e. ischemia – inadequate blood flow

10.2   Principal sources of electrons for the mitochondrial electron transport system are NADH and $FADH_2$.

10.4   The chemiosmotic coupling theory has the following principal features. As electrons pass through the electron transport chain (ETC), protons are transported from the matrix and released into the inner membrane space. As a result, an electrical potential (Ψ) and a proton gradient (ΔpH) are created across the inner membrane. The electrochemical proton gradient is sometimes referred to as the protonmotive force (Δp). Protons, which are present in the intermembrane space in great excess, can pass through the inner membrane and back into the matrix down their concentration gradient only through the proton-translocating ATP synthase.

10.5   According to the chemiosmotic theory, an intact inner mitochondrial membrane is required to maintain the proton gradient required for ATP synthesis.

10.7   The translocation of three protons is required to drive ATP synthesis. The fourth proton drives the transport of ADP and P_i.

10.8   Oxygen is widely used as an energy source because it makes possible greater energy yields from food molecules and is readily available.

10.10   a. rotor – a machine component that rotates; in ATP synthase, the ε, γ, and c_{12} subunits rotate

        b. stator – a stationary machine component. The stator in ATP synthase (subunits a, b_2, δ, α_3 and β_3) is fixed in relation to the rotating F_o component (subunits ε, γ, and c_{12}).

        c. uncoupler – a molecule that uncouples ATP synthesis from electron transport; it collapses proton gradients by transporting protons across a membrane

        d. malate-aspartate shuttle – a metabolic process in which the electrons from NADH in the cytosol are transferred to mitochondrial NAD^+

        e. ROS – reactive oxygen species – reactive derivatives of molecular oxygen

10.11   Examples c, d, and f are all reactive oxygen species. Each of these species can act as a free radical, that is, it can attack various cell components, producing such effects as enzyme inactivation, polysaccharide depolymerization, DNA breakage, and membrane destruction.

10.13   ROS damage cells by inactivating enzymes, depolymerizing polysaccharides, breaking DNA, and destroying membranes.

10.14   a. radical – a chemical species with an unpaired electron

        b. protonmotive force – the force arising from a proton gradient and a membrane potential

    c. antioxidant – a substance that prevents the oxidation of other molecules

    d. aerobic respiration – the energy-generating process that uses O_2 as a terminal electron acceptor

    e. respiratory burst – an oxygen-consuming process in scavenger cells such as macrophages in which ROS are generated and used to kill foreign or damaged cells

10.16 A genetic defect with survival value is G-6-PD deficiency. The lower NADPH and GSH concentrations in the red blood cells of G-6-PD-deficient individuals inhibit the growth of *Plasmodium* (the malarial parasite), a major killer of humans.

10.17 Although the end products of the following substances in an electron transport system are not named specifically in the text, we can speculate based upon the most reduced forms of the atom with the highest oxidation state. The nitrogen atom in nitrate (NO_3^-) has an oxidation state of +5; possible reduction products include NO_2^-, NO, and N_2, with NH_3 being the most reduced form. The ferric ion (Fe^{3+}) may reduce to ferrous ion (Fe^{2+}) or metallic iron. From carbon dioxide, CH_4 is the most reduced form of carbon. The most reduced form of sulfur, sulfide, S^{2-}, would likely be the final product of the reduction of both sulfate (SO_4^{2-}) and elemental sulfur.

10.19 Flagellar movement in respiring bacteria would slow or cease as dinitrophenol, an uncoupler, collapses the proton gradient across the organism's plasma membrane. Protons would be carried across the membrane by dinitrophenol instead of passing through ATP synthase to convert ADP to ATP. Without ATP from respiration, flagellar movement would depend upon the generation of ATP from glycolysis, a much less efficient process. The energy from oxidized nutrients would not be captured by ATP synthesis, but would dissipate as heat.

10.20 Reaction equations that illustrate the production of ROS from electrons leaking from the electron transport system include:

$$O_2 + e^- \rightarrow O_2^{\bar{\cdot}} \text{ (superoxide)}$$

$$2H^+ + 2O_2^{\bar{\cdot}} \rightarrow O_2 + H_2O_2 \text{ (peroxide)}$$

$$Fe^{2+} + H_2O_2 \rightarrow Fe^{3+} + OH^- + \cdot OH \text{ (hydroxyl radical)}$$

$$2O_2^{\bar{\cdot}} + 2H^+ \rightarrow H_2O_2 + {}^1O_2 \text{ (singlet oxygen)}$$

$$2 \text{ ROOH} \rightarrow 2 \text{ ROH} + {}^1O_2$$

10.22 Uncoupling agents disrupt phosphorylation by reducing (or collapsing) the proton gradient across the inner mitochondrial membrane. Uncoupling proteins such as UCP1 form a proton channel across this membrane, while other uncoupling agents such as dinitrophenol carry protons across the membrane. Both types of uncoupler provide a mechanism for protons to move across the membrane in a way that bypasses ATPase. The energy captured by the electron transport system and used to build the proton gradient is dissipated as heat instead of being used to do work, i.e., to synthesize ATP. UCP1 is activated by fatty acids that are produced by the hydrolysis of fats in brown adipose tissue. Fat hydrolysis is initiated by a cascade mechanism, which is triggered by the neurotransmitter norepinephrine.

10.23 The net reaction of the electron transport system is highly exergonic. Using a series of oxidations instead of one reaction allows a more controlled, more efficient capture of this energy, avoiding the liberation of a great deal of heat.

10.25 When added to actively respiring mitochondria, cyanide inhibits cytochrome oxidase in Complex IV, resulting in the accumulation of oxygen and H^+ on the matrix side, and reduced cytochrome c in the intermembrane space.

10.26 The minimum voltage drop for individual electron transfer events in the mitochondrial electron transport system that is necessary for ATP synthesis can be ascertained by referring to Figure 10.8. Note that of the three steps in the oxidation of NADH that account for the synthesis of an ATP molecule, the minimum voltage drop (+0.18V) occurs between UQH_2 and cyt c.

CHAPTER 10: SOLUTIONS TO THOUGHT QUESTIONS

10.28 Once a labeled acetate molecule is converted to acetyl-CoA it is processed through the citric acid cycle (Figure 9.8 on p. 317). Because of the symmetrical structure of the intermediate succinate, $^{14}CO_2$ is not released until two or more turns of the cycle. The number of moles of ATP produced from 1 mol of acetate is calculated as follows: Each turn of the cycle produces 10 ATP. Assuming that the release of $^{14}CO_2$ requires two turns of the cycle, a total of 20 ATP are generated. Because 2 ATP are required to convert acetate to acetyl-CoA, the net production of ATP is 18 mol.

10.29 The oxidation of one mole of ethanol to acetyl-CoA produces two moles of NADH. The conversion of acetate to carbon dioxide and water through the citric acid cycle produces 3 NADH, 1 $FADH_2$, and 1 GTP. Assuming that the aspartate-malate shuttle is in operation, each cytoplasmic NADH yields 2.25 ATP for a total of 4.5 ATP. Each mitochondrial NADH yields 2.5 ATP for a total of 7.5 ATP. Each molecule of $FADH_2$ yields 1.5 ATP for a total of 1.5 ATP. Each GTP yields 0.75 ATP for a total of 0.75 ATP. The total ATP produced by the oxidation of ethanol is therefore 14.25 ATP.

10.31 Dinitrophenol collapses the proton gradient across the mitochondrial inner membrane. The energy normally used to drive ATP synthesis is lost as heat.

10.32 Other ions such as sodium or potassium moving across the membrane would disrupt the electrical potential and reduce or eliminate oxidative phosphorylation.

10.34 In order for the system to operate efficiently energy should be released in stages. Each relatively large decrease in potential corresponds to the energy required to generate an ATP molecule. If the energy is released all at once much of the energy would be released as heat and little ATP would be produced. The gradual change in voltage of the cytochromes makes this gradual release of energy possible.

10.35 Complex I transfers protons from NADH to UQ. Inhibition of this process will prevent the oxidation of NADH to NAD^+. However, NADH will accumulate and the NADH/NAD ratio will increase. UQH_2 will be oxidized by the remainder of the cytochrome system thus generating more UQ. As a result the UQ/UQH_2 ratio will also increase.

10.37 If the cytochrome complexes were not embedded in the mitochondrial inner membrane the proton gradient could not be established and ATP synthesis would not occur.

10.38 Rotenone inhibits complex I which oxidizes NADH. NADH is provided by the citric acid cycle with acetyl-CoA, a derivative of pyruvate, as the substrate. Succinate, the substrate for complex II, enters the series below the point where rotenone inhibits.

10.40 Uncouplers do not cause a buildup of ETS intermediates. Inhibitors would cause a buildup of reduced intermediates above the point of inhibition and oxidized intermediates below the point of inhibition.

10.41 Dehydroascorbic acid is a diketo acid. Glutathione reduces the carbonyls to regenerate ascorbic acid.

10.43 The inactivation of aconitase is probably due to the oxidation of one of the iron atoms. As long as there is an oxidizing condition aconitase remains inactive. When ROS levels drop the iron atom is reduced by an intracellular reducing agent and the complex reforms.

10.44 If the MnSOD gene is inactive, superoxide levels will rise and aconitase will be inactivated. This will tend to inhibit the citric acid cycle, which consumes acetyl-CoA derived from pyruvate. If acetyl-CoA is not oxidized it will be shunted into fatty acid synthesis and fatty deposits will build up both in the liver and muscle.

10.46 The carbon radicals generated by the attack of oxygen on an amino acid can dimerize to produce a crosslink between two amino acids.

$$2R'CHCONH \rightarrow RCH(CONH)\text{-}CH(CONH)R$$

11 Lipids and Membranes

Brief Outline of Key Terms and Concepts

11.1 LIPID CLASSES

FATTY ACIDS
Long-chain carboxylic acids; components of triacylglycerols, membrane-bound lipids, and acylated membrane proteins.
CIS-ISOMERS VS. TRANS-ISOMERS
ESSENTIAL VS. NONESSENTIAL FATTY ACIDS
OMEGA-3 AND OMEGA-6 FATTY ACIDS

THE EICOSANOIDS
AUTOCRINE REGULATORS derived from arachidonic acid or EPA
PROSTAGLANDIN   THROMBOXANE
LEUKOTRIENE

TRIACYLGLYCEROLS (NEUTRAL FATS) consist of glycerol esterified to three fatty acids.
MONOUNSATURATED; POLYUNSATURATED
In both animals and plants they are a rich energy source.

WAX ESTERS

PHOSPHOLIPIDS are amphipathic molecules, with a POLAR HEAD GROUP and long, nonpolar hydrocarbon chains. Functions include: membrane components, emulsifying agents, and surface active agents.
Phospholipid types: **PHOSPHOGLYCERIDES** and **SPHINGOMYELINS**.

SPHINGOLIPIDS are important membrane components of animals and plants, contain a complex long-chain amino alcohol (either **SPHINGOSINE** or **PHYTOSPHINGOSINE**). The core of each sphingolipid is **CERAMIDE**, a fatty acid amide derivative of the alcohol molecule. **GLYCOLIPIDS** are derivatives of ceramide that possess a carbohydrate component.
Example: **GPI ANCHORS**

SPHINGOLIPID STORAGE DISEASES

ISOPRENOIDS have repeating units derived from isopentenyl pyrophosphate.

TERPENES	**MIXED TERPENOIDS**
CAROTENOIDS	**PRENYLATION**
STEROIDS	

LIPOPROTEINS
Plasma lipoproteins transport lipids through the bloodstream. Lipoproteins are classified by density (very low density, intermediate density, low density, and high density): **CHYLOMICRONS, VLDL, IDL, LDL, AND HDL**.

11.2 MEMBRANES

MEMBRANE STRUCTURE
LIPID BILAYER, FLUID MOSAIC MODEL, LIPID RAFTS
Lipid bilayers consist of **PHOSPHOLIPIDS** and other **AMPHIPATHIC** lipid molecules.
PERIPHERAL PROTEIN; INTEGRAL PROTEIN

MEMBRANE FUNCTION
Membrane transport mechanisms:
PASSIVE TRANSPORT: does not require energy; solutes cross the membrane down (or with) their concentration gradient; Types of passive transport: **SIMPLE DIFFUSION, FACILITATED DIFFUSION**
ACTIVE TRANSPORT: requires energy (directly or indirectly from ATP hydrolysis or another energy source); solutes cross the membrane against their concentration gradient.
CYSTIC FIBROSIS; CYSTIC FIBROSIS TRANSMEM-BRANE CONDUCTANCE REGULATOR (CTFR)

ADDITIONAL TERMS
HETEROKARYON
AQUAPORINS
NEPHROGENIC DIABETES INSIPIDIS

11.1 LIPID CLASSES

FATTY ACIDS

Fatty acids have a long hydrocarbon chain (12-20 carbons) and one carboxylic acid. They are components of lipid molecules, primarily triacylglycerols and membrane-bound lipids. Fatty acids are AMPHIPHILIC, meaning they contain both hydrophobic and hydrophilic ends.

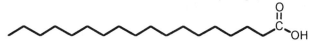

SATURATED VS. UNSATURATED

If the hydrocarbon chain contains at least one alkene, the fatty acid is UNSATURATED (not completely filled with hydrogens). If the hydrocarbon chain contains no double bonds, it's SATURATED (completely filled with hydrogens). MONOUNSATURATED fatty acids have one double bond, and POLYUNSATURATED fatty acids have two or more double bonds. Double bonds can be oxidized.

MELTING POINTS AND FLUIDITY OF A FATTY ACID; *CIS*- VS. *TRANS*-ISOMERS

The lower the melting point, the more fluid a fatty acid (or lipid) is likely to be at a given temperature. What determines whether the fatty acid is a solid or liquid at a given temperature? 1) number of carbons 2) number and type of double bonds.

The longer the hydrocarbon chain, the higher the melting point (that is, it takes more energy to put those long chains in motion).

The greater the number of double bonds, the lower the melting point. A *cis*-alkene causes a "kink" in the chain that prevents *cis*-unsaturated fatty acids from packing as closely as saturated fatty acids do. So, it takes less energy to melt *cis*-unsaturated fatty acids (than a saturated fatty acid with the same number of carbons), and they have lower melt points. (*Trans*-unsaturated fatty acids are similar in conformation to saturated fatty acids.)

Example: Compare the melting points of these fatty acids. Why does stearic acid have the highest melting point, and linoleic acid have the lowest?

myristic acid (14 carbons)	$CH_3(CH_2)_{12}COOH$	54°C
palmitic acid (16 carbons)	$CH_3(CH_2)_{14}COOH$	63°C
stearic acid (18 carbons)	$CH_3(CH_2)_{16}COOH$	70°C
oleic acid	$CH_3(CH_2)_7CH=CH(CH_2)_7COOH$	4°C
linoleic acid	$CH_3(CH_2)_4CH=CHCH_2CH=CH(CH_2)_7COOH$	−12°C

Naturally occurring fatty acids are typically:
unbranched, even number of carbon atoms, if unsaturated: *cis*-isomeric form

DESIGNATION:

(total number of carbons) : (number of double bonds)$^{\Delta(\text{location of double bond})}$

Location of double bond:

Counting from the carboxyl carbon, the first carbon that begins the double bond

Alternatively: The carbon furthest from the carboxyl carbon is the omega(ω) carbon. When one of the double bonds begin 3 or 6 carbons from the end, its designation is ω-3 or ω-6. OMEGA-3 FATTY ACIDS and OMEGA-6 FATTY ACIDS are now believed to promote cardiovascular health.

Example: Linoleic acid, $18:2^{\Delta 9,12}$, has 18 total carbons and two double bonds, one located between carbons 9-10 and the second located between carbons 12-13. The second double bond is located at the sixth carbon from the last, or carbon, so it may be designated $18:2\omega$-6.

NONESSENTIAL VS. ESSENTIAL FATTY ACIDS

NONESSENTIAL FATTY ACIDS can be synthesized by the body.

ESSENTIAL FATTY ACIDS must be obtained from the diet (sources include some vegetable oils, nuts, seeds). Examples: linoleic and linolenic acids; metabolic precursors; symptoms of dietary deficiency include: dermatitis, poor wound healing, reduced resistance to infection, alopecia, thrombocytopenia.

CHEMICAL REACTIONS OF FATTY ACIDS

- esterification: carboxylic acid + alcohol = ester (reversible)
- hydrogenation of unsaturated fatty acids = saturated fatty acids
- oxidation of unsaturated fatty acids (Chapter 12)
- protein acylation – fatty acid groups (ACYL GROUPS) covalently attach to proteins to create ACYLATED PROTEINS

Example: Fatty acids undergo reactions in the body in order to be transported. Esterification to serum proteins allows transport from fat cells to body cells. Acyl transfer reactions allow fatty acids to enter cells.

EICOSANOIDS

FUNCTION: AUTOCRINE REGULATORS (rather than hormones, since they tend to be active in the cell in which they're produced)

NAMING: type of eicosinoid + type of modification + number of double bonds
Type of eicosinoid: PG = PROSTAGLANDIN   TX = THROMBOXANE   LT = LEUKOTRIENE
Type of modification: A = –OH, ether ring   B = two –OH groups

PROSTAGLANDINS
STRUCTURE: contain a cyclopentane ring, –OH groups at C-11 and C-15.
E-series: have a carbonyl at C-9
F-series: have an –OH at C-9
2-series: derived from arachidonic acid; most important group for humans
3-series: derived from EPA (eicosapentaenoic acid)

FUNCTION: involved in wide range of regulatory functions, including involvement in inflammation, reproduction, and digestion.

THROMBOXANES contain a cyclic ether, and are derivatives of arachidonic acid (TXA_2, TXB_2 in platelets) or EPA (TXA_3, TXB_3 in polymorphonuclear leukocytes).

LEUKOTRIENES
STRUCTURE: linear (noncyclic); synthesis begins with peroxidation reaction (lipoxygenase); contains a thioether near the site of this peroxidation.

EXAMPLES of leukotrienes: LTC_4, LTD_4, LTE_4 are components of SRS-A – SLOW-REACTING SUBSTANCE OF ANAPHYLAXIS (an unusually severe allergic reaction). LTB_4 is a potent CHEMOTACTIC AGENT, or CHEMOATTRACTANT; it attracts white blood cells to damaged tissue.

TRIACYLGLYCEROLS – ESTERS OF GLYCEROL WITH THREE FATTY ACIDS
– also called NEUTRAL FATS (they are not charged)

STRUCTURE: All three hydroxyl groups of glycerol have a fatty acid attached via an ester bond. For each ester bond that's formed, a water molecule is removed.

glycerol   fatty acid   Remember: To be fatty acids, the R groups   triacylglycerol
                                      must be *long* hydrocarbon chains.

FATS VS. OILS: AT ROOM TEMPERATURE, FATS ARE SOLID, OILS ARE LIQUID. WHY?
FATS have more saturated fatty acids, and OILS have more *cis*-unsaturated fatty acids. PARTIAL HYDROGENATION: commercial hydrogenation of double bonds in oils converts them to fats.

FUNCTIONS OF TRIACYLGLYCEROLS:

* To store and transport fatty acids

  Triacylglycerols store energy more efficiently than glycogen because:

  1. hydrophobic; coalesce into compact anhydrous droplets in adipocytes; take up 1/8 the volume of glycogen (because glycogen binds a substantial amount of water)

  2. when degraded (oxidized), triacylglycerols release more energy per gram than glycogen

* Insulation; poor conductor of heat, prevents heat loss

* Water-repellent – secreted by specialized glands to make fur or feathers water-repellent

* Plants – important energy reserve in fruits (avocados and olives) and seeds (peanut, corn, palm, safflower, soybean)

* Soapmaking (SAPONIFICATION): SOAP = sodium or potassium salts of fatty acid anions; EMULSIFYING AGENTS; how soap works

WAX ESTERS

Wax esters are complex mixtures that consist primarily of esters from long-chain fatty acids and long-chain alcohols, and may also contain hydrocarbons, alcohols, fatty acids, aldehydes, sterols.

FUNCTION: Protective coatings on leaves, stems, fruits and animal skin, fur.

Examples: carnauba wax (melissyl cerotate, a 26-carbon carboxylic acid esterified to a 30-carbon alcohol) and beeswax (triacontyl hexadecanoate)

PHOSPHOLIPIDS: PHOSPHOGLYCERIDES AND SPHINGOMYELINS

STRUCTURE: **AMPHIPATHIC:** hydrophilic (polar head group) and hydrophobic (long hydrocarbon chains); polar head group contains a phosphoryl group

In water, phospholipids spontaneously rearrange into micelles and/or bilayers.

Other types of polar or charged molecules can be attached to the phosphate group via a phosphodiester bond linkage.

PHOSPHATIDIC ACID is the simplest phospholipid, and consists of:

Glycerol-3-phosphate esterified to

two fatty acids, typically with 16-20 carbons each and:

- R1 = saturated fatty acid esterified to C-1
- R2 = unsaturated fatty acid esterified to C-2

Phosphatidic acid

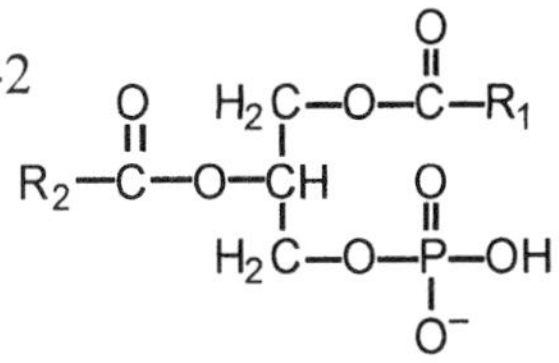

FUNCTIONS: Structural components of all cell membranes

EMULSIFYING AGENTS

SURFACE ACTIVE AGENTS or **SURFACTANTS** (lower the surface tension of a liquid so it can spread out over a surface)

intracellular signal transduction; Example: phosphatidylinositol-4,5-bisphosphate (PIP$_2$) phosphatidylinositol cycle

component of **GPI** (glycosyl phosphatidylinositol) **ANCHORS** – attach proteins to the external surface of the plasma membrane by embedding the phosphatidylinositol fatty acid chains in the membrane, and connecting to the carboxy-terminal end of a protein through a linker component

TYPES: **PHOSPHOGLYCERIDES:** glycerol, fatty acids, phosphate, and an alcohol

SPHINGOMYELINS: sphingosine, fatty acids, phosphate, and an alcohol

PHOSPHOGLYCERIDES:

FAMILY NAME	STRUCTURE
PHOSPHATIDYLETHANOLAMINE (PE)	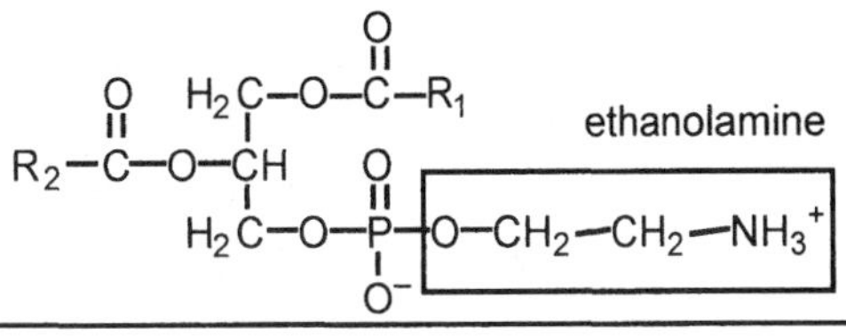 ethanolamine
PHOSPHATIDYLCHOLINE (PC) (LECITHIN)	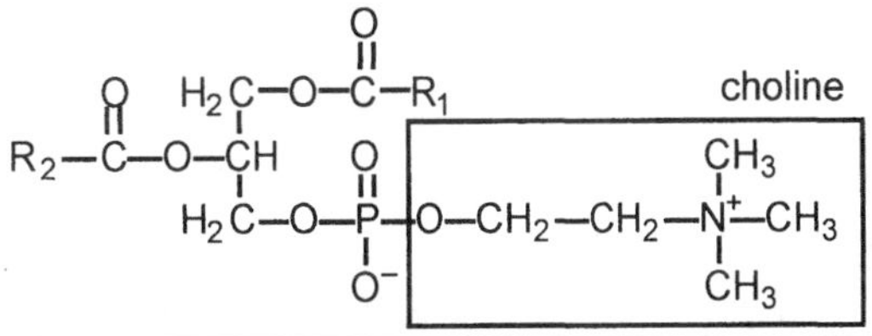 choline
PHOSPHATIDYLSERINE	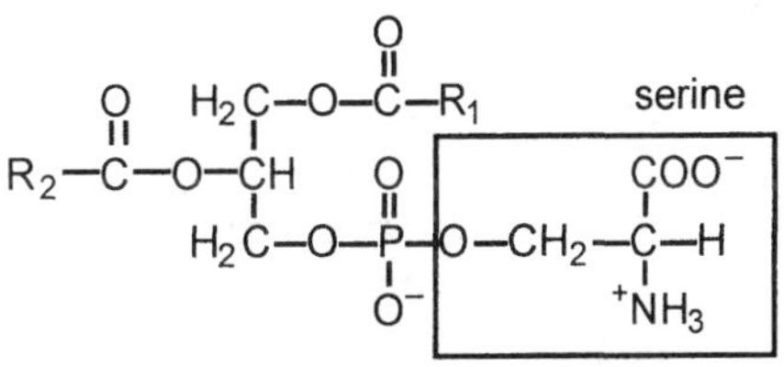 serine

(continued on next page)

PHOSPHOGLYCERIDES, *CONTINUED*

PHOSPHATIDYLGLYCEROL	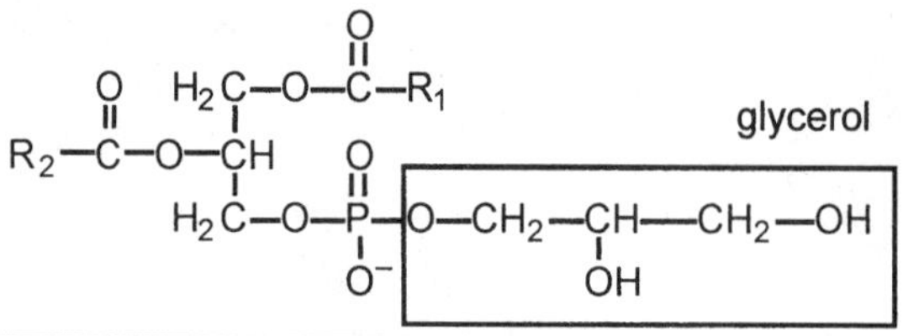 glycerol
PHOSPHATIDYLINOSITOL a prominent component of GPI anchors	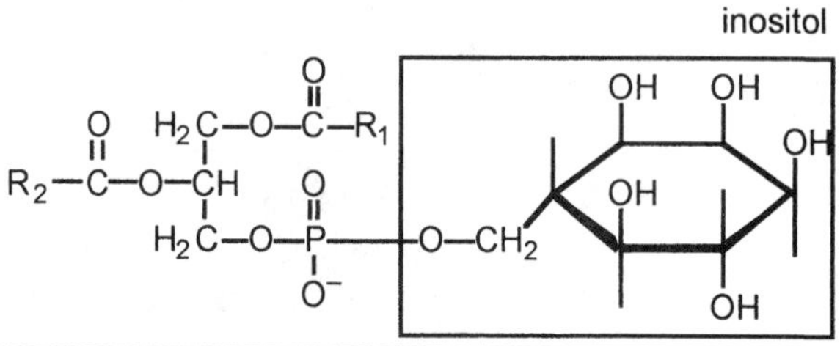 inositol

SPHINGOLIPIDS: ANOTHER LIPID COMPONENT OF BIOLOGICAL MEMBRANES

SPHINGOSINE is the "backbone" of sphingolipids (in animals; phytosphingosine is in plants).	Sphingosine
CERAMIDES have a fatty acid attached to the amino group of sphingosine.	Ceramide
SPHINGOMYELIN • a ceramide with a phosphorylcholine or phosphorylethanolamine attached to the last –OH group • an important brain lipid • located in myelin sheath of nerve cells (and in other cell membranes) • insulates nerve cells to allow for rapid transmission of nerve impulses	Sphingomyelin

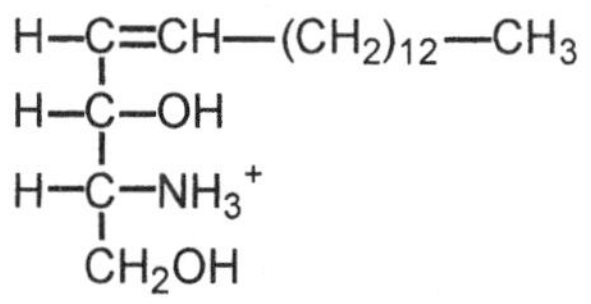

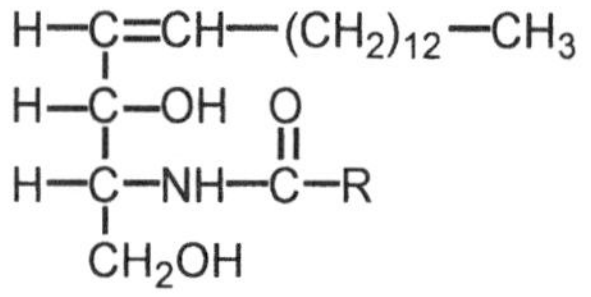

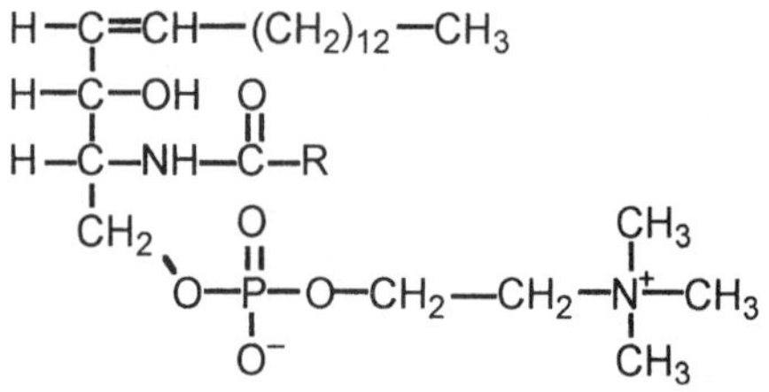

164

GLYCOLIPIDS (sugar + lipid) (also **GLYCOSPHINGOLIPIDS**) – monosaccharide, disaccharide, or oligosaccharide *O*-glycosidic linkage to ceramide; contain no phosphoryl group

CEREBROSIDES – polar head group is a monosaccharide (nonionic)

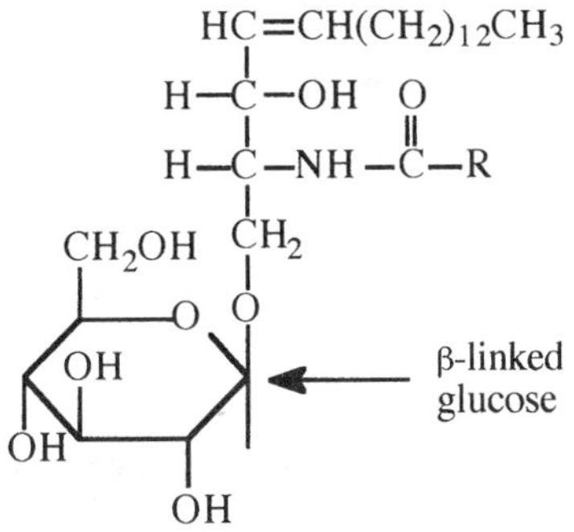

Glucocerebroside 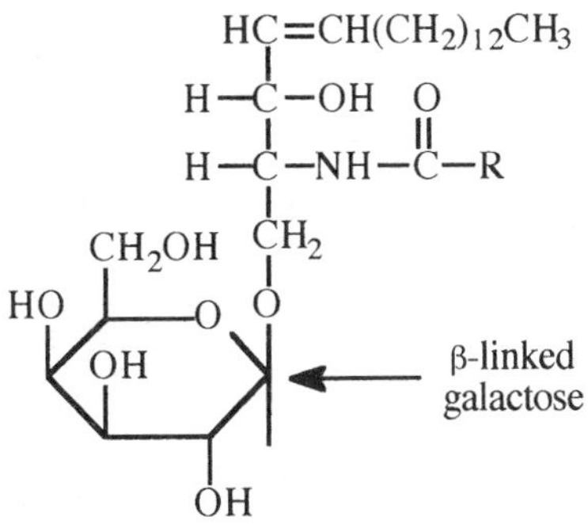   Galactocerebroside

GALACTOCEREBROSIDES – found in cell membranes of the brain
SULFATIDES – sulfated cerebroside, negatively charged at pH 7

GANGLIOSIDES – sphingolipids with oligosaccharide groups with at least one sialic acid residue; Names: subscript letter (M, D, or T = 1, 2, or 3 sialic acid residues) and numbers (the sequence of sugars that are attached to ceramide)

SPHINGOLIPID STORAGE DISEASES

Sphingolipidoses (enzymes that metabolize sphingolipids): An enzyme deficiency results in the accumulation of a specific sphingolipid; most of these diseases are fatal.

Examples: Tay-Sachs disease, Gaucher's disease, Krabbe's disease, Niemann-Pick disease. Tay-Sachs disease is caused by a deficiency of β-hexosaminidase A, which degrades G_{M2}, a ganglioside.

ISOPRENOIDS: TERPENES AND STEROIDS

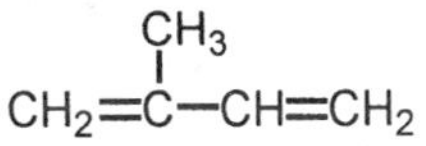

isoprene 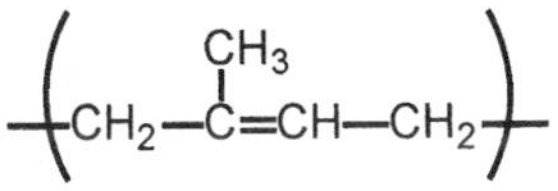   isoprene unit or prenyl group 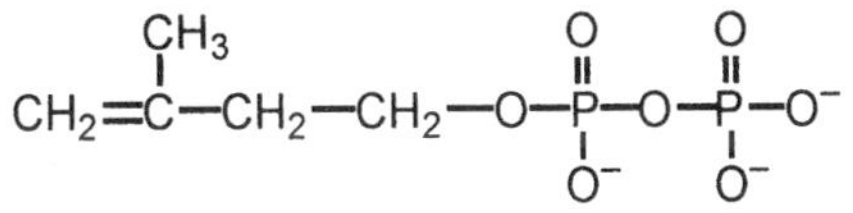   isopentenyl pyrophosphate

ISOPRENOIDS contain repeating **ISOPRENE** units, and are synthesized from acetyl-CoA (not isoprene), beginning with the formation of isopentenyl pyrophosphate.

TERPENES – classified according to number of isoprene residues; found in essential oils of plants.

# ISOPRENE UNITS	TYPE
2	monoterpene
3	sesquiterpene
4	diterpene
6	triterpene
8	tetraterpene

CAROTENOIDS – orange pigments in most plants; the only tetraterpenes (examples: CAROTENES, xanthophylls)

MIXED TERPENOIDS – nonterpene components attached to isoprenoid groups (PRENYL or ISOPRENYL groups) *Examples*: vitamins E and K, ubiquinone, some cytokinins

PROTEIN PRENYLATION – prenyl groups covalently attached to proteins; function isn't clear; typically farnesyl and geranylgeranyl groups

STEROIDS – derivatives of triterpenes, found in all eukaryotes, few bacteria

STEROID STRUCTURE:
4 fused rings; may contain double bonds and various substituents (OH, C=O, alkyl groups); sterols contain an OH

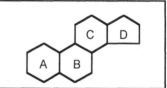

Examples: **CHOLESTEROL**: functions as an animal cell membrane component and as a precursor to steroid hormones, vitamin D and bile salts (emulsifying agents that aid in the digestion of fats in the small intestine); cholesterol usually stored within cells as a fatty acid ester

Many hormones and vitamins are steroids. Testosterone, progesterone, and cortisol are examples of steroid hormones.

Example of a steroid derivative:
Cardiac glycosides increase the force of cardiac muscle contraction; digitoxin is a glycoside in digitalis, an extract from the foxglove plant; used to treat congestive heart failure; toxic in higher than therapeutic doses; inhibits Na^+-K^+ ATPase.

LIPOPROTEINS TRANSPORT LIPID MOLECULES THROUGH THE BLOODSTREAM
Molecular complexes found in mammalian blood plasma; also contains lipid-soluble antioxidants (α-tocopherol, carotenoids)

APOLIPOPROTEINS (or **APOPROTEINS**) are the protein components of lipoproteins (that is, the lipid portion is *A*bsent). Five classes of apolipoproteins: A, B, C, D, E

CLASSIFIED ACCORDING TO DENSITY
Higher lipid content = lower density; higher protein content = higher density
As the lower-density lipids are removed (and the proteins aren't), what remains has a higher percentage of protein that has a higher density.

LIPOPROTEIN	FUNCTION
CHYLOMICRONS (large, extremely low density)	Transports dietary triacylglycerols and cholesteryl esters from intestine to muscle and adipose tissues
VERY LOW DENSITY LIPOPROTEINS (VLDL) (produced in the liver)	Transports lipids to tissues
INTERMEDIATE-DENSITY LIPOPROTEINS (IDL) (VLDLs after some depletion of lipid components)	Continue to transport lipids to tissues; may be removed by the liver
LOW-DENSITY LIPOPROTEINS (LDL) (converted from lipid-depleted VLDLs)	Transports cholesterol to tissues; engulfed by cells after binding to LDL receptors
HIGH-DENSITY LIPOPROTEIN (HDL) (produced in the liver)	Scavenges excess cholesterol from cell membranes. (see top of next page)

166

A plasma enzyme transfers a fatty acid residue from lecithin to cholesterol, forming a cholesteryl ester that's transported by HDL to the liver, which converts most of it to bile acids.

11.2 MEMBRANES

FLUID MOSAIC MODEL – proteins float within a LIPID BILAYER; the types and proportion of proteins and lipids depend on the type of cell (or organelle).

MEMBRANE STRUCTURE

MEMBRANE LIPIDS

1. MEMBRANE FLUIDITY is determined by percentage of unsaturated fatty acids in phospholipids; more unsaturated chains = greater fluidity.
   Cholesterol moderates fluidity because it has both rigid and flexible components. Lateral diffusion is rapid; flipping across the membrane is rare.
   HETEROKARYON – formed by fusion of cells from two different species; evidence that proteins move freely in the lipid bilayer

2. SELECTIVE PERMEABILITY: A bilayer is a good barrier to charged molecules like Na^+, K^+, Cl^-, and most polar molecules. Nonpolar substances diffuse through the lipid bilayer down their concentration gradients. Transport across the membrane is controlled by carrier proteins or protein channels.

3. SELF-SEALING CAPABILITY: Resealing small breaks is spontaneous and immediate; repairing larger tears is an energy-requiring, Ca^{2+} dependent process.

4. ASYMMETRY: different components face the interior vs. the exterior of the cell

MEMBRANE PROTEINS

CLASSIFICATION
- by function: structural, enzymes, hormone receptors, transport, signal transduction, immunological identity
- by structural relationship to membrane: INTEGRAL vs. PERIPHERAL

INTEGRAL PROTEINS in red blood cell membrane:
Glycophorin – 60% carbohydrate (by weight) includes the ABO and MN blood group antigens, function is unknown
Anion channel protein (band 3) – CO_2 transport in blood; HCO_3^- exchanged with Cl^- (chloride shift)

PERIPHERAL PROTEINS in red blood cell membrane: spectrin, ankyrin, band 4.1
Preserves biconcave cell shape – allows rapid diffusion of O_2
Links cytoskeleton and membrane
Anion channel protein linked to ankyrin linked to spectrin linked to band 4.1 linked to actin filaments (a cytoskeleton component); band 4.1 also binds to glycophorin

MEMBRANE MICRODOMAINS
– nonuniform distribution of membrane components; example: lipid rafts

LIPID RAFTS:
- specialized microdomains in the external leaflet of eukaryotic plasma membranes
- have a higher degree of order and less fluidity than surrounding membrane regions

- components are primarily cholesterol, sphingolipids, and certain membrane proteins (GPI-anchored proteins, doubly acylated tyrosine kinases, etc.)

MEMBRANE FUNCTION

MEMBRANE TRANSPORT OF NUTRIENTS, WASTES, IONS, ETC.

PASSIVE TRANSPORT – transport down a concentration gradient with no energy input needed

SIMPLE DIFFUSION – spontaneous transport down a concentration gradient; the greater the gradient, the faster the rate; small nonpolar molecules (example: diffusion of gases such as O_2 and CO_2)

FACILITATED DIFFUSION – special CHANNELS or CARRIERS increase the rate at which certain solutes (large or charged molecules) move down their concentration gradients

CHANNELS – tunnel-like transmembrane proteins, designed for a specific solute, often chemically or voltage-regulated or "gated"

CHEMICALLY REGULATED CHANNELS – open or close in response to a specific chemical signal (example: chemically gated Na^+ channel)

VOLTAGE-REGULATED CHANNELS

Membrane potential – electrical gradient across a membrane: decrease = DEPOLARIZATION; reestablishment = REPOLARIZATION (example: voltage-gated Na^+ or K^+ channel)

CARRIERS OR PASSIVE TRANSPORTERS – solute binds to carrier on one side, causing a conformational change in the carrier that results in translocation across the membrane, carrier releases solute (example: red blood cell glucose transporter)

ACTIVE TRANSPORT needs energy to transport against a concentration gradient

PRIMARY ACTIVE TRANSPORT – ATP hydrolysis provides the energy; Example: Na^+-K^+ pump (or Na^+-K^+ ATPase pump)

SECONDARY ACTIVE TRANSPORT – primary active transport generates a concentration gradient that is used to move substances across membranes

IMPAIRED MEMBRANE TRANSPORT MECHANISMS

Example: CYSTIC FIBROSIS (CF) is a fatal autosomal recessive disease caused by a missing or defective plasma membrane glycoprotein, CFTR.

CF – most common fatal genetic disorder in Caucasians; affects lungs and other organs; marked by lung disease and pancreatic insufficiency

CFTR – CYSTIC FIBROSIS TRANSMEMBRANE CONDUCTANCE REGULATOR

- Cl^- channel in epithelial cells
- ABC transporter (contains a polypeptide segment called an **ATP-b**inding cassette
- failure of CFTR results in retention of Cl^- within cells; osmotic pressure causes excessive water uptake, resulting in a thick mucus layer or other secretion.

TRANSPORT MECHANISMS:

TYPE:	TRANSPORT:
UNIPORTERS	1 solute
SYMPORTERS	2 different solutes simultaneously in the same direction
ANTIPORTERS	2 different solutes simultaneously in opposite directions

MEMBRANE RECEPTORS

- Ligand-receptor binding results in a conformational change which causes a programmed response

- Functions: allow cells to monitor and respond to changes in the environment; intracellular communication; cell-cell recognition or adhesion

- **RECEPTOR-MEDIATED TRANSPORT** – receptor and ligand in coated pits are engulfed by endocytosis. *Example*: LDL receptor-mediated endocytosis

---

AFTER STUDYING THIS CHAPTER, YOU SHOULD BE ABLE TO:

- Predict relative melt points of fatty acids, given their names or structures.

- Draw a fatty acid given its designation. Example: $18:3^{\Delta 9, 12, 15}$

- Draw a triacylglycerol given the fatty acids' names, structures, or designations.

- Draw a phospholipid or sphingomyelin given the fatty acids and the alcohol.

- Determine the terpene class; locate isoprene units in terpenes.

- Given the structure of a lipid, determine its class.

- Describe membrane structure and membrane fluidity.

- Transport across a membrane: Differentiate between types of transport. Identify the type of transport given the molecule to be transported.

- Differentiate between types and functions of lipoproteins.

Use this space to note any additional objectives provided by your instructor.

CHAPTER 11: SOLUTIONS TO REVIEW QUESTIONS

11.1   a. lipids – naturally occurring substances that dissolve in nonpolar solvents

       b. autocrine regulator – a hormone-like molecule that is active in the cell in which it is produced

       c. amphipathic – molecules that contain both hydrophobic and hydrophilic groups

       d. sesquiterpene – a terpene that contains three isoprene units

       e. lipid bilayer – the basic structural feature of biological membranes

11.2   a. prenylation – the covalent attachment of isoprenoid groups to a protein

       b. fluidity – a measure of the resistance of membrane components to movement

       c. chylomicron – large lipoprotein complex with extremely low density

       d. voltage-gated channel – a membrane channel that is opened by changes in membrane voltage

       e. terpene – a large group of molecules that are composed of isoprene units and are found primarily in the essential oils of plants

11.4   a. CFTR – cystic fibrosis transmembrane conductance regulator – the plasma membrane chloride channel in epithelial cells

       b. aquaporin – a water channel protein in a biological membrane

       c. GPI anchor – glycosyl phosphatidylinositol anchor – attaches proteins to the external surface of the plasma membrane by embedding the phosphatidylinositol fatty acid chains in the membrane, and connecting to the carboxy-terminal end of a protein through a linker component

       d. essential oil – volatile hydrophobic liquid mixtures that are extracted from plants, fruits, or flowers and that have characteristic odors

       e. mixed terpenoid – a biomolecule consisting of nonterpene components attached to isoprenoid groups; examples include ubiquinone and vitamins E and K

11.5   a. intermediate density lipoprotein – a type of lipoprotein that contains a component ratio of protein to lipid that lies between those of low-density lipoproteins and high-density lipoproteins

       b. fluid mosaic model – the currently accepted model of cell membranes in which the membrane is a lipid bilayer with integral proteins buried in the lipid, and peripheral proteins more loosely attached to the membrane surface

       c. lipid raft – specialized microdomains in the external leaflet of eukaryotic plasma membranes; lipid rafts have a higher degree of order and less fluidity than surrounding membrane regions; components are primarily cholesterol, sphingolipids, and certain membrane proteins

       d. SREBP – sterol regulatory element-binding proteins – membrane-bound ER proteins that activate genes involved in lipid metabolism, including the gene for LDL receptors; located in liver cells

       e. *trans* fatty acid – a fatty acid that contains a *trans*-alkene

11.7   a. triacylglycerol   d. unsaturated fatty acid

       b. steroid   e. phosphatidyl choline

       c. wax ester   f. sphingolipid

11.8   Plasma lipoproteins improve the solubility of hydrophobic lipid molecules as they are transported in the bloodstream to the organs.

11.10   Unsaturated fatty acid content increases fluidity; cholesterol decreases fluidity.

11.11   Both a and c are true. (Ionophores are discussed on page 348.)

11.13   When acetylcholine binds to the acetylcholine receptor complex in the nerve cell membrane, sodium ions flow into the nerve cell and a smaller number of potassium ions flow out. During the repolarization phase, potassium ions flow out of the cell through voltage-regulated potassium channels.

11.14   Eicosanoids are derived from arachidonic acid. Medical conditions in which it is advantageous to suppress the synthesis of eicosanoids are anaphylaxis, allergies, pain, the inflammation caused by injury, and fever.

11.16   The indicated compounds are classified as follows:

a. monoterpene   d. polyterpene

b. monoterpene   e. diterpene

c. sesquiterpene   f. triterpene

11.17   Triacylglycerol functions include energy storage, insulation, and shock absorption. Their highly reduced, long hydrocarbon chains are a very efficient storage of energy; in addition, their hydrophobicity results in compact storage in adipocytes. The relatively low heat conductivity helps to prevent heat loss and serves organisms as insulation.

11.19   To increase a cell's resistance to mechanical stress, increase the content of cholesterol and cardiolipin (two phospholipids linked by a glycerol) in the cell membrane.

11.20   HDL scavenges free cholesterol and transports it to the liver where it is converted to bile acids and excreted. This reduces total cholesterol in the serum and helps prevent plaque formation.

11.22   The sodium gradient created by the Na^+-K^+-ATPase in the plasma membrane of kidney tubule cells transports the glucose. This is secondary active transport.

11.23   In facilitated diffusion, polar, charged, or large molecules that normally cannot penetrate the cell membrane diffuses across the membrane through protein channels or carriers that "facilitate" this diffusion. Because facilitated diffusion occurs with (or down) a concentration gradient, this is a spontaneous process, so a coupled reaction that provides energy (such as the hydrolysis of ATP) is not needed. The transport of glucose across red blood cell membranes is an example of facilitated diffusion by a carrier protein, and anion channels in red blood cell membranes are examples of protein channels. (Note that the transport of glucose across the plasma membrane of kidney tubule cells is an example of secondary active transport, a distinctly different transport mechanism. See Review question 11.22.)

11.25   a. simple diffusion   c. and d. facilitated diffusion
            b. diffusion through aquaporins   e. active transport or gated channel

11.26   The outer layer of lipoproteins consist of a single layer of phospholipids and proteins. The hydrophobic hydrocarbon chains of the phospholipids face inward, towards the neutral lipids contained within. The hydrophilic group of the phospholipids face outward and are solvated by water molecules, allowing the lipoprotein to dissolve in the bloodstream.

11.28 Detergents disrupt membranes and extract membrane proteins by solubilizing both their hydrophobic and their hydrophilic components, effectively dispersing them throughout the aqueous solvent.

11.29 To form bilayers, molecules need to be amphipathic. Although triacylglycerols have three polar ester bonds, their three hydrocarbon chains are sufficiently long to cause these molecules to be nonpolar overall. Their hydrophobic nature causes them to coalesce into droplets rather than forming bilayers.

11.31 Branched chain fatty acids prevent regular packing of the long hydrocarbon chains. As with *cis*-unsaturated fatty acids, this would require less energy to disrupt the chain packing, and would lower the melting point of a fatty acid mixture.

11.32 The hydrophobic hydrocarbon chains in lecithin (phosphatidylcholine) associate with the hydrocarbon chains in the oil molecules, and the polar phosphate and choline groups are solvated by water, thus keeping the oil dispersed throughout the mixture.

CHAPTER 11: SOLUTIONS TO THOUGHT QUESTIONS

11.34 As time progresses, the antigens of the heterokaryon will intermingle. This suggests that the membrane is fluid and the antigens, as well as other components of the cell membrane, can move freely within the lipid bilayer.

11.35 Most of the cholesterol in plaque results from the ingestion of LDL by the foam cells that line the arteries. High blood plasma LDL therefore promotes atherosclerosis. Because the coronary arteries are narrow, they are especially prone to occlusion by atherosclerotic plaque.

11.37 For a phospholipid to move from one side of the bilayer to the other, the polar head must move through the hydrophobic portion of the phospholipid membrane. This process requires a significant amount of energy and is therefore relatively slow.

11.38 The hooves and lungs are subjected to much lower temperatures than the rest of the body. At these low temperatures, the membrane must be modified so that the membranes remain fluid. This can be done by increasing the unsaturation of the nonpolar tails of the membrane phospholipids.

11.40 Many transmembrane and peripheral proteins are attached to the cytoskeleton and therefore are not free to move in the phospholipid bilayer.

11.41 Both carbohydrates and proteins contain large numbers of atoms capable of hydrogen bonding (oxygen and nitrogen). In the presence of water these materials would either dissolve or swell. Waxes on the other hand are composed of hydrophobic molecules that are resistant to the penetration of water from the leaf interior. A relatively thick wax layer prevents insect penetration.

11.43 The steroids are lipid soluble molecules and would tend to dissolve in adipose tissue. During a diet the fat content of this tissue is reduced and the dissolved steroid molecules are released. This causes an overall increase in blood steroid levels.

11.44 Boric acid is a hard crystalline solid. One of the ways that this abrasive molecule kills insects is that when it contacts the exoskeleton the crystals cut into the wax coat. These gaps allow water to escape and the insect dies of dehydration. (Boric acid is also an internal poison in insects. When insects that are covered with boric acid powder groom themselves, they ingest the boric acid crystals. Boric acid interferes with the digestion of food, thereby causing starvation.)

12 Lipid Metabolism

Brief Outline of Key Terms and Concepts

12.1 FATTY ACIDS AND TRIACYLGLYCEROLS
LIPOLYSIS VS. LIPOGENESIS

BILE SALTS; CHYLOMICRON REMNANTS; HORMONE-SENSITIVE LIPASE; FATTY ACID–BINDING PROTEIN

FATTY ACID DEGRADATION VIA β-OXIDATION
CARNITINE is needed to transport acyl-CoA into the mitochondrion, where β-oxidation oxidizes fatty acids in a series of 4 reactions. The fourth step is a THIOLYTIC CLEAVAGE (addition of CoASH with the breakage of the bond between the α- and β-carbons), giving an acetyl-CoA and an acyl-CoA that's two carbons shorter. β- oxidation repeats until only acetyl-CoA is left.

THE COMPLETE OXIDATION OF A FATTY ACID
The number of ATP from each β-oxidation cycle(followed by ETC and oxidative phosphorylation) is approximately:
- 1.5 ATP/FADH$_2$ x number of β-oxidation cycles
- 2.5 ATP/NADH x number of β-oxidation cycles
- 10 ATP/acetyl CoA x total acetyl-CoA
- = Total ATP produced during β-oxidation
- –2 ATP equivalents to form the fatty acyl-CoA
- = Total ATP produced from a fatty acid

β-OXIDATION IN PEROXISOMES shorten very-long-chain fatty acids.

THE KETONE BODIES (acetoacetate, β-hydroxybutyrate, acetone) are produced from excess acetyl-CoA from β-oxidation. KETOGENESIS; KETOSIS

FATTY ACID OXIDATION: DOUBLE BONDS, ODD CHAINS
Additional reactions are needed to oxidize fatty acids that are unsaturated, have an odd number of carbons, and/or are branched (α-OXIDATION).

FATTY ACID BIOSYNTHESIS
In animals, fatty acids are synthesized in the cytoplasm from acetyl-CoA, beginning with the carboxylation of acetyl-CoA by ACC to form malonyl-CoA, which is converted to malonyl-ACP (ACYL CARRIER PROTEIN). ACC, a key enzyme in fatty acid metabolism, is regulated by allosteric modulators and phosphorylation. The remaining reactions of fatty acid synthesis take place on the fatty acid synthase multienzyme complex.

FATTY ACID ELONGATION AND DESATURATION
are carried out by mitochondrial and ER enzymes.

EICOSANOID METABOLISM

REGULATION OF FATTY ACID METABOLISM
Short term: allosteric modulators (malonyl-CoA inhibits CAT-I), covalent modification (AMPK-catalyzed phosphorylation of ACC1 and glycerol-3-phosphate acyltransferase), and hormones (e.g., insulin, glucagon, and epinephrine). Long term regulation: gene expression triggered by TRANSCRIPTION FACTORS (e.g., the SREBPs and the PPARs) and certain hormones.

12.2 MEMBRANE LIPID METABOLISM
PHOSPHOLIPID METABOLISM

After phospholipids have been synthesized at the interface of the SER and the cytoplasm, they are often "remodeled"; that is, their fatty acid composition is adjusted. TURNOVER (i.e., the degradation and replacement) of phospholipids is rapid and is mediated by the phospholipases.

SPHINGOLIPID METABOLISM: Synthesis begins with the production of ceramide. Ceramide reacts with phosphatidylcholine to form sphingomyelin, or UDP-glucose (or UDP-galactose) to form glucosylcerebroside (or galactocerebroside). Reaction of galactocerebrosides with the sulfate donor PAPS (3′-phosphoadenosine-5′-phosphosulfate) produces the sulfatides.

12.3 ISOPRENOID METABOLISM
CHOLESTEROL METABOLISM

Cholesterol is the precursor for all steroid hormones and the bile salts.

CHOLESTEROL SYNTHESIS
Three phases of cholesterol synthesis: acetyl-CoA to HMG-CoA, HMG-CoA to squalene, and squalene to cholesterol. The rate-limiting step, the reduction of HMG-CoA to form mevalonate, is catalyzed by HMGR.

CHOLESTEROL DEGRADATION: primarily by conversion to BILE SALTS that emulsify dietary fat.

CHOLESTEROL HOMEOSTASIS: via intricate mechanisms that regulate cholesterol biosynthesis, LDL receptor activity, and bile acid biosynthesis.

STEROID HORMONE SYNTHESIS
STEROL CARRIER PROTEIN ; CONJUGATION REACTION
GLUCOCORTICOID; MINERALOCORTICOID; PREGNENOLONE

BIOCHEMISTRY IN PERSPECTIVE:

ATHEROSCLEROSIS: ATHEROMA; CHEMOKINES; PLAQUE

BIOTRANSFORMATION: PHASE I AND PHASE II REACTIONS; CYTOCHROME P$_{450}$ SYSTEM; MONOOXYGENASE; CONJUGATION REACTIONS; DETOXICATION; DETOXIFICATION

12.1 FATTY ACIDS AND TRIACYLGLYCEROLS

COMPARISON OF LIPOLYSIS AND LIPOGENESIS	
LIPOLYSIS (degradation/catabolism) triacylglycerol + 3H$_2$O → 3 fatty acids + glycerol	**LIPOGENESIS** (synthesis/anabolism) 3 fatty acids + glycerol → triacylglycerol + 3H$_2$O
energy reserves are low and energy (ATP) is needed: fasting, exercise, in response to stress	energy reserves are high and ATP levels are fine
mobilizes fat stores to fatty acids + glycerol	stores dietary fat in adipocytes
Produce energy – ATP – by oxidizing fatty acids.	Store energy by synthesizing triacylglycerols from excess nutrients and metabolites

HOW IS DIETARY FAT (TRIACYLGLYCEROLS) TRANSPORTED TO THE CELLS TO BE METABOLIZED?

1. **BILE SALTS** emulsify (solubilize) fats in the lumen of the small intestine because they're amphipathic, with detergent-like properties. Bile salts are made by the liver, stored in the gall bladder, and secreted into the small intestine.

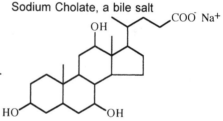

Sodium Cholate, a bile salt

2. **PANCREATIC LIPASE** (and other intestinal lipases) hydrolyzes triacylglycerols to free two fatty acids, leaving a monoacylglycerol.

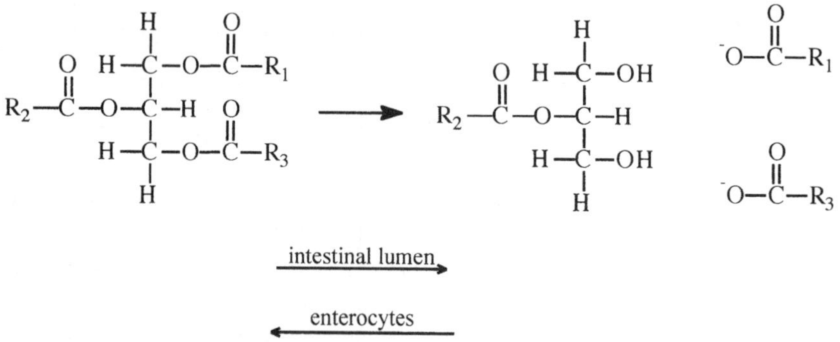

intestinal lumen →

← *enterocytes*

3. Bile salts form mixed micelles with these and with other fats. Micelles are then taken up into enterocytes (intestinal wall cells), and reconverted to triacylglycerols.

4. These triacylglycerols are combined with cholesterol, phospholipids and protein to form **CHYLOMICRONS**, transported out to the lymph and then into the bloodstream for distribution to cells. (Chylomicrons and other lipoproteins are described on pp. 392-4 of your text.)

The use of fatty acids for energy depends on the tissue.
Fatty acids are: – a major source of energy in muscle tissue (cardiac and skeletal)
                    – used minimally in nervous tissue.
During fasting, many tissues use fatty acids or ketone bodies for energy.

Target Tissue: primarily muscle cells (cardiac and skeletal) and adipocytes; also lactating mammary gland cells. All of these cells synthesize lipoprotein lipase, transferred to the endothelial surface of capillaries

5. Lipoprotein lipase converts triacylglycerols in chylomicrons (and VLDLs) to fatty acids and glycerol. Glycerol must be transported back to the liver to be metabolized.

6. In the liver:
    *   Glycerol $\xrightarrow{\text{glycerol kinase}}$ Glycerol-3-phosphate, which is used in the synthesis of triacylglycerols, phospholipids, or glucose.
    *   Chylomicron remnants are removed from the blood via receptor-mediated endocytosis.
    *   VLDLs are synthesized.

FATTY ACID DEGRADATION (IN THE MITOCHONDRIAL MATRIX)

1. **Activation** by acyl-CoA synthetase (in the outer mitochondrial membrane)
2. **Transport** (through the inner membrane into the matrix)
3. **β-oxidation**

ACTIVATION BY ACYL-CoA SYNTHETASE

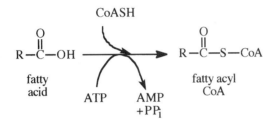

The ATP used to drive this reaction might be thought of as an "activation fee." The production of AMP means that two ATP equivalents were used. PP_i hydrolysis helps to drive the reaction to completion.

TRANSPORT: CARNITINE TRANSPORTS ACYL-CoA INTO THE MATRIX
Why is this necessary?
Fatty acid catabolism occurs inside the mitochondrial matrix. Fatty acyl-CoA cannot cross the inner membrane, but carnitine can. Carnitine forms an ester bond with the fatty acyl group and carries it into the matrix (with the help of a carrier protein).

Consider the acyl-carnitine structure. Can you see why it needs a carrier protein to get through the membrane?

Carnitine

$$CH_3-\overset{+}{N}{}^{\pm}CH_2-CH-CH_2-COO^-$$

forms ester bond with fatty acid

The figure on the next page shows the transport of palmitoyl-CoA through the inner membrane. CoASH doesn't pass through the membrane, so the cell maintains separate cytoplasmic and mitochondrial pools of CoASH.

CARNITINE TRANSPORT OF ACYL-CoA ACROSS THE MITOCHONDRIAL MEMBRANE

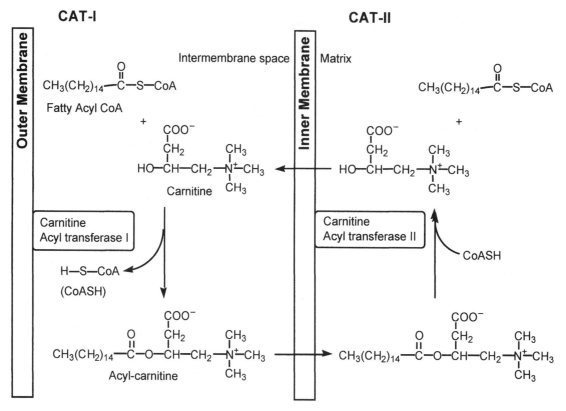

Author's note: I thought it worthwhile to see a specific example. It's so convenient to use "R" for the long hydrocarbon chain that it can be easy to forget just how long these molecules are.   --PD

THE RATE OF β-OXIDATION IN

TISSUE:	DEPENDS UPON:
MUSCLE	– current energy requirements. – substrate availability (fatty acid concentration in the blood).
LIVER	– the rate of fatty acid transport into mitochondria. CAT-I is strongly inhibited by malonyl-CoA (the product of fatty acid synthesis' first committed step).
	– blood glucose levels. Increased blood glucose levels promote fatty acid synthesis. The product of the first committed step is malonyl-CoA. Malonyl-CoA inhibition of β-oxidation prevents fatty acid synthesis and oxidation from occurring simultaneously.

176

β-Oxidation

OF A SATURATED FATTY ACYL-CoA WITH AN EVEN NUMBER OF CARBONS

Additional steps and enzymes are needed if the fatty acid is
unsaturated, branched and/or has an odd number of carbons.

The following four steps repeat until the entire fatty acid is converted to acetyl-CoA molecules. Fatty acyl-CoA	CH_3—$(CH_2)_n$—C—C—C—S—CoA (β, α positions) FAD → FADH$_2$
1. Acyl-CoA dehydrogenase Oxidation (removes 2e⁻ and 2H⁺) Forms a *trans*-double bond between the α- and β-carbons. *trans*-α,β-Enoyl-CoA	CH_3—$(CH_2)_n$—C=C—C—S—CoA H_2O →
2. Enoyl-CoA hydrase Hydration (adds water) Forms –OH at β-carbon L-β-Hydroxyacyl-CoA (3-Hydroxyacyl-CoA)	CH_3—$(CH_2)_n$—C—CH_2—C—S—CoA (OH) NAD⁺ → NADH + H⁺
3. 3-Hydroxyacyl-CoA dehydrogenase Oxidation at the β-carbon to form C=O Now we officially have β-oxidation! (Recall that NAD⁺ transfers 2e⁻ in the form of a hydride.) β-Ketoacyl-CoA	CH_3—$(CH_2)_n$—C—CH_2—C—S—CoA CoASH →
4. Thiolase (β-ketoacyl-CoA thiolase) Thiolytic cleavage adds a CoASH and cleaves a bond to release acetyl-CoA. Acyl-CoA	Acetyl-CoA CH_3—C—S—CoA CH_3—$(CH_2)_n$—C—S—CoA

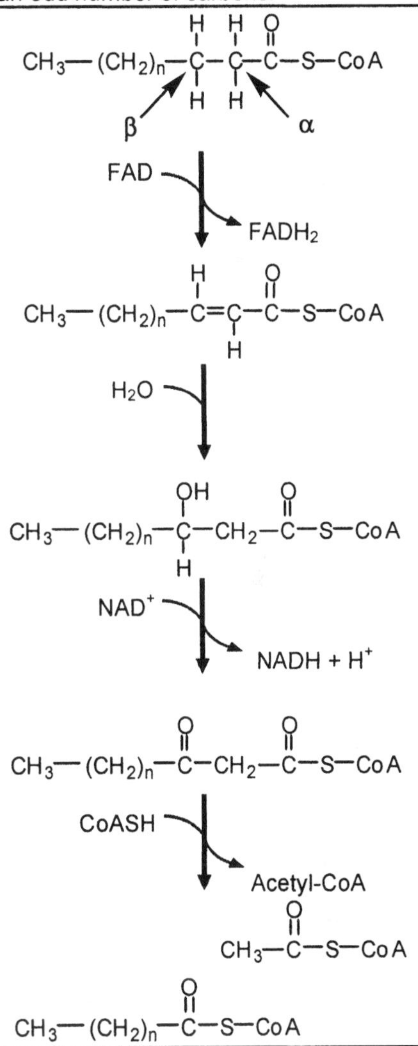

THE COMPLETE OXIDATION OF A FATTY ACID

TO CALCULATE THE TOTAL ATP PRODUCED PER FATTY ACID:

1. Determine the number of cycles of β-oxidation required. Each cycle produces:
   1 FADH$_2$, 1 NADH, and 1 acetyl-CoA; the last cycle produces 2 acetyl-CoA. [The last cycle splits a 4-carbon acyl-CoA into 2 acetyl-CoA, so palmitate (16 carbons) produces 8 acetyl-CoA in 7 cycles (*not* 8).]

2. List out the total number of FADH$_2$, NADH, and acetyl-CoA produced.
   [Palmitate: 7 FADH$_2$, 7 NADH, 8 acetyl-CoA]

3. Multiply each molecule above by the number of ATP it produces from the ETC:
   1.5 ATP per FADH$_2$, 2.5 ATP per NADH, and 10 ATP per acetyl-CoA.

177

[The acetyl-CoA molecule can enter the citric acid cycle to give 3 NADH, 1 FADH$_2$, and 1 GTP. So, [(3 NADH x 2.5 ATP/NADH) + (1 FADH$_2$ x 1.5 ATP/FADH$_2$) + 1 ATP] = 10 ATP per acetyl-CoA.]

[Palmitate: (1.5 x 7 FADH$_2$) + (2.5 x 7 NADH) + (10 x 8 acetyl-CoA) = 108 ATP!]

4. Subtract 2 ATP from the total ATP to arrive at the final answer. Those 2 ATP were the "activation fee" – the energy cost to activate the fatty acid. It counts as *two* ATP because an AMP, not an ADP, was formed.

5. Make further adjustments if the fatty acid was unsaturated and/or branched and/or had an odd number of carbons. (Each of these are described later in the chapter.)

Palmitic acid (16 carbons) is described on pp. 423-4 of your text, and stearic acid (18 carbons) is Question 12.5 (on p. 424).

Try calculating how many NADH, FADH$_2$, and ATP molecules can be synthesized from 1 molecule of arachidic acid (20 carbons, saturated).

Solution: 1. Nine cycles of β-oxidation is required, producing 9 FADH$_2$, 9 NADH, and 10 acetyl-CoA.
2. (9 x 1.5) + (9 x 2.5) + (10 x 10) = 13.5 + 22.5 + 100 = 136 ATP
3. 136 – 2 = 134 ATP total from one arachidic acid. Wow!

β-OXIDATION IN PEROXISOMES *SHORTENS VERY LONG-CHAIN FATTY ACIDS*

Peroxisomal enzymes differ somewhat from the mitochondrial enzymes. FADH$_2$ produced in the first step donates its electrons directly to O$_2$ (instead of UQ) to produce H$_2$O$_2$, which is converted to H$_2$O by catalase. The last enzyme has a low affinity for medium-chain acyl-CoA molecules, these are transported to mitochondria to continue oxidation.

KETONE BODIES

KETOGENESIS: Excess acetyl-CoA molecules are converted into **KETONE BODIES** in liver cells, in the mitochondrial matrix.

KETONE BODIES:

acetoacetate   β-hydroxybutyrate   acetone

Ketone bodies are a result of excess acetyl-CoA from β-oxidation:

Intake: high lipid, low carb diet (and/or not enough oxaloacetate)

Starvation (body consumes fats)

Diabetes (problems with carbohydrate catabolism)

KETOGENESIS:

2 Acetyl-CoA

$$CH_3-\overset{O}{\underset{\|}{C}}-S-CoA$$

Acetoacetyl-CoA thiolase
Condensation of 2 acetyl-CoA
Reverse of the last step of β-oxidation

$$CH_3-\overset{O}{\underset{\|}{C}}-S-CoA \quad \searrow \quad CoASH$$

Acetoacetyl-CoA

$$CH_3-\overset{O}{\underset{\|}{C}}-CH_2-\overset{O}{\underset{\|}{C}}-S-CoA$$

HMG-CoA synthase
The acetyl-CoA added here will be regenerated in the next step.

$$CH_3-\overset{O}{\underset{\|}{C}}-S-CoA + H_2O \quad \searrow \quad CoASH$$

HMG-CoA
(β-Hydroxy-β-methylglutaryl-CoA)

$$^-O-\overset{O}{\underset{\|}{C}}-CH_2-\underset{\underset{CH_3}{|}}{\overset{OH}{\underset{|}{C}}}-CH_2-\overset{O}{\underset{\|}{C}}-S-CoA$$

HMG-CoA lyase
Compare to the reaction with citrate synthase.

$$CH_3-\overset{O}{\underset{\|}{C}}-S-CoA$$
Acetyl CoA

Acetoacetate

$$CH_3-\overset{O}{\underset{\|}{C}}-CH_2-\overset{O}{\underset{\|}{C}}-O^-$$

$$\searrow NADH + H^+$$
$$\searrow NAD^+$$

β-Hydroxybutyrate dehydrogenase

β-Hydroxybutyrate

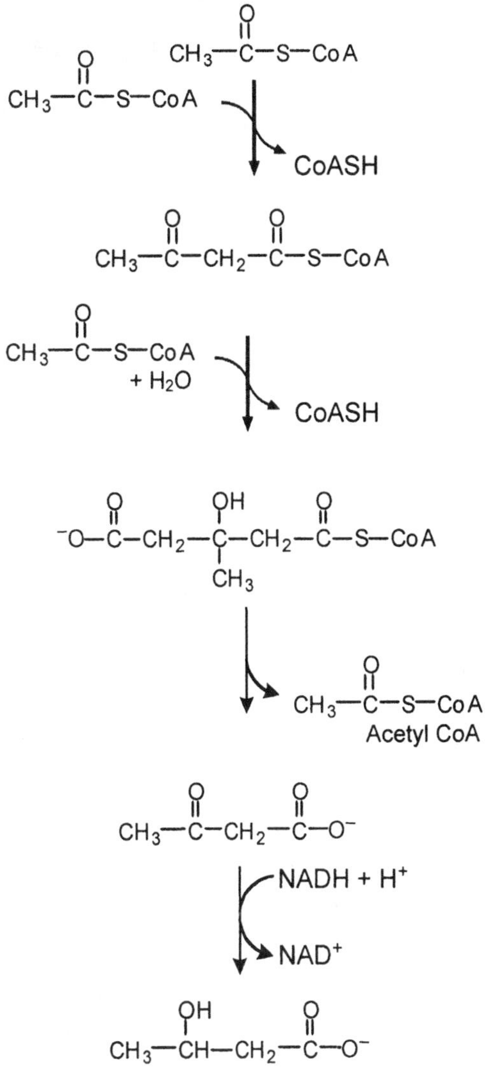

$$CH_3-\underset{\underset{OH}{|}}{CH}-CH_2-\overset{O}{\underset{\|}{C}}-O^-$$

Ketone bodies are used for energy by many tissues, most notably cardiac and skeletal muscle. The brain prefers glucose for its fuel, but can use ketone bodies when it's forced to (i.e., during starvation or no-carb diets). Ketone bodies are made in the liver, but the liver does not use them as fuel.

KETOSIS occurs when acetoacetate is made faster than it can be used. Under these conditions, acetoacetate spontaneously degrades (no enzyme is needed) to acetone, which is exhaled.

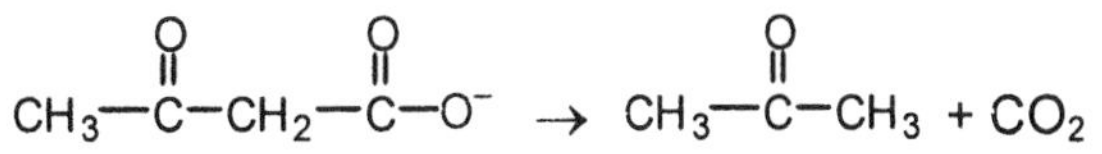

$$CH_3-\overset{O}{\underset{\|}{C}}-CH_2-\overset{O}{\underset{\|}{C}}-O^- \rightarrow CH_3-\overset{O}{\underset{\|}{C}}-CH_3 + CO_2$$
acetoacetate   acetone

179

In target tissues, acetoacetate is converted back to two molecules of acetyl-CoA:

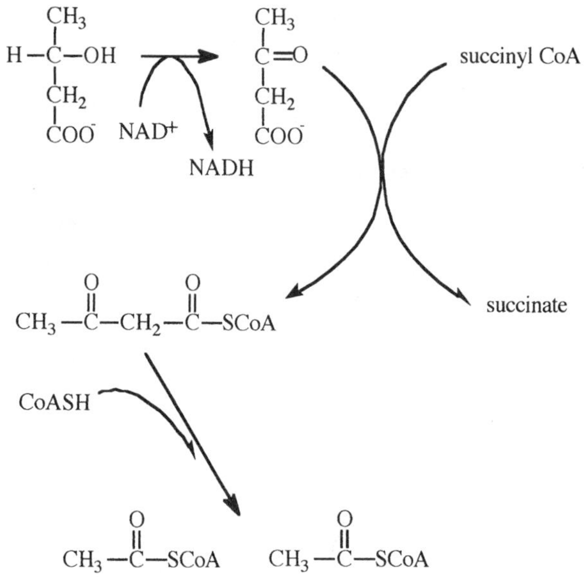

FATTY ACID OXIDATION: DOUBLE BONDS, ODD CHAINS, AND BRANCHES

UNSATURATED FATTY ACID OXIDATION

β-Oxidation proceeds normally until the alkene (usually *cis-*) is in the β,γ-position.

Enoyl-CoA isomerase catalyzes a *cis*-β,γ-enoyl-CoA to a *trans*-α,β-enoyl-CoA. β-Oxidation continues.

One less FADH₂ is produced because the double bond is already present. FAD isn't needed to oxidize the acyl-CoA to produce a double bond.

ODD CHAIN FATTY ACID OXIDATION

β-Oxidation proceeds normally, but the products of the last cycle are acetyl-CoA and propionyl-CoA, which is carboxylated and isomerized to produce succinyl-CoA. The total energy produced depends upon the degradation pathway of succinyl-CoA. Additional enzymes needed for odd chain fatty acid oxidation are: **propionyl-CoA carboxylase**, **methylmalonyl-CoA racemase**, and **methylmalonyl-CoA mutase**.

α-OXIDATION OF BRANCHED-CHAIN FATTY ACIDS

Fact: a β-methyl group blocks β-oxidation, but an α-methyl does not. α-Oxidation essentially oxidizes the α-carbon and removes the carboxyl group so that the product fatty acid has an α-methyl instead of a β-methyl (and one less carbon than the original). β-oxidation can then proceed normally because the remaining branches occur in α-positions.

Phytanic acid is a common dietary branched-chain fatty acid.

FATTY ACID BIOSYNTHESIS TAKES PLACE IN THE CYTOPLASM

Fatty acid biosynthesis cannot be a simple reversal of β-oxidation because of thermodynamic considerations. (Both pathways must be spontaneous ($-\Delta G$) to go forward, but a simple reversal of any reaction also reverses the sign of ΔG.)

Fatty acids are typically made from glucose:

glucose → pyruvate $\xrightarrow{\text{transport into mitochondrion}}$

pyruvate → acetyl-CoA → citrate $\xrightarrow{\text{transport to the cytoplasm}}$

citrate → acetyl-CoA → fatty acids

Lipogenic tissues: liver, adipose tissue, lactating mammary gland.

1. Citrate transports acetyl-CoA from the mitochondrion to the cytoplasm.

    Problem: Acetyl-CoA can't cross the mitochondrial membrane (where it's produced) so it must be transported to the cytoplasm to be used in fatty acid synthesis.

    Solution: Acetyl-CoA + oxaloacetate → citrate. Citrate crosses the membrane. In the cytoplasm: citrate → acetyl-CoA + oxaloacetate.

    Recall that this synthesis of citrate is the first reaction of the citric acid cycle. Why would citrate leave the citric acid cycle in the mitochondrion and venture out into the cytoplasm? Remember that the citric acid cycle exists to produce energy. If energy isn't needed by the cell, citrate will accumulate.

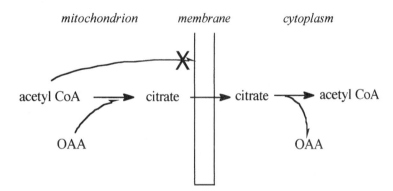

2. **ACETYL-COA CARBOXYLASE (ACC)** carboxylates acetyl-CoA to form **MALONYL-COA**. ACC requires biotin, a CO_2 carrier. Eukaryotic ACC has three domains: **BCCP, BC**, and **CT** (**BIOTIN CARBOXYL CARRIER PROTEIN, BIOTIN CARBOXYLASE**, and **CARBOXYLTRANSFERASE**, respectively.)
    Remember this step! Both ACC and malonyl-CoA are important regulators.

3. The malonyl group is transferred from CoA to **ACP (ACYL CARRIER PROTEIN)**. Malonyl-ACP serves as the carbon donor for fatty acid biosynthesis.

4. A second acetyl group is transferred from acetyl-CoA to synthase to form acetyl-S-synthase.

5. Synthase condenses malonyl-ACP with acetyl-S-synthase to form acetoacetyl-ACP. CO_2 is removed as well. This decarboxylation drives the process forward.

6. The reduction/dehydration/reduction reactions have the same intermediates as the β-oxidation of fatty acids, but use different enzymes.

7. The product butyryl-ACP transfers its acyl group to synthase, which condenses with another malonyl-ACP (again, with loss of CO_2). The process repeats until the fatty acid is synthesized (up to 16 carbons total).

SUMMARY OF FATTY ACID BIOSYNTHESIS:

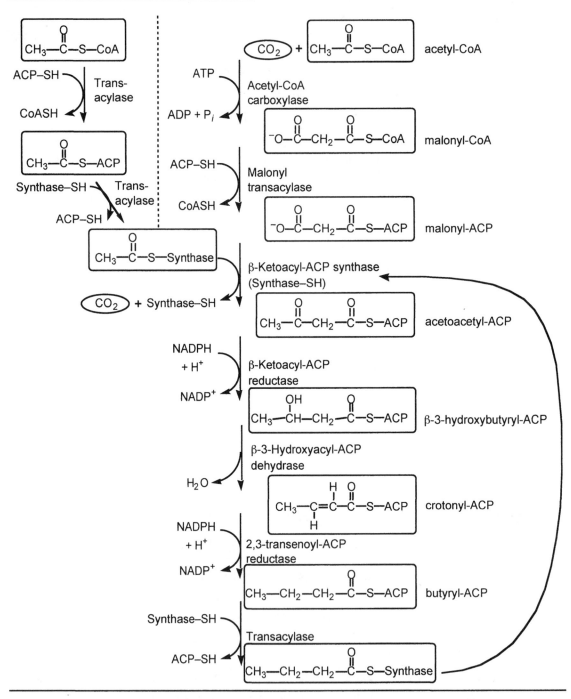

Note: *Fatty acid fatty acid synthase is a multienzyme complex and a dimer, so two fatty acid molecules are synthesized at the same time. (See Figure 12.17 on page 434 of your text.)*

THE OVERALL REACTION TO SYNTHESIZE A FATTY ACID WITH 16 CARBONS:

8 acetyl-CoA + 14 NADPH + 14 H$^+$ + 7 ATP →

palmitate + 14 NADP$^+$ + 7 ADP + 7 P$_i$ + 8 CoASH + 6 H$_2$O

β-OXIDATION OF FATTY ACIDS	*vs.*	FATTY ACID SYNTHESIS
• Location: mitochondria		• Location: cytoplasm (chloroplasts/plants)
• CoA is the acyl carrier.		• ACP is the acyl carrier.
• FAD and NAD$^+$ are electron acceptors.		• NADPH is the electron donor.
• FADH$_2$ and NADH are formed.		• NADPH is consumed; NAD$^+$ is formed.
• Acetyl-CoA is the 2-carbon-unit product.		• Malonyl-CoA is the 2-carbon-unit donor.
• Enzymes differ.		• Fatty acid synthase is a dimeric multi-enzyme complex. (See Figure 12.17, p. 434 of your text.)

β-OXIDATION

Fatty acyl-CoA (C$_{n+2}$)

Acyl-CoA dehydrogenase ⟶ FAD ⟶ FADH$_2$

trans-α,β-Enoyl-CoA

Enoyl-CoA hydrase ⟶ H$_2$O

β-Hydroxyacyl-CoA

β-Hydroxyacyl-CoA dehydrogenase ⟶ NAD$^+$ ⟶ NADH + H$^+$

β-Ketoacyl-CoA

Thiolase ⟶ CoASH ⟶ Acetyl-CoA

Fatty acyl-CoA (C$_n$)

FATTY ACID SYNTHESIS

Fatty acyl-ACP (C$_{n+2}$)

NADP$^+$ ⟵ *trans*-α,β-Enoyl-ACP reductase ⟵ H$^+$ + NADPH

trans-α,β-Enoyl-ACP

H$_2$O ⟵ β-Hydroxyacyl-ACP dehydrase

β-Hydroxyacyl-ACP

NADP$^+$ ⟵ β-Ketoacyl-ACP reductase ⟵ H$^+$ + NADPH

β-Ketoacyl-ACP

CO$_2$ + Synthase-SH ⟵ β-Ketoacyl-ACP synthase ⟵ Malonyl-ACP

Fatty acyl-S-synthase

ACP-SH ⟵ Transacylase ⟵ Synthase-SH

Fatty acyl-ACP (C$_n$)

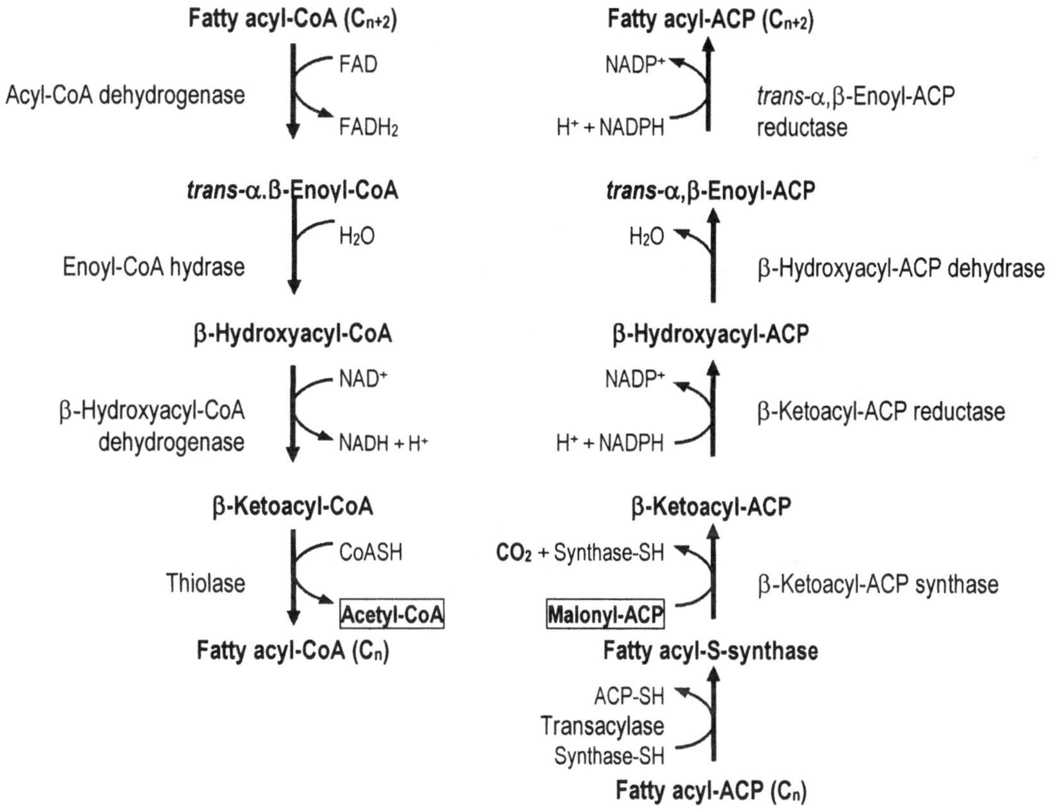

FATTY ACID ELONGATION AND DESATURATION – PRIMARILY BY ER ENZYMES

ER enzymes synthesize fatty acids longer than palmitate (16 carbons) and desaturates fatty acids (removes hydrogens to add double bonds) when these fatty acids aren't obtained by the diet.

Why is this important? Cells need unsaturated (and perhaps longer) fatty acids to maintain proper membrane fluidity, and as precursors for fatty acid derivatives (such as: eicosanoids; cerebrosides and sulfatides in myelin).

EICOSANOID METABOLISM

The first step in eicosanoid synthesis is the release of arachidonic acid from C-2 of glycerol in membrane phosphoglyceride molecules. Cyclooxygenase converts arachidonic acid to PGG_2, a precursor of the prostaglandins and the thromboxanes. The lipoxygenases convert arachidonic acid to the precursors of the leukotrienes.

REGULATION OF FATTY ACID METABOLISM IN MAMMALS: SHORT-TERM VS. LONG-TERM

SHORT-TERM REGULATION OF FATTY ACID METABOLISM:

COVALENT MODIFICATION

ACC1[1] catalyzes the first step of fatty acid synthesis to store energy (and/or to lower glucose levels). ACC1 dimers aggregate to form polymers.

> Inactive: ACC1 monomer; phosphorylated
> Active: ACC1 polymer (NO phosphate)

Phosphorylation/dephosphorylation:

> Phosphorylated-ACC1 = **inactive;** it will not polymerize.
> Inactive ACC1 halts fatty acid synthesis (prevents energy storage). Phosphorylation (depolymerization): AMPK (activated by upstream AMPK kinases and AMP), PKA (glucagon-stimulated), and palmitoyl-CoA accumulation.

> Dephosphorylation of ACC1 = **active**; allows ACC1 to polymerize. Dephosphorylation: PP2A – phosphoprotein phosphatase 2A – (activated by insulin, inhibited by glucagon and epinephrine)

Glycerol-3-phosphate acyltransferase phosphorylation is catalyzed by AMPK.

ALLOSTERIC REGULATION; SUBSTRATES AND PRODUCTS

Citrate stimulates **ACC**, the first step of fatty acid synthesis (its product is malonyl-CoA). (Excess citrate indicates that the citric acid cycle isn't needed, energy level is good, and fatty acid synthesis can begin to store excess energy.)

Malonyl-CoA inhibits **CAT-I**, which is required to transport fatty acids into the mitochondrion for β-oxidation. (CAT-I is carnitine acyltransferase I).
> Why malonyl-CoA? Remember that the first step of fatty acid synthesis makes malonyl-CoA. When that first step is inhibited, levels of malonyl-CoA drop. Since it's not there to act as inhibitor, the acyl-CoA can enter the mitochondrion to be degraded by β-oxidation. So, inhibiting the first step of fatty acid synthesis propels β-oxidation.

Palmitoyl-CoA inhibits fatty acid synthesis by causing the depolymerization of ACC. Palmitoyl-CoA also inhibits the pentose phosphate pathway.

HORMONES
INSULIN promotes lipogenesis and inhibits lipolysis.
GLUCAGON and **EPINEPHRINE** inhibit fatty acid synthesis.
> Both increase hydrolysis of triacylglycerols so fatty acids enter the blood.

---

[1] Mammals have two forms of ACC: ACC1 is cytoplasmic, in lipogenic tissues. ACC2 is mitochondrial, in oxidative tissues such as cardiac and skeletal muscles, where the product, malonyl-CoA, inhibits CAT-I.

Both inhibit PP2A, and glucagon activates PKA, resulting in inactive, phosphorylated ACC1.

LONG-TERM REGULATION OF FATTY ACID METABOLISM:

- occurs in response to fluctuating nutrient availability and energy demand.
- involves changes in gene expression triggered by transcription factors (e.g., the **SREBPs** and the **PPARs**), and certain hormones.

For an excellent summary, refer to Figure 12.22, Regulation of Intracellular Fatty Acid Metabolism, on p. 441 of your text.

12.2 MEMBRANE LIPID METABOLISM: PHOSPHOLIPIDS AND SPHINGOLIPIDS

PHOSPHOLIPID METABOLISM

Phospholipid turnover (degradation and replacement mediated by the phospholipases) is rapid.

PHOSPHOLIPID SYNTHESIS

Location: at the interface of the SER and the cytoplasm

Once choline (or ethanolamine) has entered a cell, it is phosphorylated and converted to a CDP derivative, which reacts with a diacylglycerol to form phosphatidylcholine (or phosphatidylethanolamine).

REMODELING: a process that allows a cell to adjust the fluidity of its membranes. Unsaturated fatty acids replace the original fatty acids that were incorporated during synthesis.

PHOSPHOLIPID DEGRADATION is catalyzed by several phospholipases.

SPHINGOLIPID METABOLISM

SPHINGOLIPID SYNTHESIS BEGINS WITH THE PRODUCTION OF CERAMIDE:

Palmitoyl-CoA combines with serine to form 3–ketosphinganine, which is reduced by NADPH to form sphinganine, which is converted to ceramide in a two-step process involving acyl-CoA and $FADH_2$. Ceramide can then be used to form sphingomyelin, glucosylceramide, or galactosylceramide.

SPHINGOLIPID DEGRADATION occurs in lysosomes by specific hydrolytic enzymes.

12.3 ISOPRENOID METABOLISM

Isoprenoids occur in all eukaryotes.

First phase of isoprenoid synthesis: the synthesis of **ISOPENTYL PYROPHOSPHATE**
Isopentyl pyrophosphate synthesis appears to be identical in all of the species in which this process has been investigated (despite the huge diversity of isoprenoids produced).

CHOLESTEROL METABOLISM

Location: within the ER membrane; also in the cytoplasm, at or near the ER.

CHOLESTEROL SYNTHESIS occurs primarily in the liver.

Three multi-step phases convert:
1. 2 acetyl-CoA to HMG-CoA (β-hydroxy-β-methylglutaryl-CoA)
2. HMG-CoA $\xrightarrow{\text{HMGR}}$ mevalonate → → → squalene
   HMGR is the rate-limiting enzyme for cholesterol synthesis.
3. squalene to cholesterol.

A sterol carrier protein binds squalene and its intermediate to lanosterol[2], which binds to a second carrier protein.

CHOLESTEROL DEGRADATION
Cholesterol is degraded and eliminated primarily by conversion to **BILE SALTS**.
(Bile salts facilitate the emulsification and absorption of dietary fat.)

SYNTHESIS OF BILE ACIDS (OR BILE SALTS)
The rate-limiting reaction:

$$\text{cholesterol} \xrightarrow{\text{cholesterol-7-hydroxylase}} 7\text{-}\alpha\text{-hydrocholesterol}$$

Subsequent reactions to form cholic acid and deoxycholic acid:
- rearrangement and reduction of double bond at C-5
- introduction of an additional –OH group
- reduction of the 3-keto group to a 3-α-hydroxyl group

ER enzymes catalyze **CONJUGATION REACTIONS** that convert cholic acid and deoxycholic acid to bile salts.

CHOLESTEROL HOMEOSTASIS
To maintain cholesterol homeostasis, regulate:
cholesterol synthesis, LDL receptor activity, and bile acid synthesis.

TO REGULATE CHOLESTEROL SYNTHESIS, REGULATE **HMGR**:
HMGR – the rate-limiting enzyme for cholesterol synthesis – catalyzes the formation of mevalonate.
Phosphorylated HMGR is **inactive**, and is promoted by:
high AMP levels activate AMPK-catalyzed phosphorylation of HMGR
cAMP – activates PKA, which phosphorylates and activates phosphoprotein phosphatase inhibitor 1 (PPI-1), which inhibits phosphoprotein phosphatase, which prevents HMGR from becoming active by dephosphorylation. (Phew!) (Recall cAMP regulation by hormones.)
Endocytosis of LDL receptors release cholesterol that inhibits HMGR by negative feedback.

GENE EXPRESSION: mediated by sterols
SREBP2 – ER membrane protein; stimulates expression of genes that code for: cholesterol biosynthesis enzymes, LDL receptor, three NADPH synthesis enzymes (G-6-PD, 6-phosphogluconate-dehydrogenase, malic enzyme).

Refer to Figure 12.36 (p. 457) in your text for an excellent summary figure of SREBP2 regulation, which involves:
SCAP – SREBP cleavage-activating protein; binds SREBP2
SSD – sterol-sensing domain
Insig – insulin-induced gene
NF-1 – nuclear factor-1
CREB – cAMP response element binding protein
ACAT – acyl-CoA-cholesterol acyltransferase

STEROID HORMONE SYNTHESIS

Cholesterol is the precursor for all steroid hormones.

$$\text{Cholesterol} \xrightarrow{\text{Desmolase (mitochondria)}} \text{Pregnenolone}$$

---

[2] The C-18 methyl shifts from C-8 to C-13 in the conversion of squalene-2,3-epoxide to lanosterol.

$$\text{Pregnenolone} \xrightarrow{\text{(in the ER)}} \text{Progesterone}$$

Pregnenolone and progesterone are the precursors for all other steroid hormones.

BIOCHEMISTRY IN PERSPECTIVE

ATHEROSCLEROSIS

BIOTRANSFORMATION

AFTER STUDYING THIS CHAPTER, YOU SHOULD BE ABLE TO:

- State which pathway(s) (lipogenesis, lipolysis, β-oxidation, or fatty acid synthesis) will occur under given conditions (high or low energy needs, levels of: ATP, AMP, cAMP, NADH, NAD^+, and blood glucose).

- Compare and contrast opposing pathways, including differences in intermediates, enzymes, location, transport mechanisms (if needed) and regulatory mechanisms.

- Write the β-oxidation pathway (including structures and enzymes) of a given fatty acid.

- Determine how many ATP equivalents are obtained from the complete oxidation of a given fatty acid.

- Describe how the oxidation of a fatty acid differs if it has a double bond, branches, or an odd number of carbons.

- Describe the regulation mechanisms of triacylglycerol, fatty acid, and/or cholesterol metabolism. Include how the location of these pathways impact their regulation.

- Describe a specific transport mechanism required for a given pathway to proceed. For example, describe the role of carnitine in fatty acid metabolism.

- Explain the effects of insulin and glucagon on the pathways studied in this chapter.

- State the key enzymes in a given pathway that are either rate-limiting enzymes or highly regulated.

- Describe the roles of SREBPs and PPARs in lipid metabolism.

Use this space to note additional objectives provided by your instructor:

CHAPTER 12: SOLUTIONS TO REVIEW QUESTIONS

12.1   a. *de novo* – biosynthesis from new materials

          b. oil bodies – structures in plants that store triacylglycerols

          c. β-oxidation – the main pathway for the degradation of fatty acids in which two-carbon fragments in the form of acetyl-CoA are removed from the carboxyl end of fatty acids

          d. turnover – the rate at which molecules are degraded and replaced

          e. thiolytic cleavage – cleavage by thiolase between the α- and β-carbons of a β-ketoacid during β-oxidation to produce a molecule of acetyl-CoA and a new acyl-CoA

12.2   a. carnitine – an amino acid that is used to transport acyl-CoA molecules into the mitochondria.

          b. sterol carrier protein – a cytoplasmic molecule that binds squalene during cholesterol synthesis

          c. thrombin – a proteolytic enzyme that converts fibrinogen to fibrin

          d. thiolase – an enzyme that catalyzes a thiolytic cleavage in the final reaction in the β-oxidation cycle to form an acetyl-CoA product

          e. desmolase – an enzyme that catalyzes the conversion of cholesterol to pregnenolone, which is the first reaction in steroid hormone synthesis

12.4   Three differences between fatty acid synthesis and β-oxidation are the following:
        (1) The two pathways take place in different cell compartments. Synthesis occurs in the cytoplasm and β-oxidation is within the mitochondria.
        (2) The intermediates of fatty acid synthesis and β-oxidation are linked through thioester linkages to ACP and CoASH respectively.
        (3) The electron carrier for fatty acid synthesis is NADPH while those for β-oxidation are NADH and $FADH_2$.

12.5   a. autoantibodies – molecules that bind to surface antigens on a patient's own cells as if they were foreign

          b. ketone bodies – acetone, acetoacetate, or β-hydroxybutyrate; produced in the liver from acetyl-CoA

          c. conjugation reaction – a reaction that converts a molecule into a derivative that contains a water-soluble functional group in order to improve its solubility

          d. ACP – acyl carrier protein – a component of fatty acid synthase; a protein that contains a phosphopantetheine prosthetic group to which the acyl intermediates in fatty acid synthesis link via a thioester bond; analogous to CoASH in β-oxidation

          e. AMPK – 5′-AMP-activated protein kinase – a trimeric enzyme with catalytic and regulatory subunits; allosterically activated by AMP

12.7   Steroids are compounds containing the following ring system:

The term *steroid* is often used to designate all compounds containing this ring system. It is more accurately reserved for those derivatives that contain carbonyl groups. Sterols have a similar structure but also contain hydroxyl groups.

12.8   a. hypertriglyceridemia – high levels of triglycerides in the blood

        b. SREBP 1 – sterol regulatory element binding proteins – includes SREBP-1a and SREBP-1c; regulate the expression of genes involved in fatty acid metabolism

        c. SREBP 2 – sterol regulatory element binding proteins – regulates the expression of genes involved in cholesterol metabolism

        d. chemokines – signal molecules that attract fully activated macrophages to an injured site, e.g., an atheroma

        e. atheroma – soft masses that accumulate within arterial walls as a result of the accumulation of LDLs and the subsequent inflammatory response (see plaque)

12.10 The products of the oxidation of the molecule in question 9 (3-methylhexanoic acid) are two molecules of propionyl CoA, two NADH, one $FADH_2$ and one AMP (which corresponds to a loss of two ATP molecules). Assuming that the propionyl-CoA does not enter the citric acid cycle via its conversion to succinyl-CoA, the total ATP molecules produced are: 5 (from 2NADH) + 1.5 (from $FADH_2$) – 2 (from the conversion of 2-methyl-pentanoic acid to its acyl-CoA activated form) = 4.5 ATP.

The propionyl-CoA may be further oxidized for energy via its conversion to succinyl-CoA with its subsequent entry into the citric acid cycle and the electron transport chain. Listed below is a tally of the ATP molecules produced and their source reaction(s) or pathway. (Conversion of GTP to ATP is assumed.) "Oxalo-acetate to oxaloacetate" covers one full turn of the citric acid cycle. "Oxaloacetate to succinyl-CoA" is a partial turn of the citric acid cycle, stopping at succinyl-CoA.

propionyl-CoA to succinyl-CoA:	+1 CO_2	–1 ATP		
succinyl-CoA to oxaloacetate:		+1 ATP	+1 $FADH_2$	+1 NADH
oxaloacetate to oxaloacetate:	–2 CO_2	+1 ATP	+1 $FADH_2$	+3 NADH
oxaloacetate to succinyl-CoA:	–2 CO_2			+2 NADH
total (propionyl-CoA → 3CO2):	–3 CO_2	+2 ATP	+2 $FADH_2$	+6 NADH
x 2 propionyl-CoA				
total (2 propionyl-CoA → 6 CO_2): –6 CO_2		**+4 ATP**	+4 $FADH_2$	+12 NADH
From the electron transport chain:		(x1.5 ATP/$FADH_2$)		(x2.5 ATP/NADH)
total ATP from propionyl-CoA and ETC:		**+4 ATP**	**+6 ATP**	**+30 ATP**
		= 40 ATP		

Finally, add the 4.5 ATP from the conversion of 3-methylhexanoic acid to two molecules of propionic acid. Disregarding inorganic products, the complete oxidation of 3-methylhexanoic acid via α-oxidation, β-oxidation, the citric acid cycle, and the ETC yields 7 CO_2 and **44.5 ATP**.

12.11 Insulin facilitates transport into adipocytes and stimulates fatty acid synthesis and triacylglycerol synthesis. It prevents lipolysis by inhibiting protein kinase.

12.13 a. plaque – atherosclerotic lesions – the build-up of damaged endothelial cells, macrophages, and smooth muscle cells in the intima (inner lining) of arterial walls; also forms a fibrous cap to separate the damaged tissue from healthy tissue

        b. SAM – S-adenosylmethionine – a methyl donor used to methylate molecules such as phosphatidylethanolamine to form phosphatidylcholine

        c. PAPS – 3′-phosphoadenosine-5′-phosphosulfate – a high-energy sulfate donor

   d. HMGR – HMG-CoA reductase – the rate-limiting enzyme in cholesterol
       synthesis, located on the cytoplasmic face of the ER

   e. allyl group – an organic functional group consisting of a methylene group and an
       alkene:

$$-CH_2-CH=CH_2$$

12.14 The hydrophobic portions of the molecule are the long hydrocarbon tails of the
       molecule. The hydrophilic portion of the molecule is the phosphate ester functional
       group. The hydrocarbon tails are within the bilayer and the phosphate head group is
       on the surface of the membrane.

12.16 The indicated bond is cleaved by glucocerebrosidase.

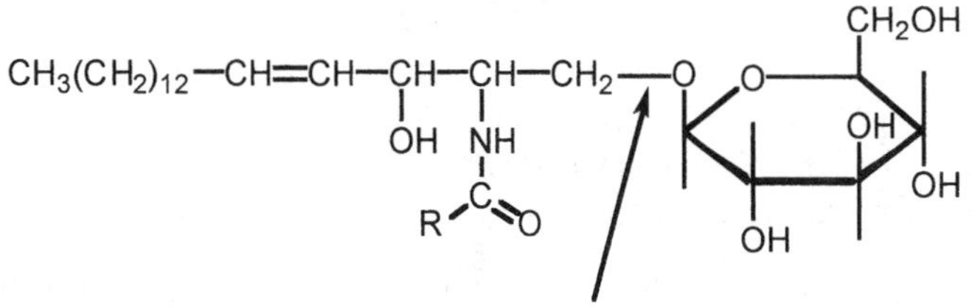

12.17 As described (see solution to In-Chapter Question 12.5) each stearic acid molecule
       generates 120 ATP. Consequently, the three separate stearate products of the
       hydrolysis of tristearin yield 360 ATP. The glycerol product is transported to the
       liver where it is used in gluconeogenesis.

12.19 a. detoxication – the conversion of a toxic molecule to a more soluble (and usually
           less toxic) product so that it may be excreted

       b. detoxification – the correction of a state of toxicity, as in the removal or
           metabolism of a specific toxin from an organism

       c. epoxide – a highly reactive ether in which one oxygen atom is bound to two
           carbon atoms in a three-membered ring

       d. monooxygenase – enzymes that catalyze redox reactions in which one molecule
           of oxygen is reduced per substrate molecule; one oxygen atom appears in the
           product of interest and the other oxygen atom forms H_2O

       e. cytochrome P_{450} – cyt P_{450} – a family of microsomal monooxygenases that
           catalyze the oxidation of many endogenous and exogenous substances, are
           involved in the detoxication of many xenobiotics, and require NADPH as an
           external reductant

12.20 Review Figure 12.33 on p 455. The bile salts are emulsifying agents that facilitate fat
       digestion.

12.22 The functions of each substance are as follows:

       a. phospholipid exchange protein – to bind phospholipids in one membrane and
           transfer them to another membrane; phospholipid exchange proteins are water-
           soluble

       b. sterol carrier protein – to bind squalene during cholesterol synthesis in the
           cytoplasm

       c. pregnenolone – with progesterone, to serve as a precursor for all other steroid
           hormones

       d. glucocorticoid – to promote carbohydrate, protein, and fat metabolism;
           glucocorticoids are hormones

e. hormone sensitive lipase (HSL) – triacylglycerol lipase – to catalyze lipolysis of triacylglycerols to fatty acids and glycerol; HSL is activated by phosphorylation via a phosphorylation cascade initiated by cAMP, which is produced in response to hormone-receptor interactions

12.23 The components of the cytochrome P_{450} electron transport system are: 1) cytochrome P_{450}, which contains heme, and 2) NADPH-cytochrome P_{450} reductase, a flavoenzyme that contains both FAD and FMN. Cyt P_{450} catalyzes oxidation reactions involving a wide variety of hydrophobic substrates. A hydroxyl group appears in each reaction. The electrons required for each Cyt P_{450}-catalyzed reaction are donated by Cyt P_{450} reductase.

12.25 In the small intestine, triacylglycerols mix with bile salts and are emulsified (solubilized). The size of the triacylglycerols and their emulsification with bile salts prevent them from crossing the enterocytes' cell membranes.

12.26 The inability of peroxisomal thiolase to bind medium-chain acyl-CoA requires the final three-to-six β-oxidation cycles of every fatty acid to take place in one organelle, the mitochondria. This is an advantage to the cell because it allows for tighter regulation of the β-oxidation of fatty acids to form acetyl-CoA molecules. Efficiency is also increased since acetyl-CoA produced in mitochondria may enter the citric acid cycle without needing to be transported across membranes. Similarly, the NADH and $FADH_2$ may enter the ETC directly. Otherwise, there would be an energy cost associated with transporting acetyl-CoA from the peroxisomes to the mitochondria. The maximum number of ATP could not be realized from a fatty acid molecule in a peroxisome, since the NADH and $FADH_2$ produced there would not be able to enter the mitochondria to enter the ETC. A further consideration is the additional regulation by compartmentation made possible by the requirement of β-oxidation to be completed in mitochondria. For example, to prevent fatty acid synthesis from occurring at the same time as β-oxidation in the liver, the transport of fatty acids into mitochondria is inhibited when fatty acid synthesis is taking place in the cytoplasm. If β-oxidation could go to completion in peroxisomes, it would negate this control mechanism.

12.28 Fatty acid biosynthesis occurs in the cytoplasm, whereas its catabolism takes place in mitochondria and peroxisomes so that these pathways may be regulated independently and energy-wasting futile cycles may be prevented.

12.29 NADH donates electrons to an electron transport system composed of cytochrome b_5 reductase (a flavoprotein) and cytochrome b_5 (which contains heme). An oxygen-dependent desaturase uses these electrons (from NADH) to activate O_2. The activated O_2 oxidizes the fatty acyl-CoA to create an alkene in its hydrocarbon chain, and is reduced to form two molecules of water.

12.31 In a long hydrocarbon chain, a *cis*-alkene introduces a kink that partially disrupts the otherwise close packing. *Cis*-fatty acids have lower melting points than saturated fatty acids because less heat energy is required to fully disrupt the packing, which must occur for the melting process to take place. More heat energy is required to disrupt the closer packing in saturated chains, so the melting point is higher for saturated fatty acids. In lipid bilayers that contain a relatively high proportion of *cis*-alkenes in their long hydrocarbon chains, the amphipathic nature of the phospholipids helps to prevent the bilayers from melting *per se*. However, the greater number of *cis*-alkenes does cause an increased degree of fluidity within the membrane bilayer, resulting in more facile lateral motion within the bilayer.

12.32 Cholesterol is used as a stiffening agent in animal cell membranes because its four fused rings are all-*trans*, making for a rigid and relatively flat structure. In addition to adding physical support to the membrane structure, this rigid fused-ring system induces a higher degree of order in nearby hydrocarbon chains as well.

12.34 The rate-limiting enzyme in cholesterol synthesis is HMGR (HMG-CoA reductase), which catalyzes the formation of mevalonate. HMGR is inactivated when it is phosphorylated via mechanisms that are initiated by AMP or cAMP. Physical exercise depletes energy stores, increasing cellular levels of AMP and cAMP. As a result, exercise lowers cholesterol by indirectly inhibiting its synthesis via AMP and cAMP. Specifically, high cellular [AMP] activates AMPK, and cAMP stimulates a signal transduction mechanism that activates RK. Both AMPK and RK phosphorylate and thus deactivate HMGR. cAMP also stimulates the phosphorylation of PPI-1, which inhibits phosphoprotein phosphatase, thus preventing dephosphorylation, i.e. preventing activation, so that HMGR remains in its inactive phosphorylated state. Also, cAMP is regulated by hormones such as glucagon and epinephrine, both of which increase during exercise.

12.35 Refer to Figure 11.16, which illustrates the numbered structure of cholesterol, and copy that structure onto a piece of paper. Each group of carbon atoms that originated as an intact mevalonate unit should be circled as listed below:
- C-2, C-3, C-4 (two methyl groups were lost in the synthesis of 7-dehydrocholesterol from lanosterol)
- C-6, C-5, C-10, C-1, C-19
- C-7, C-8, C-9, C-11 (A methyl shift and/or a demethylation occurred, leaving this mevalonate unit with only four carbons)
- C-15, C-14, C-13, C-12 (C-18 also originated in a mevalonate unit, but is a result of a methyl shift that occurred in the reaction that converted squalene-2,3-epoxide to lanosterol.)
- C16, C-17, C-20, C-21, C-22
- C-23, C-24, C-25, C-26, C-27

12.37 Glycerol (generated from the lipolysis of adipocyte triacylglycerols) is released into the blood and transported to the liver, where glycerol kinase converts it to glycerol-3-phosphate, which is a substrate for the synthesis of glucose, phospholipids, and triacylglycerols.

12.38 *Trans*-fatty acids, like saturated fatty acids, are able to pack much more closely than *cis*-fatty acids and have higher melting points as a result. The incorporation of *trans*-fatty acids into phospholipids would result in cell membranes with a higher degree of order and rigidity that is, perhaps, not as predictable as the rigidity that is induced by incorporating cholesterol into the bilayer. The reactivity of the alkenes may also pose a problem, as their oxidation would change the polarity and the properties of lipids that were synthesized from *trans*-fatty acids.

12.40 Of the four reactions in β-oxidation, only reactions 1 and 3 are oxidation reactions:

1) Acyl-CoA + FAD $\rightarrow$ *trans*-α,β-Enoyl-CoA + FADH$_2$

3) L-β-Hydroxyacyl-CoA + NAD$^+$ $\rightarrow$ β-Ketoacyl-CoA + NADH + H$^+$

CHAPTER 12: SOLUTIONS TO THOUGHT QUESTIONS

12.41 The reaction in which HMG-CoA reductase reduces HMG to form mevalonate is a rate-limiting step in cholesterol synthesis. Statins, inhibitors of this enzyme, act to lower blood levels of cholesterol in patients.

12.43 Carnitine is required for the transport of fatty acids into mitochondria where they are oxidized to generate energy. When carnitine levels are low, fat metabolism is impaired. Although glucose metabolism accelerates, an energy deficit occurs. In addition, accumulating acyl-CoA molecules become substrates for competing processes such as peroxisomal β-oxidation and triacylglycerol synthesis.

12.44 a. Hydrophobic interactions are probable between the enzyme and the lipid in the micelle.

        b. For phospholipases to be drawn into the micelle, they must have a hydrophobic surface.

12.46 Although regular eating is not a panacea, it does provide sufficient carbohydrate to act as a fuel to sustain vital metabolic processes.

12.47 Membrane phospholipids are synthesized on the cytoplasmic side of SER membrane. Because the polar head groups of phospholipid molecules make transport across the hydrophobic core of a membrane an unlikely event, a translocation mechanism is used to transfer phospholipids across the membrane to ensure balanced growth. Choline-containing phospholipids are found in high concentration on the lumenal side of ER membrane because a prominent phospholipid translocator protein called flippase preferentially transfers this class of molecule.

12.49 In the oxidation of butyric acid by the β-oxidation pathway, 1 mole each of $FADH_2$ and NADH and two moles of acetyl-CoA are produced.

12.50 The reaction for the oxidation of one mole of palmitic acid is:

$$CH_3(CH_2)_{14}COOH + 23\ O_2 \rightarrow 16\ CO_2 + 16\ H_2O$$

One mole of palmitic acid produces a yield of 16 moles of water molecules. From this number must be subtracted 8 moles of water of which 7 moles are used in the hydration reaction in each round of the β-oxidation spiral and 1 mole is used for the hydrolysis of pyrophosphate, the reaction that drives the activation of the palmitic acid molecule to completion. The net reaction, therefore, yields a total of 8 molecules of metabolic water.

12.52 Cholesterol is the precursor of the bile salts. If they are not reabsorbed new bile salts are synthesized from the cholesterol pool. This has the result of lowering cholesterol stores in liver, thereby lowering serum cholesterol.

12.53 Fatty acids are assembled from two carbon units, hence most biological fatty acids have an even number of carbon atoms. In many animals, fatty acids are synthesized by fatty acid synthase which ends the process at C-16. Although there is a robust capacity to elongate fatty acids, the most abundant fatty acids are C16 and C18, the product of one round of elongation process.

12.55 The ^{14}C label will appear as indicated:
Note: If ^{14}C-labeled acetyl-CoA is also used to make the β-ketobutyryl-CoA, then a ^{14}C label would appear on the internal alkene carbon in addition to the –CH_2–O– group.

13 Photosynthesis

Brief Outline of Key Terms and Concepts

OVERVIEW

Incorporating CO_2 into organic molecules requires energy and reducing power. In photosynthesis, both these requirements are provided by a complex process driven by light energy. **REACTION CENTER; PHOTOSYSTEM**

13.1 CHLOROPHYLL AND CHLOROPLASTS

The light energy absorbed by chromophores causes electrons to move to higher energy levels. In photosynthesis, absorption of energy from light drives electron flow.
CHLOROPHYLL; CAROTENOID CHROMOPHORE; ANTENNA PIGMENT

Photosynthesis occurs in **CHLOROPLASTS**, which consist of: an outer membrane, an inner membrane (encloses the **STROMA**), and the **THYLAKOID MEMBRANE**, which forms an intricate series of flattened vesicles called **GRANA. STROMAL LAMELLA** connect grana, and the **THYLAKOID LUMEN** is the space within the thylakoid membrane.

THE WORKING UNITS OF PHOTOSYNTHESIS:
  PHOTOSYSTEM I (PSI);
  PHOTOSYSTEM II (PSII);
  CYTOCHROME b_6f COMPLEX;
  ATP SYNTHASE (CF_0CF_1ATP SYNTHASE)

13.2 LIGHT

Electrons absorb energy at specific wavelengths and become excited. Electrons release energy (return to ground) by: fluorescence; resonance energy transfer; oxidation-reduction; or radiationless decay. This energy drives photosynthesis.

13.3 LIGHT REACTIONS

During the light reactions, H_2O is oxidized to O_2, and the ATP and NADPH required to drive carbon fixation are produced. The **Z SCHEME** is a mechanism that connects PSI and PSII in series. Light-driven photosynthesis begins with PSII.

PHOTOSYSTEM II: O_2 GENERATION
  OXYGEN-EVOLVING COMPLEX
  WATER-OXIDIZING CLOCK

PHOTOSYSTEM I AND NADPH SYNTHESIS

NONCYCLIC ELECTRON TRANSPORT:
Electrons from H_2O are transferred from photosystem II to photosystem I to NADP with the production of O_2 and NADPH.
CYCLIC ELECTRON TRANSPORT involves only PSI and generates additional ATP but no NADPH. During the light-independent reactions, CO_2 is incorporated into organic molecules. The first stable product of carbon fixation is glycerate-3-phosphate.

PHOTOPHOSPHORYLATION
Pumping 8 H^+ by the cyt b_6f complex (as a result of absorbing 4 photons) yields 2 ATP.

13.4 LIGHT-INDEPENDENT REACTIONS

THE CALVIN CYCLE is a series of light-independent reactions that incorporates CO_2 into organic molecules such as starch and sucrose (both important energy sources). Three phases of the Calvin cycle reactions are: **CARBON FIXATION,** reduction, and regeneration. (Sucrose is also used to translocate fixed carbon throughout the plant.)

PHOTORESPIRATION is a wasteful process in which photosynthesizing cells consume O_2 and release CO_2. Its role in plant metabolism is not understood.

ALTERNATIVES TO C3 METABOLISM
  C4 METABOLISM; C4 PLANTS can suppress photorespiration.
  CRASSULACEAN ACID METABOLISM

13.5 REGULATION OF PHOTOSYNTHESIS

LIGHT CONTROL OF PHOTOSYNTHESIS
Light is the principal regulator of photosynthesis, and affects the activities of regulatory enzymes such as rubisco by means of indirect mechanisms, which include changes in pH, $[Mg^{2+}]$, the **FERREDOXIN-THIOREDOXIN SYSTEM**, and **PHYTOCHROME**.

RIBULOSE-L,5-BISPHOSPHATE CARBOXYLASE (RUBISCO), the most important enzyme in photosynthesis is highly regulated by: allosteric effectors, covalent modification (carbamoylation of an active site lysine residue is required for optimal activity), and genetic control (light-activated synthesis of rubisco).

PHOTOSYNTHESIS: WHAT ARE THE ULTIMATE GOALS?

- to convert light energy into chemical energy (ATP)

- to produce reducing equivalents (NADPH)

- to convert CO_2 into sugars: **CARBON FIXATION**

LIGHT ENERGY drives the production of **ATP** and **NADPH**. The ATP and NADPH produced are used to convert **CO_2** into **SUGARS**.

STUDY HINTS

To make sense out of photosynthesis, we need to understand and integrate all of the following topics, and each topic contains new terms and new concepts. The good news is that you're already familiar with a fair amount of this, and there are a number of similarities between systems that you've studied before (especially the electron transport system) and those involved in photosynthesis.

- Chloroplast structure
- How light excites electrons, and what those excited electrons can do
- Molecular structures and complexes that are receptive to light (chromophores)
- Electron transfer and the use of the energy released as a result
- ATP as energy currency in the cell; NADPH as reducing power
- Metabolism of carbohydrate molecules and regulation of pathways

STAY FOCUSED ON THE GOAL of each part of photosynthesis that you're studying, and how it relates to the ultimate goals listed above. A stumbling block to avoid is to become so mired in the details of one section that you lose sight of its importance. Understand the big picture first, then tackle the specifics.

Example: The *Z Scheme* is just another way of looking at photosynthesis. Compare Figure 13.12, which outlines the Z scheme, to Figure 13.16, which includes a general schematic of the light reactions of photosynthesis.

CHLOROPLAST STRUCTURE: Knowing the location (i.e., stroma, thylakoid lumen, membrane) and how the parts of the chloroplast fit together will really help to give a clear picture of the overall process, and will also help to clarify how the different photosystems and the dark reactions fit together. Keep a diagram of the chloroplast in front of you, and when terms such as "thylakoid membrane" and "stroma" pop up, you'll have an immediate visual. Figure 13.2 on page 471 is great. Be sure to include *appressed* and *nonappressed* in your diagram.

Example: Since the ATP synthase complex works by allowing H^+ to pass through its pore (and as such, is driven by an $[H^+]$ gradient, just like mitochondrial ATP synthase), it's significant *where* each H^+ is used or released by a particular reaction.

ABBREVIATIONS: Many of the molecules, complexes, and systems have long names, and most of them have abbreviations. To stay afloat in this alphabet soup, keep a running list of abbreviations that you don't know right off the top of your head, and keep this list in front of you as well. *(Examples include PSII, PQ, Yz, MSP, and LHCII.)*

Some molecules have more than one abbreviation. It's valuable to be aware of all of the abbreviations (since different sources/texts have different preferences), but don't let them frustrate the learning process. Have a good concise list on hand. Example: Ferredoxin-NADP oxidoreductase is referred to as both FP and FNR, and pheophytin a is shown as both Ph and Phe a.

A PICTURE IS WORTH A THOUSAND WORDS: Make your own diagrams, using the figures in the text as a guide. The drawing and writing process is a powerful learning tool. These figures in your text are particularly good visual summaries:

Figure 13.16	***Membrane Organization of the Light Reactions in Chloroplasts: the Noncyclic Electron Transport Chain and the ATP Synthase Complex.*** (p.484)
Figure 13.17	***The Cyclic Electron Transport Pathway.*** This figure shows the interactions of light with photosystems I and II, their connection, and the roles of the cyt b_6F complex, PQ, and ATP synthase, on location in and around the thylakoid membrane. (p.484)
Figure 13.2	***Chloroplast structure.*** Be sure to differentiate between appressed and nonappressed thylakoid membrane in your diagrams. (p. 471)
Figures 13.12	***More Details of the Z Scheme.*** (p. 481)
Figure 13.18	***The Calvin Cycle***: using CO_2, ATP, and NADH to make a 3-carbon sugar. What is the net equation, and what happens at each phase? Note that parts of this cycle resembles the pentose phosphate pathway. (p. 487)
Figure 13.19	***Photorespiration.*** (p.490)
Figures 13.20-21	***C4 Metabolism. Crassulacean Acid Metabolism.*** (p. 491-2)

OVERVIEW

Light energy is converted into chemical energy at **reaction centers**. Each reaction center is a complex of light-absorbing pigments and electron transfer proteins.

13.1 CHLOROPHYLL AND CHLOROPLASTS

SPECIALIZED PIGMENT MOLECULES ABSORB LIGHT ENERGY.

Chromophores that absorb visible light typically have extended chains of conjugated alkenes, like the carotenoids. (β-carotene contains 11 conjugated alkenes.) In contrast, chlorophylls *a* and *b* and pheophytin *a* have a highly conjugated porphyrin ring system that absorbs visible light. Their long, nonpolar phytol chain embeds (and anchors) in the cell membrane (and contains only one alkene).

Chlorophyll *a*, chlorophyll *b*, and **pheophytin *a*:**
Chlorophylls *a* and *b* have a Mg atom in the porphyrin ring; pheophytin *a* has 2 H atoms. On the porphyrin ring II, Chlorophyll *a* and pheophytin *a* have $-CH_3$; chlorophyll *b* has –CHO at the same site.

Lutein and **β-carotene** are the most abundant carotenoids in thylakoid membranes.

CHLOROPLASTS: WHERE PHOTOSYNTHESIS OCCURS IN PLANTS

CHLOROPLAST STRUCTURE *(See Figure 13.2, p. 471 of your text.)*
Chloroplasts have three membranes:

1. an outer membrane that's highly permeable
2. an inner membrane that regulates transport into and out of the chloroplast

   STROMA – the space inside the inner membrane (similar to mitochondrial matrix); contains enzymes (for light-independent reactions and starch synthesis), DNA, and ribosomes

3. **THYLAKOID MEMBRANE** – a third membrane that forms an intricate series of flattened vesicles called **GRANA.** Light-dependent reactions of photosynthesis take place within the thylakoid membrane.

   GRANA– stacks of several flattened vesicles (*granum* is singular)

APPRESSED THYLAKOID MEMBRANE – adjacent layers of membrane that fit closely together within each granum; location of most PSII

NONAPPRESSED THYLAKOID MEMBRANE – directly exposed to the stroma; location of most PSI complexes

THYLAKOID LUMEN (SPACE) – internal compartment created by the formation of grana

STROMAL LAMELLA – thylakoid membrane that interconnects grana

THE WORKING UNITS OF PHOTOSYNTHESIS:

PSI – PHOTOSYSTEM I
Function: energizes and transfers e^- that are eventually donated to $NADP^+$.
Location: primarily in nonappressed thylakoid membrane

PsaA, PsaB – subunits and the largest polypeptides in PSI
PsaAB dimer contains: P700, A_0, A_1, F_4

P700 (SPECIAL PAIR) – a special pair of chlorophyll *a* molecules that absorb light at 700 nm

A_0, A_1, F_4 – a series of $1e^-$ carriers

A_0 a specific chlorophyll *a* molecule;

A_1 (phylloquinone, Q, vitamin K_1) – similar in structure to ubiquinone

F_X a 4Fe-4S center

F_A, F_B – two 4Fe-4S centers in an adjacent protein

ANTENNA PIGMENTS – absorb light energy and transfer it to the reaction center; include: additional chlorophyll *a* molecules, chlorophyll *b,* carotenoids

LHCI – LIGHT HARVESTING COMPLEX I – associated with PSI

PSII – PHOTOSYSTEM II
Function: oxidize H_2O to O_2 and donate energized e^- to electron carriers that eventually reduce photosystem I.
Location: primarily in appressed thylakoid membrane (not exposed to stroma)

PSII is a large membrane-spanning protein-pigment complex with at least 23 components. Its reaction center contains: D_1/D_2 dimer, cytochrome b_{559}, P680 (a special pair of chlorophyll *a* molecules that absorb light at 700 nm; and several hundred antenna pigments.

O_2-evolving component contains:
MSP (manganese-stabilizing protein)
Mn_3CaO_4 (a cubelike cluster linked to another Mn by a mono-μ-oxo bridge)
Tyrosine residue (Tyr^{161}, or Y_z) – located on D_1

Proteins: CP43, CP47
Also associated with PSII:
pheophytin a
plastoquinone (PQ) – similar to ubiquinone; two forms: Q_A, Q_B

LIGHT HARVESTING COMPLEX II (LHCII) – transmembrane protein and a trimer of Lhcb1, Lhcb2, Lhcb3 (light harvesting complex proteins). Each binds 12-14 chlorophyll a and chlorophyll b molecules and several carotenoids; major component of thylakoid membrane; detachable

PQH_2 – reduced plastoquinone – maintains balanced absorption of energy between PSI and PSII via a PQ H_2 –dependent phosphorylation of LHCII, which (as always) causes a conformational change, which (surprisingly) causes LHCII to detach so it

can wander over to PSI. When PQH_2 drops, LHCII loses its phosphoryl group and locks back with PSII.

CYTOCHROME b_6f COMPLEX is similar to cytochrome bc_1 complex in mitochondria. Location: throughout the thylakoid membrane.

ATP SYNTHASE = CF_0CF_1ATP SYNTHASE

Structurally and functionally similar to mitochondrial ATP synthase (see Figure 13.6, p. 475). ATP synthase phosphorylates ADP, driven by the transmembrane proton gradient (produced during light-driven electron transport).
Location: thylakoid membrane that's in direct contact with the stroma.

13.2 LIGHT HAS PROPERTIES OF BOTH WAVES AND PARTICLES

VISIBLE LIGHT:

violet		*red*
400 nm	to	700 nm
shorter wavelength		longer wavelength
higher frequency		lower frequency
higher energy		*lower energy*

WAVE PROPERTIES OF LIGHT: LIGHT ACTS LIKE WAVES	PARTICLE PROPERTIES OF LIGHT: PHOTONS LIGHT ACTS LIKE PARTICLES
$\lambda = c/v$ λ = wavelength, v = frequency, c = speed of light	$\varepsilon = hv$ ε = energy of a photon h = Planck's constant

Molecules absorb light at specific energies that correspond to specific wavelengths. (Complex molecules can absorb light at several wavelengths.)

Chromophores absorb light. When electrons in the chromophores absorb light, they become excited (move from the ground state to higher energy levels).

HOW AN EXCITED ELECTRON CAN RETURN TO ITS GROUND STATE:

1. **FLUORESCENCE.** Excited electron first relaxes to a lower vibrational (energy) state, then goes back to ground state and emits a photon (of lower energy than original photon absorbed).

2.* **RESONANCE ENERGY TRANSFER.** Excitation energy is transferred to a neighboring molecule via resonance

3.* **OXIDATION-REDUCTION.** Transfer of an excited electron to a neighboring molecule makes it a strong reducing agent. The electron returns to its ground state by reducing another molecule.

4. **RADIATIONLESS DECAY.** Excitation energy is given off as heat

*These two methods are the most important to photosynthesis.

LIGHT REACTIONS: LIGHT EXCITES ELECTRONS THAT ARE USED TO SYNTHESIZE ATP AND NADPH

OVERALL REACTION: H_2O IS OXIDIZED TO O_2, $NADP^+$ IS REDUCED TO NADPH

$$2 \text{ NADP}^+ + 2 \text{ H}_2\text{O} \rightleftharpoons 2 \text{ NADPH} + \text{O}_2 + 2 \text{ H}^+ \qquad \Delta E^{\circ\prime} = -1.136 \text{ V}$$

For each mole of O_2 generated:

at least 438 kJ of free energy ($\Delta G°'$) is needed

at least 8 photons are absorbed to provide 1360 kJ of energy

SUMMARY OF LIGHT REACTIONS

Light is used in photosystem II to excite electrons to a higher energy state. As the electrons are transferred down an energy gradient, protons (H^+) are pumped across the membrane to generate a proton gradient, which is used for ATP synthesis.

The electrons from photosystem II can then be transferred to photosystem I. More light is needed to excite the electrons to an energy state that's high enough to reduce $NADP^+$ to NADPH.

Water serves as a source of electrons when electrons are transferred from photosystem II to photosystem I to $NADP^+$. When H_2O gives up its electrons, the oxygen in H_2O is oxidized to O_2.

Z SCHEME: ELECTRON TRANSFER FROM H_2O TO $NADP^+$

The Z scheme shows where electrons go during photosynthesis. If this path of electrons is shown on a graph where the y axis is energy, then part of this path resembles the letter Z.

Light drives photosynthesis, and the Z scheme shows where and why this is true. From the energy differences between molecules within the photosystems, it's clear that a boost of energy – in the form of a photon of light – is needed at PSII and at PSI to give the electrons enough energy to continue through their path to $NADP^+$.

Overall, ELECTRONS COME FROM H_2O and REDUCE $NADP^+$ TO FORM NADPH.

PATH OF ELECTRONS THROUGH THE PHOTOSYSTEMS IN THE PRESENCE OF LIGHT:

H_2O

    PSII (photosystem II)

        PQ (plastoquinone)

            cyt b$_6$f (cytochrome b$_6$f complex)

                PC (plastocyanin)

                    PSI (photosystem I)

                        Fd (ferredoxin)

                            FNR (ferredoxin-NADP oxidoreductase)

                           $NADP^+$

Here's another way to think about this: Begin with the excitation of photosystem I, which results in the transfer of electrons, ultimately to $NADP^+$ to form NADPH. After the electrons are released from PSI, how are they replaced?

P700 in PSI absorbs a photon of light and releases an energized electron. That electron is replaced by an electron from PSII (through PQ, cyt b$_6$f, and PC, as shown above). The PSII electron needs to be excited by a photon as well, in order to have enough energy to be transferred. How is the PSII electron replaced? *Water* gives up electrons to form O_2 and H^+.

PHOTOSYSTEM II AND OXYGEN GENERATION

The function of photosystem II is to oxidize water molecules and donate energized electrons to electron carriers which eventually reduce photosystem I. As the electrons are donated to the electron carriers, protons are pumped out of the thylakoid and ultimately used for ATP synthesis. The water-oxidizing clock is the mechanism by which H_2O is converted into O_2.

REACTIONS OF PHOTOSYSTEM II:

Light energy excites a P680 electron, giving it a large potential energy. These energized electrons are transferred to a series of electron transport carriers that are analogous to the mitochondrial electron transport chain. As the electrons are transferred, protons are pumped from the stroma into the thylakoid lumen. This creates a proton gradient that is used to synthesized ATP.

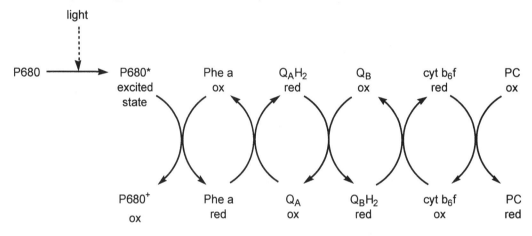

red = reduced state; the carrier marked 'red' has the electrons ready to pass to the next carrier
ox = oxidized state; the carrier marked 'ox' is ready to accept electrons

The first reaction (first set of arrows pointing down) of P680* is:

$$P680^* + Phe\ a \rightarrow P680^+ + Phe\ a$$
excited state   ox   ox   red

The second reaction (second set of arrows, pointing up) is:

$$Phe\ a + Q_A \rightarrow Phe\ a + Q_AH_2$$
red   ox   ox   red

THE REDUCED PLASTOCYANIN (PC) TRANSFERS ITS ELECTRONS TO P700 IN PSI.

WHAT HAPPENS TO THE OXIDIZED P680?
These electrons are replaced when H_2O is oxidized.

PQ TRANSFERS ELECTRONS FROM PSII TO cyt b₆f AND PUMPS H⁺ TO THE LUMEN.
PQ is plastoquinone, an electron carrier within the thylakoid membrane

$$PQ + 2e^- \text{ (from PSII)} + 2H^+ \text{ (from stroma)} \rightarrow PQH_2$$
$$PQH_2 \rightarrow PQ + 2e^- \text{ (to cyt b}_6\text{f)} + 2H^+ \text{ (to the thylakoid lumen)}$$

CYTOCHROME b₆f COMPLEX TRANSFERS ELECTRONS FROM PQ TO PC
An Fe-S site on the complex transfers electrons from PQ to PC.

PC – plastocyanin – a water-soluble, copper-containing protein in the thylakoid lumen.

PHOTOSYSTEM I USES ELECTRONS FROM PC TO REDUCE NADP$^+$ → NADPH

The ultimate goal of PSI: to reduce NADP$^+$ to NADPH, which requires 2e$^-$.

The path of the electrons:

1. from PC (in the lumen) to P700 (in the membrane)

2. reduced P700 absorbs a photon of light that excites and energizes this electron

3. the excited electron is transferred through this series of electron carriers in PSI, in the thylakoid membrane:

   P700* → A$_0$ → Q → F$_X$ → F$_A$, F$_B$
   chlorophyll a → phylloquinone → Fe-S proteins

The next electron acceptor is ferredoxin.
Ferredoxin is mobile, water-soluble, and in the stroma.

FERREDOXIN TRANSFERS ITS ELECTRON FROM PSI

EITHER: **TO FNR (TO MAKE NADPH)**

This is the **NONCYCLIC ELECTRON TRANSPORT PATHWAY.**
FNR (ferredoxin-NADP oxidoreductase) uses two electrons (and a stromal H$^+$) to reduce NADP$^+$ to NADPH.

OR: **TO PQ (TO PUMP H$^+$ FOR ATP SYNTHESIS)**

This is the **CYCLIC ELECTRON TRANSPORT PATHWAY.**

Electrons are cycled *back* to PQ (in the thylakoid membrane), which also takes two H$^+$ from the stroma to form PQH$_2$. Remember that PQH$_2$ then gives its electrons to cyt b$_6$f and its two H$^+$ go to the lumen.

So, energized electrons were used to pump additional H$^+$ across the thylakoid membrane into the lumen *instead of making NADPH*.

The extra H$^+$ in the lumen produces additional ATP.

The cyclic electron transport pathway drives ATP synthesis without making NADPH. This only occurs in some species (e.g., algae), when the NADPH/NADP$^+$ ratio is high.

ATP SYNTHASE (CF$_0$CF$_1$ATP SYNTHASE)

ATP synthase phosphorylates ADP, driven by the transmembrane proton gradient (produced during light-driven electron transport), and is located in the thylakoid membrane, in direct contact with the stroma.

PHOTOPHOSPHORYLATION is the light-driven synthesis of ATP from ADP + P$_i$.

LIGHT-INDEPENDENT REACTIONS

THE CALVIN CYCLE[1] INCORPORATES CO_2 INTO CARBOHYDRATE MOLECULES

Many Calvin cycle reactions resemble pentose phosphate pathway reactions.

PHASES OF THE CALVIN CYCLE:

1. **CARBON FIXATION** by the enzyme rubisco (ribulose-1,5-bisphosphate carboxylase) in C3 plants: CO_2 reacts with ribulose-1,5-bisphosphate to form 2 glycerate-3-phosphates (per CO_2). This reaction requires ATP.

2. **REDUCTION OF GLYCERATE-3-PHOSPHATE** by NADPH to form glyceraldehyde-3-phosphate

3. **REGENERATION OF RIBULOSE-1,5-BISPHOSPHATE**: Of every six glyceraldehyde-3-phosphates formed, five are regenerated to form three molecules of ribulose-1,5-bisphosphate.

CALVIN CYCLE REACTION SUMMARY:

3 ribulose-1,5-bisphosphate + 3 CO_2 $\rightarrow$	6 glycerate-3-phosphate
6 glycerate-3-phosphate + 6 ATP + 6 NADPH $\rightarrow$	6 glyceraldehyde-3-phosphate + 6 ADP + 6 $NADP^+$ + 6 P_i
5 glyceraldehyde-3-phosphate + 3 ATP $\rightarrow$	3 ribulose-1,5-bisphosphate + 3ADP + 2P_i

CALVIN CYCLE NET EQUATION:

3 CO_2 + 6 NADPH + 9 ATP $\rightarrow$	glyceraldehyde-3-phosphate + 6 $NADP^+$ + 9 ADP + 8 P_i

PHOTORESPIRATION CONSUMES O_2 AND EVOLVES CO_2

Photorespiration is a wasteful light-dependent process that undermines photosynthesis. In photorespiration, O_2 is consumed and CO_2 is evolved by plant cells while actively engaged in photosynthesis.

The rate of photorespiration depends on the concentrations of CO_2 and O_2, and its function is unknown. Plants that use C4 metabolism and crassulacean acid metabolism (CAM) can counteract the photorespiration process. (See further notes, below.)

REGULATION OF PHOTOSYNTHESIS: RUBISCO CONTROL; EFFECT OF LIGHT

WHY REGULATE PHOTOSYNTHESIS?
- to increase the rate of photosynthesis when light is available,
- to make sure that when the rate of photosynthesis is high, the rates of CO_2 fixation and **sucrose synthesis** is also high.

PHOTOSYNTHESIS DEPENDS ON TEMPERATURE, CELLULAR CO_2 CONCENTRATION, AND LIGHT.
Light: activates certain photosynthetic enzymes, and
deactivates some of the enzymes that degrade sugars.

---

[1] The Calvin cycle is also known as: the dark reactions, light-independent reactions, the reductive pentose phosphate cycle (RPP cycle) and the photosynthetic carbon reduction cycle (PCR cycle.)

The key regulatory enzyme in photosynthesis is **RIBULOSE-1,5-BISPHOSPHATE CARBOXYLASE (RUBISCO)**. Rubisco catalyzes the fixation of CO_2, which is relatively slow. The best way to increase rubisco's rate is to increase the number of copies of rubisco. Genes that code for rubisco are activated by an increase in light intensity (and appears to involve phytochrome).

Rubisco is also controlled by covalent modification (carbamoylation). The rate of carbamoylation depends on the CO_2 concentration and an alkaline pH.

LIGHT AFFECTS ENZYMES BY INDIRECT MECHANISMS:

1. **pH** of the stroma increases during photosynthesis (when H^+ is pumped out of the stroma into the thylakoid lumen). This higher pH (~ 8) activates rubisco and other enzymes.

2. **Mg^{2+}** moves across the membrane into the stroma during the light reactions; This increase in stromal [Mg^{2+}] activates several photosynthetic enzymes (including rubisco).

3. **FERREDOXIN-THIOREDOXIN SYSTEM**

4. **PHYTOCHROME**

ALTERNATIVES TO C3 METABOLISM: C4 METABOLISM AND CAM (CRASSULACEAN ACID METABOLISM)

C4 plants have evolved mechanisms to reduce water loss, and photorespiration. C4 metabolism is a mechanism for assimilating CO_2.

When the plant opens its stomata at night, CO_2 enters and reacts to form oxaloacetate. When light is available, photosynthesis makes ATP and NADPH, and CO_2 is released in the bundle sheath cells and converted into sugar. Because the concentration of CO_2 within bundle sheath cells is significantly higher than that of O_2, photorespiration is drastically reduced.

In C4 plants, the light reactions occur in the mesophyll cells, and the Calvin cycle reactions (CO_2 fixation) occur in the bundle sheath cells. This division of labor lets C4 plants concentrate CO_2 in the bundle sheath cells at the expense of ATP hydrolysis.

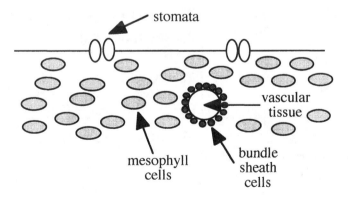

ADVANTAGES OF C4 METABOLISM

- Minimizes photorespiration by increasing CO_2 concentration in cells that have rubisco.

- Reduces the plant's need for water. Stomata must open to allow CO_2 to enter and O_2 to exit the leaf. But water is also lost when the stomata are open. C4 plants can effectively concentrate CO_2, so that the stomata do not need to be open as often. Water loss in C4 plants is only 10-30% compared to C3 plants.

MESOPHYLL CELLS[2]
IN DIRECT CONTACT WITH AIR
LACK RUBISCO
LIGHT REACTIONS MAKE **ATP, NADPH**

BUNDLE SHEATH CELLS
IN CONTACT WITH VASCULAR TISSUE
HAVE RUBISCO
CALVIN CYCLE REACTIONS USE CO_2 RELEASED

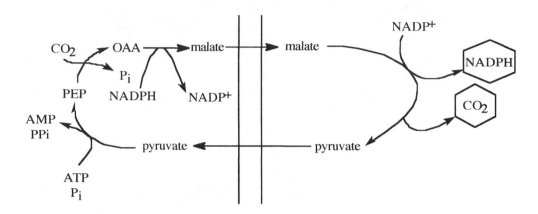

CAM (CRASSULACEAN ACID METABOLISM)

CAM is another type of photosynthetic specialization used by desert plants. These plants grow in high intensity light without very much water and, in order to survive, have evolved to the point where they are able to temporally separate carbon fixation and ATP/NADPH synthesis. The growth of CAM plants is slowed because the extent of photosynthesis is limited by how much CO_2 is stored as malate during the night.

AT NIGHT:	DURING THE DAY:
The stomata open at night when water loss would be at a minimum.	The stomata close.
CO_2 is stored as malate.	Photosystems I and II work to make ATP and NADPH.
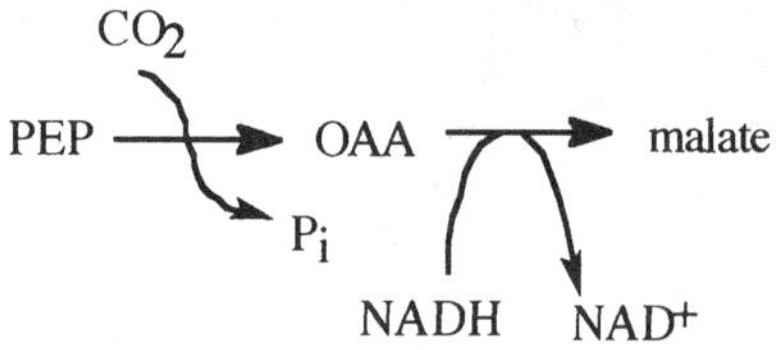	The CO_2 that was stored in malate during the night is released, and rubisco uses this CO_2 to make carbohydrates in the Calvin Cycle. 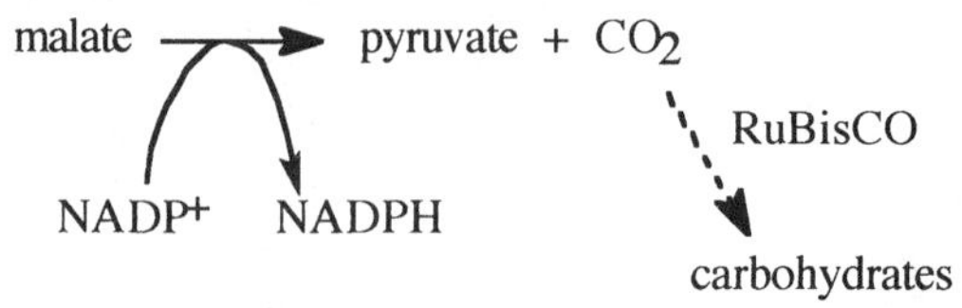

[2] In some C4 plants, oxaloacetate is converted to aspartate instead of malate. The Asp travels to the bundle sheath cells, where it's converted back to oxaloacetate, which can then release its CO_2.

AFTER STUDYING THIS CHAPTER, YOU SHOULD BE ABLE TO:

- Demonstrate an understanding of the overarching goals of photosynthesis, and how each pathway studied contributes to those goals.

- Describe the properties of light and how energy from an excited electron is captured in photosynthesis.

- Describe the structure of the chloroplast. State the locations of specific pathways, and the property of each location that is advantageous for each pathway.

- Differentiate between the light-dependent and the light-independent reactions.

- Trace electrons through the Z scheme. Identify the points where:

    o energy is captured as reducing power in NADH

    o energy is captured as ATP

    o water is oxidized to O_2

- State the role served by various components of photosystems I and II.

- Describe the similarities and differences between the electron transport system and photosynthesis, and between oxidative phosphorylation and photophosphorylation.

- Explain why photorespiration is counterproductive to photosynthesis, and describe the mechanisms (pathways) that some plants use to suppress photorespiration.

- Trace carbon and/or oxygen atoms from CO_2 through the Calvin cycle.

- State under which conditions photosynthesis will be activated or inhibited. Specifically, relate the enzyme(s) and/or pathway will be affected, the type of regulation (allosteric, covalent modification, compartmentation, genetic control) and the activating or inhibiting agent.

Use this space to note any additional objectives provided by your instructor.

CHAPTER 13: SOLUTIONS TO REVIEW QUESTIONS

13.1   a.   photosystem – a molecular complex that absorbs light and converts it to chemical energy

b.   reaction center – a pigment-protein component of a photosystem that mediates the conversion of light energy into chemical energy

c.   light reactions – a biochemical mechanism whereby the electrons energized by light are subsequently used in ATP and NADPH synthesis

d.   dark reactions – light-independent reactions in which the ATP and NADPH generated during light reactions are utilized in the synthesis of carbohydrate

13.2   The most significant contribution of early photosynthesizing organisms to the earth's environment was the conversion of a reducing atmosphere (ammonia and methane) to an oxidizing atmosphere.

13.4   Chloroplasts resemble mitochondria in the following ways: (1) they are both similar in size and structure to modern prokaryotes; (2) they both reproduce by binary fission; (3) the genetic information and protein synthesizing capability of both chloroplasts and mitochondria are similar to that of prokaryotes; (4) the ribosomes of chloroplasts and mitochondria are similar in size and function; and (5) they are both thought to have arisen from ancient free living prokaryotes.

13.5   a.   chloroplast – a chlorophyll-containing plastid found in the cells of algae and higher plants

b.   photorespiration – a wasteful process in which O_2 is consumed and CO_2 is released

c.   chlorosome – membrane-bound vesicles attached to the plasma membrane that are extremely efficient at harvesting light, collecting low energy photons and transferring them to the photosynthetic reaction centers in the plasma membrane

d.   stromal lamellae – thylakoid membrane that interconnects two grana

e.   granum – a folded portion of the thylakoid membrane

13.7   The final electron acceptor in photosynthesis is carbon dioxide, regardless of the $NADPH/NADP^+$ ratio.

13.8   During the light restrictions of photosynthesis, light-driven electron transport results in ATP and NADPH synthesis.

13.10   a. D_1/D_2 dimer – two polypeptide units that are part of the reaction center of PSII

b. CF_0 – a component of ATP synthase; a membrane-spanning protein complex that contains a proton-conducting channel and four different types of protein subunits

c.   wavelength – the distance from the crest of one wave to the crest of the next wave

d.   chromophore – a molecular component that absorbs light of a specific frequency

e.   fluorescence – a form of luminescence in which certain molecules can absorb light of one wavelength and emit light of a longer wavelength

13.11   The molecules of O_2 are generated from H_2O.

13.13   The maximum rate of photosynthesis will occur at the λ_{max} of the photosynthesizing system. This absorption maximum should match the absorption maxima of the light absorbing pigments. See the Biochemistry in the Lab box on photosynthetic studies (located on the companion website at www.oup.com/us/mckee) for a discussion of the absorption spectrum of chlorophyll and several other pigment molecules.

13.14 a. radiationless decay – occurs when an excited molecule decays to its ground state by converting the excitation energy into heat

      b. resonance energy transfer – the transfer of energy from an excited molecule to a nearby molecule, thereby exciting the second molecule

      c. Z scheme – a mechanism whereby electrons are transferred from water to $NADP^+$. This process produces the reducing agent NADPH required for fixing carbon dioxide in the light-independent reactions of photosynthesis.

      d. PSI – photosystem I – a large membrane-spanning complex that contains protein and pigments including over 200 chlorophyll; PSI energizes and transfers the electrons that eventually are donated to $NADP^+$

      e. PSII – photosystem II – a large membrane-spanning protein-pigment complex that functions to oxidize water molecules and to donate energized electrons to electron carriers that eventually reduce photosystem I.

13.16 Among the most notable metals present in photosynthesizing systems are magnesium, manganese, iron, and copper. Magnesium stabilizes the porphyrin ring in chlorophyll molecules without adding a redox element site during the reduction of water to oxygen. Manganese acts as a redox center in the transfer of electrons from the quinone electron carriers to the terminal electron acceptors in photosynthesis. Finally, copper acts as the terminal electron acceptor in PSII before the electrons are transferred to P700 in PSI.

13.17 The oxygen-evolving complex of PSII exists in five transient oxidation states (S_0 through S_4), collectively referred to as a clock. Because oxygen evolution occurs only when the S_4 state has been reached, several light bursts are required.

13.19 The Z scheme is a mechanism whereby electrons are transferred from water to $NADP^+$. This process produces the reducing agent NADPH required for fixing carbon dioxide in the light-independent reactions of photosynthesis. Removal of the electrons from water also results in the production of oxygen. As electrons flow from PSII to PSI, protons are pumped across the thylakoid membrane, a process that establishes the proton gradient that drives ATP synthesis.

13.20 Carbon dioxide fixation occurs in the stroma of the chloroplasts.

13.22 If blue wavelengths are used in addition to red ones, the rate of oxygen evolution is increased. This phenomenon is known as the Emerson enhancement effect. If photosynthesis occurs in a single photosystem, then the magnitude of the enhancement should reflect the ratios of the λ_{max} in the red and blue regions. The enhancement was much greater than predicted by this ratio. The existence of a second chromophore (and by inference a second photosystem) was therefore suggested.

13.23 The first reaction in both photosynthesis and photorespiration is catalyzed by rubisco, which has both carboxylase and oxidase activity. CO_2 and O_2 compete for the enzyme's active site. Higher levels of CO_2 enable it to out-compete O_2, and photosynthesis will occur at the expense of photorespiration.

13.25 The term triose phosphate is used to describe the molecules glyceraldehyde-3-phosphate and dihydroxyacetone phosphate. Formed during the Calvin cycle, the triose phosphates are used in plants in biosynthetic processes such as the formation of sucrose, polysaccharides, fatty acids and amino acids. For a deeper understanding of triose phosphate metabolism, read the Biochemistry in Perspective box on Starch and Sucrose Metabolism on the Companion Web Site.

13.26   Since O_2 is consumed in the process of photorespiration, and C4 plants are able to avoid photorespiration, exposure to $^{15}O_2$ should not produce any molecules that contain ^{15}O-labeled atoms. Although C4 plants utilize rubisco, which catalyzes the first reaction in both the Calvin cycle and in photorespiration, C4 plants maintain a much higher concentration of CO_2 than O_2. This higher concentration, combined with rubisco's greater affinity for CO_2, results in CO_2 outcompeting any O_2 that might be present. Should CO_2 be removed from the air surrounding the C4 plant prior to exposure to $^{15}O_2$, then it's possible that photorespiration may occur until ATP and NADH are depleted. In this case, the first products of photorespiration, glycolate-2-phosphate and glycerate-3-phosphate, would contain ^{15}O-labeled atoms.

13.28   Theoretically, plants should be able to grow in an oxygen-free atmosphere. However, since oxygen is produced by growing plants, the atmosphere would not remain oxygen-free for long.

13.29   When corn has been exposed to $^{14}CO_2$, the radioactive label will first be found in oxaloacetate, at the carbon atom located in the carboxylate group adjacent to the CH_2.

13.31   Dinitrophenol is an uncoupler molecule that destroys the proton gradient needed for ATP synthesis. Photosynthesis requires ATP. Therefore, without ATP photosynthesis comes to a halt.

CHAPTER 13: SOLUTIONS TO THOUGHT QUESTIONS

13.32   The electrons energized by light absorption are used to generate NADPH. This NADPH is used in large quantity to fix CO_2 into carbohydrate molecules. If CO_2 is not available, NADPH accumulates and the chlorophyll molecules do not have electron transfer available as a way to return to the relaxed state. One avenue available is to release the energy as a photon, i.e., to fluoresce.

13.34   Only light of a particular energy can be absorbed by photosynthetic pigments. Increasing the intensity of the light increases the number of photons present and hence can improve the rate of photosynthesis. Increasing the energy level of the light, that is, the energy of the photons, decreases the rate of photosynthesis by shifting the photons to energy levels that are not absorbed by the photosystems.

13.35   Oxidative phosphorylation and photophosphorylation use many of the same molecules in their reactions and both are linked to an electron transport system. However, chloroplasts use light energy to drive redox reactions while mitochondria use the energy of chemical bonds to drive redox reactions. In contrast to mitochondrial inner membrane, thylakoid inner membrane is permeable to magnesium and chloride ions. Therefore, the electrochemical gradient across the thylakoid membrane consists mainly of a proton gradient.

13.37   If sufficient carbon dioxide is already present to saturate all of the ribulose-1,5-bisphosphate carboxylase molecules, the presence of additional carbon dioxide molecules will not increase the rate of photosynthesis. In addition, photosynthesis is depressed by low light levels.

13.38 Chloroplasts possess DNA similar to that of modern cyanobacteria as well as prokaryotic-like protein synthesizing machinery. In addition, they multiply by binary fission as do bacteria.

13.40 Under conditions of high temperature, the carbon dioxide compensation point of C3 plants rises because the oxygenase activity of rubisco increases more rapidly than the carboxylase activity.

13.41 Because of the capacity of C4 plants to avoid the process of photorespiration, herbicides that promote photorespiration do not affect these organisms.

13.43 Chloroplasts must be relatively large to intercept the light necessary to carry out photosynthesis.

13.44 Because of its negative value for $\Delta E^{0}{}'$, hydrogen sulfide is easier to oxidize than water.

13.46 C4 plants do not have as much photorespiration. As a consequence the total energy that they expend in the fixation of carbon dioxide is less. Therefore. they are more efficient and can out compete the C3 plants for carbon.

13.47 Because of extensive photorespiration, which competes with photosynthesis, C3 plants expend more energy per gram of carbon fixed. As a result C4 plants are more efficient.

13.49 Triazine herbicides promote photorespiration.

13.50 The ratio of energies is: $\dfrac{-hc/700}{-hc/1000} = 1.48$

Therefore, 1.48 quanta of light of 1000 nm radiation is required to provide the same amount of energy as 1 quanta of light at 700 nm.

14 Nitrogen Metabolism I: Synthesis

Brief Outline of Key Terms and Concepts

OVERVIEW
Nitrogen cycle
NAA – nonessential amino acid
EAA – essential amino acid
BCAA – branched-chain amino acid
  transamination reactions
  transaminases; aminotransferases

14.1 NITROGEN FIXATION

THE NITROGEN FIXATION REACTION
NITROGENASE COMPLEX – dinitrogenase and dinitrogenase reductase (Fe protein); contains P cluster (8Fe-7S) and a MoFe cofactor; regulated by transcriptional control of nif (nitrogen fixation genes).

NITROGEN ASSIMILATION
Nitrogen assimilation is the incorporation of inorganic nitrogen compounds into organic molecules.

14.2 AMINO ACID BIOSYNTHESIS

AMINO ACID METABOLISM OVERVIEW
Amino acid pool; positive nitrogen balance; negative nitrogen balance (Kwashiorkor)
Transport of amino acids into cells

REACTIONS OF AMINO GROUPS
TRANSAMINATION REACTIONS transfer amino groups from one carbon skeleton to another; transaminases require PLP in a ping-pong reaction mechanism. Important pairs: α-ketoglutarate/Glu pair; oxaloacetate/Asp pair; pyruvate/Ala pair
REDUCTIVE AMINATION is the synthesis of amino acids by incorporating free NH_4^+ or the amide nitrogen of glutamine or asparagine into α-keto acids. Ammonium ions are also incorporated into cellular metabolites by the amination of glutamate to form glutamine.

SYNTHESIS OF THE AMINO ACIDS
Six families of amino acids:
glutamate, serine, aspartate, pyruvate, the aromatics, and histidine.

The nonessential amino acids are derived from precursor molecules available in many organisms. The essential amino acids are synthesized from metabolites produced only in plants and some microorganisms.

14.3 BIOSYNTHETIC REACTIONS INVOLVING AMINO ACIDS

ONE-CARBON METABOLISM
Important carriers of single carbon atoms:
– THF (tetrahydrofolate, the biologically active form of folic acid)
– SAM (*S*-adenosylmethionine)

GLUTATHIONE (GSH)
GSH, the most common intercellular thiol, is involved in many cellular activities. In addition to reducing sulfhydryl groups, GSH protects cells against toxins and promotes the transport of some amino acids.
γ-GLUTAMYL CYCLE; BLOOD-BRAIN BARRIER; GLUTATHIONE-S-TRANSFERASES

NEUROTRANSMITTERS
NEUROTRANSMITTERS may be excitatory or inhibitory. **BIOGENIC AMINES** are amino acid derivatives and include: γ-aminobutyric acid (GABA), catecholamines [dopamine (D), norepinephrine (NE), and epinephrine], serotonin, and histamine. DOPA is 3,4-dihydroxyphenylalanine; BH_4 is tetrahydrobiopterin; and PNMT is phenylethanolamine-*N*-methyltransferase.

ALKALOIDS
Classes of alkaloids include: tropane alkaloids (cocaine, atropine); pyridine alkaloids (nicotine); and isoquinoline alkaloids (codeine, morphine).

NUCLEOTIDES
Nucleotides are the building blocks of the nucleic acids. They also regulate metabolism and transfer energy.
PURINE and **PYRIMIDINE NUCLEOTIDES** are synthesized in both *de novo* and salvage pathways. **NUCLEOSIDES**; *syn* vs. *anti*; **IMP** (inosine-5′-monophosphate)
DEOXYRIBONUCLEOTIDES; **RNRI** (ribonuclease reductase I)

HEME
Heme has an iron-containing porphyrin ring system and is synthesized from glycine and succinyl-CoA. Protoporphyrin IX, the precursor of heme, is also a precursor of the chlorophylls.

BIOCHEMISTRY IN PERSPECTIVE:
GASOTRANSMITTERS (NO, CO, H_2S); **ONLINE:** Parkinson's Disease and Dopamine; Pb Poisoning

OVERVIEW

OUR SOURCE OF NITROGEN: AMINO ACIDS

BCAA Branched chain amino acids — Leu, Ile, Val

EAA <u>Essential amino acids</u> must be provided by the diet. — Ile, Leu, Lys, Met, Phe, Thr, Trp, Val

NAA <u>Nonessential amino acids</u> can be synthesized from available metabolites. — Ala, Arg*, Asn, Asp, Cys, Glu, Gln, Gly, His*, Pro, Ser, Tyr

*essential for infants

> *Hint:*
>
> *Now would be a great time to brush up on the amino acid R groups! It'll be a tremendous help to you if their structures instantly pop into your mind as you study.*

14.1 NITROGEN FIXATION

THE NITROGEN FIXATION REACTION

Nitrogen Fixation ($N_2 \rightarrow 2NH_3$) is a reduction that requires energy: at least 16 ATP/N_2.

NITROGENOUS COMPLEX includes: (*must be anaerobic – protected from O_2*)

Dinitrogenase (or Fe-Mo protein): $N_2 + 8H^+ + 8e^- \rightarrow 2NH_3 + H_2$

Dinitrogenase reductase (or Fe protein): binds ATP; ATP hydrolyzes to ADP and P_i; this causes conformational changes that help the e^- transfer to dinitrogenase

Path of electrons:

NADH (or NADPH) → ferredoxin → dinitrogenase reductase → dinitrogenase

NH_3 then travels from the bacteria to the host cell to be used in glutamine synthesis.

Nitrogen fixation allows many plants and animals to synthesize many *N*-containing biomolecules such as proteins and nucleic acids. Only a few prokaryotes can fix nitrogen.

NITROGEN ASSIMILATION

Plants receive their nitrogen via symbiotic relationships with N_2-fixing prokaryotes or by absorbing NH_3 and NO_3^- synthesized by soil bacteria (or provided by artificial fertilizers). Animals take in organic nitrogen mainly as amino acids. The liver determines the fate of ingested amino acids.

14.2 AMINO ACID BIOSYNTHESIS

AMINO ACID METABOLISM OVERVIEW

Amino acid pool are the amino acids available for metabolic processes. Amino acids from the degradation of dietary and tissue proteins enter the pool; excreted nitrogenous end products such as urea and uric acid leave the pool. Amino acids enter cells via membrane-bound transport proteins; some are Na^+-transport-dependent.

NITROGEN BALANCE: NITROGEN INTAKE = NITROGEN LOSS

Positive nitrogen balance: nitrogen intake > nitrogen loss. Protein synthesis exceeds degradation (e.g., in growing children, pregnant women, recuperating patients).

Negative nitrogen balance: nitrogen intake < nitrogen loss. Nitrogen can't be replenished fast enough; leads to malnutrition (Kwashiorkor)

REACTIONS OF AMINO GROUPS

TRANSAMINATION REACTIONS: THE TRANSFER OF AMINO GROUPS MAKES THE SYNTHESIS OF NEW AMINO ACIDS POSSIBLE

Amino groups are transferred **from an α-amino acid to an α-keto acid.**
(Remember that "α-acid" places these groups *right next to* a carboxylic acid.)
Because transamination reactions are readily reversible, they play an important role in both the synthesis and degradation of the amino acids.

Transamination Example: Remember the alanine cycle?

| pyruvate | glutamate | | alanine | α-ketoglutarate |

$$\text{pyruvate} + \text{glutamate} \underset{\text{Alanine transaminase}}{\rightleftharpoons} \text{alanine} + \text{α-ketoglutarate}$$

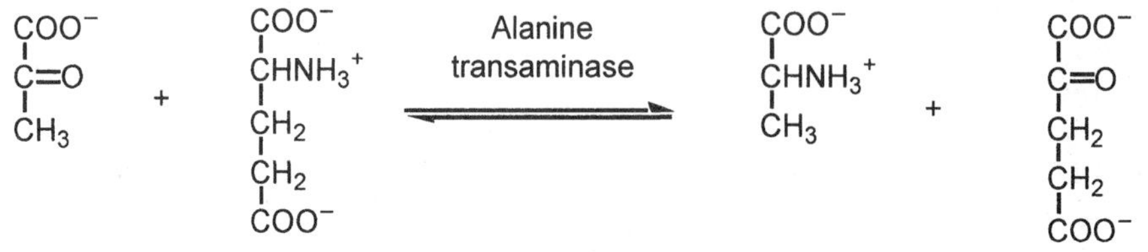

AMINOTRANSFERASES:

Aminotransferases require the coenzyme pyridoxal-5′-phosphate (PLP) (from pyridoxine / vitamin B_6) that forms a Schiff base (–C=N–) between its aldehyde C and the N of the amino acid's $-NH_3^+$.

Two types of aminotransferases:
* Specific for the type of α-amino acid that donates the α-amino group
* Specific for the α-keto acid that accepts the α-amino group.

BIMOLECULAR PING-PONG MECHANISM: The first substrate has to leave

before the second one can enter.
1. An amino acid enters, leaves its amino group behind (with the PLP), and leaves as an α-keto acid.
2. A different α-keto acid enters, the reverse reactions occur for it to take the amino group (from the PLP), and it leaves as an amino acid.

IMPORTANT TRANSAMINATION PAIRS

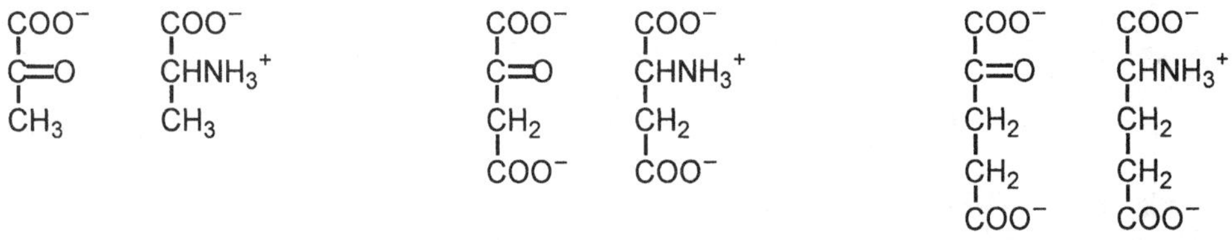

pyruvate/alanine   oxaloacetate/aspartate   α-ketoglutarate/glutamate

Note that α-ketoglutarate and oxaloacetate are citric acid cycle intermediates.

DIRECT INCORPORATION OF AMMONIUM IONS INTO ORGANIC MOLECULES

REDUCTIVE AMINATION OF α-KETO ACIDS

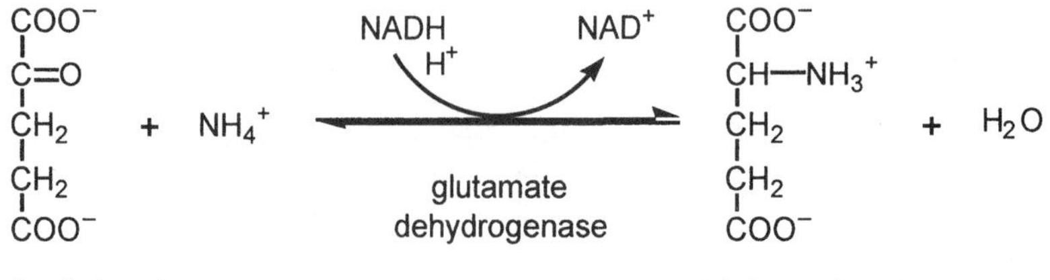

α-ketoglutarate   glutamate

(This reaction typically functions to get rid of NH_4^+, but can also be reversed if excess ammonia is present.)

FORMATION OF THE AMIDES OF ASP AND GLU (TO MAKE ASN AND GLN)

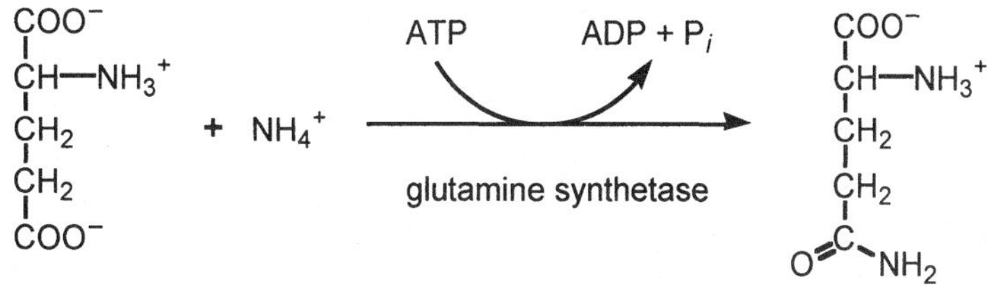

glutamate (Glu)   glutamine (Gln)

The brain uses this reaction to get rid of NH_4^+. Plants couple the reaction above with the following reaction, using two electrons from ferredoxin or NADPH, for a net of 1 Glu per NH_4^+.

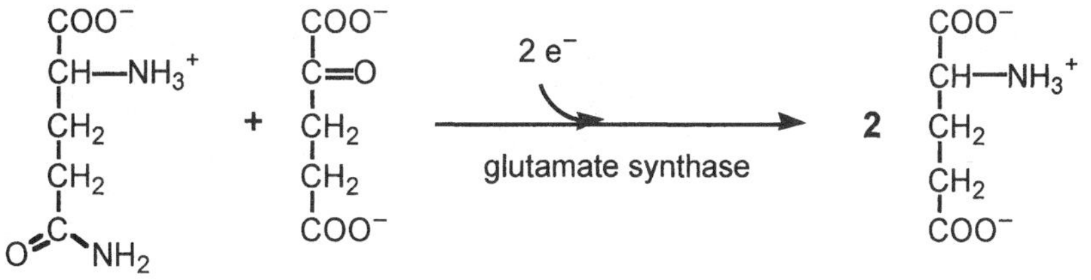

Gln   α-ketoglutarate   2 Glu

SYNTHESIS OF THE AMINO ACIDS

To be able to synthesize an amino acid, its α-keto acid precursor must be independently synthesized by the organism. In *de novo* pathways, amino acids are synthesized from metabolic intermediates (and not only by transamination).

AMINO ACIDS IN THE SAME FAMILY ARE SYNTHESIZED FROM A COMMON PRECURSOR.

FAMILY	PRECURSOR (PARENT)	AMINO ACIDS
Glutamate family	α-Ketoglutarate	Glutamate
		Glutamine
		Proline (cyclized derivative of Glu)
		Arginine
Serine family	Glycerate-3-phosphate	Serine
		Glycine
		Cysteine
Aspartate family	Oxaloacetate (OAA)	Aspartate (from OAA + Glu transamination)
		Asparagine (from Asp + Gln transamination)
		Lysine
		Methionine
		Threonine
		Isoleucine*

213

FAMILY	PRECURSOR (PARENT)	AMINO ACIDS
Pyruvate family	Pyruvate	Alanine (from pyruvate + Glu transamination) Valine Leucine Isoleucine*
Aromatic family	Phosphoenolpyruvate Erythrose-4-phosphate (shikimate pathway; chorismate intermediate)	Phenylalanine Tyrosine (from chorismate OR from hydroxylation of Phe) Tryptophan
Histidine	PRPP (phosphoribosyl-pyrophosphate)	Histidine

* Isoleucine is listed in two families because the carbons in isoleucine are derived from both pyruvate and oxaloacetate.

As an exercise, go through Figures 14.7–14.14 in the text to be sure you can follow how a given amino acid is made from each parent molecule. Make your own diagrams of each family's reactions, and trace specific atoms through the pathways.

14.3 BIOSYNTHETIC REACTIONS INVOLVING AMINO ACIDS

ONE-CARBON METABOLISM: THE TRANSFER OF ONE-CARBON GROUPS

ONE-CARBON GROUPS AND THEIR RELATIVE OXIDATION LEVELS:

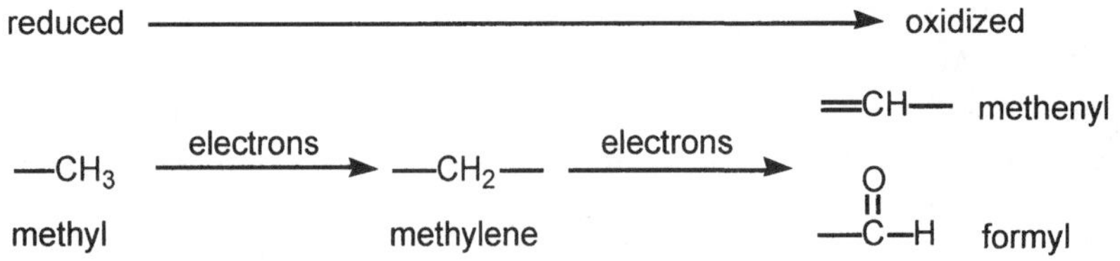

reduced ⟶ oxidized

$$-CH_3 \xrightarrow{\text{electrons}} -CH_2- \xrightarrow{\text{electrons}} \begin{array}{l} =CH- \quad \text{methenyl} \\[4pt] \overset{\displaystyle O}{\underset{}{-}}\!\! \end{array}$$

methyl   methylene

=CH— methenyl

$$-\overset{\overset{\textstyle O}{\|}}{C}-H \quad \text{formyl}$$

IMPORTANT CARRIERS/DONORS OF ONE-CARBON GROUPS:

TETRAHYDROFOLATE (THF) is the biologically active form of **FOLIC ACID**. Once the additional C is attached to THF, the oxidation state can be interconverted (see below). Sources of C groups include Gly, Ser, His, and formate (from Trp).

S-ADENOSYLMETHIONINE (SAM) contains an activated methyl thioether group that transfers its methyl group.

FOLIC ACID

TETRAHYDROFOLATE (THF)

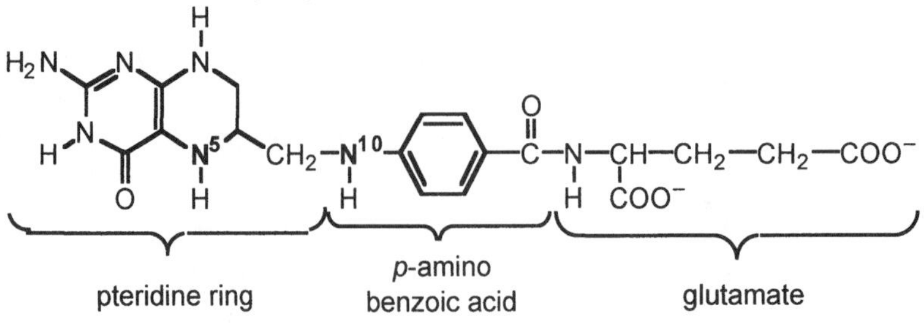

pteridine ring   *p*-amino benzoic acid   glutamate

THF: INTERCONVERSION OF ONE-CARBON GROUP OXIDATION STATES

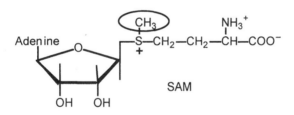

N^5,N^{10}-Methylene THF

N^5,N^{10}-Methenyl THF

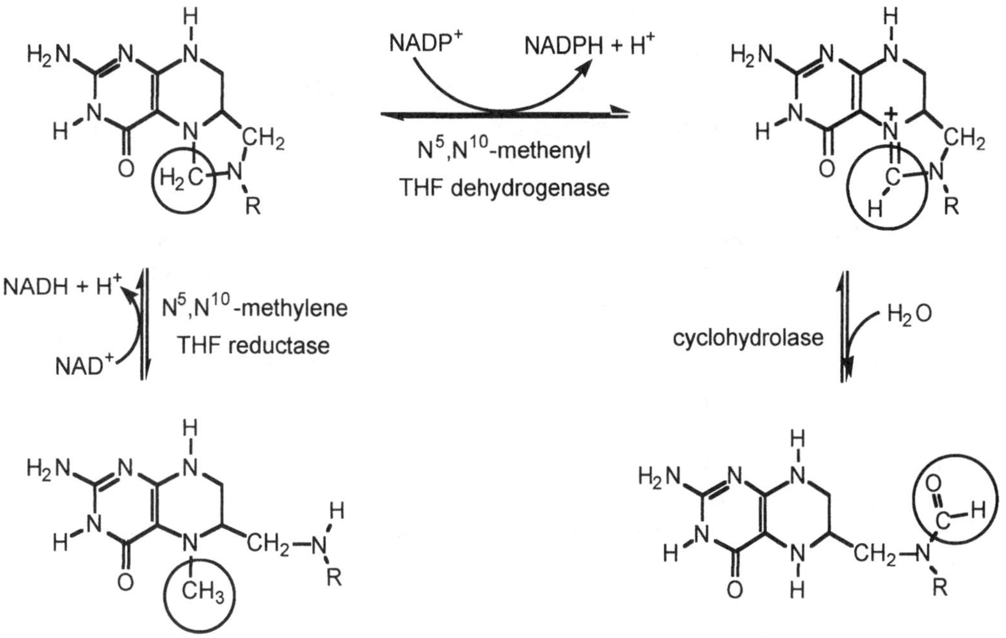

NADP$^+$ NADPH + H$^+$

N^5,N^{10}-methenyl
THF dehydrogenase

NADH + H$^+$ N^5,N^{10}-methylene
   THF reductase
NAD$^+$

cyclohydrolase   H$_2$O

N^5-Methyl THF

N^{10}-Formyl THF

S-ADENOSYLMETHIONINE (SAM)

Adenine   CH$_3$ NH$_3^+$
S—CH$_2$—CH$_2$—CH—COO$^-$

OH   OH   SAM

**SAM = S-ADENOSINE (FROM ATP) +
METHIONINE**
Methyl transferases can transfer the
circled –CH$_3$ to an acceptor. Without
the circled CH$_3$, this molecule is SAH,
S-adenosylhomocysteine, which can
be used to regenerate methionine.

GLUTATHIONE: AN IMPORTANT REDUCING AGENT FORMED FROM GLU, CYS, & GLY

Important functions of GSH (glutathione):

- reduces molecules in various biosyntheses
  (DNA, RNA, certain eicosanoids, etc.)
- protects cells from radiation, O$_2$ toxicity, and
  environmental toxins
- promotes amino acid transport

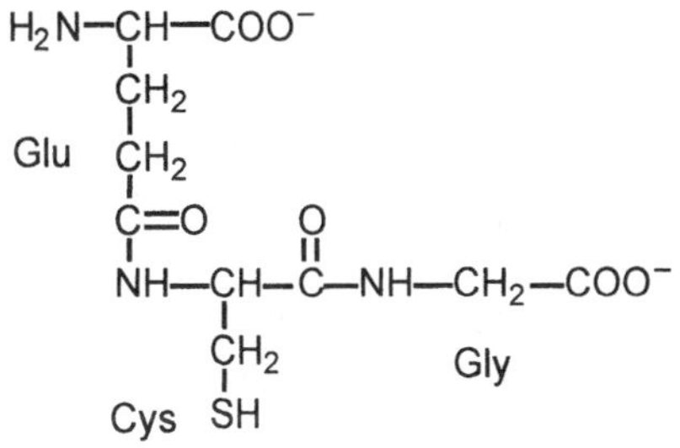

H$_2$N—CH—COO$^-$
    CH$_2$
Glu   CH$_2$
    C=O   O
    NH—CH—C—NH—CH$_2$—COO$^-$
       CH$_2$ Gly
  Cys   SH

GSH TRANSPORT OUT OF CELLS FUNCTIONS TO:

- transfer sulfur (cysteine) between cells,
- protect the plasma membrane from oxidative damage,
- provide for active transport of several amino acids (in the brain, intestine,
  pancreas, liver, and kidney)

– form γ-glutamyl amino acid derivatives in a process that begins with GSH transfer to membrane bound γ-glutamyl transpeptidases. This initiates the γ-glutamyl cycle in brain, intestine, pancreas, liver, and kidney cells.

Functions of the γ-glutamyl cycle:

– maintain cellular GSH levels,

– stimulate the Na^+-dependent transport of amino acids

– contribute to the regulation of amino acid transport across the blood-brain barrier

GLUTATHIONE-S-TRANSFERASES

GSH helps to protect cells from environmental toxins by forming GSH conjugates with these foreign molecules in order to prepare them for excretion. GSH conjugate formation may be spontaneous or catalyzed by GSH-S-transferases (**LIGANDINS**). Before their excretion in urine, GSH conjugates are usually converted to mercapturic acids by a series of reactions initiated by γ-glutamyl-transpeptidases.

NEUROTRANSMITTERS

Neurotransmitters are signal molecules released from neurons. Excitatory neurotransmitters (e.g., Glu, acetylcholine) open Na^+ channels. Inhibitory neurotransmitters (e.g., Gly) open Cl^- channels, inhibiting the formation of an action potential. **BIOGENIC AMINES** are amino acid derivatives and include:

GABA, γ-AMINOBUTYRIC ACID, is produced when glutamate decarboxylase decarboxylates glutamate. (Isn't it great when the enzyme does exactly what it says it does?)

THE CATECHOLAMINES, derivatives of Tyr, include dopamine (D), norepinephrine (NE, noradrenaline), and epinephrine (adrenaline). The enzyme tyrosine hydroxylase catalyzes the rate-limiting first step of catecholamine synthesis, producing L-DOPA (3,4-dihydroxyphenylalanine).

SYNTHESIS OF CATECHOLAMINES: <u>Type of Reaction:</u>

Tyrosine + O_2 + BH_4 $\xrightarrow{\text{Tyrosine hydroxylase}}$ H_2O + BH_2 + L-DOPA   Hydroxylation
(Rate-limiting reaction)
BH_4 (tetrahydrobiopterin) activates O_2.

L-DOPA + H^+ $\xrightarrow{\text{DOPA decarboxylase}}$ CO_2 + Dopamine   Decarboxylation
requires pyridoxal phosphate

Dopamine + O_2 $\xrightarrow[\text{Ascorbic Acid}]{\text{Dopamine-}\beta\text{-hydroxylase,}}$ H_2O + Norepinephrine   Hydroxylation

Norepinephrine + SAM $\xrightarrow{\text{PNMT}}$ SAH + Epinephrine   Methylation
PNMT is phenylethanolamine-N-methyltransferase.

SEROTONIN is a derivative of tryptophan (hydroxylated then decarboxylated). Note the similarities between the syntheses of dopamine and serotonin.

Tryptophan + O_2 + BH_4 $\xrightarrow{\text{Tryptophan hydroxylase}}$ H_2O + BH_2 + 5-Hydroxytryptophan

5-Hydroxytryptophan + H^+ $\xrightarrow[\text{decarboxylase}]{\text{5-Hydroxytryptophan}}$ CO_2 + Serotonin
requires pyridoxal phosphate

HISTAMINE

Histidine + H^+ $\xrightarrow{\text{Histidine decarboxylase}}$ CO_2 + Histamine
requires pyridoxal phosphate

ALKALOIDS

Classes of alkaloids include: **TROPANE ALKALOIDS** (cocaine, atropine); **PYRIDINE ALKALOIDS** (nicotine); and **ISOQUINOLINE ALKALOIDS** (codeine, morphine).

NUCLEOTIDES: STRUCTURE AND BIOSYNTHESIS

First, let's look at the purine and pyrimidine bases. The general structures are:

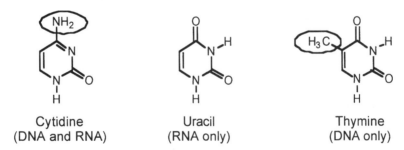

PURINE STRUCTURES

Adenine   Hypoxanthine   Guanine

Hypoxanthine is not a component of DNA or RNA, but it is the base in IMP, inosine-5′-monophosphate, an intermediate in purine biosynthesis. To convert hypoxanthine to either adenine or guanine, the circled amino groups must replace either a C=O or an H. (Note that adenine also has an extra alkene in the six-membered ring.)

PYRIMIDINE STRUCTURES

Cytidine   Uracil   Thymine
(DNA and RNA)   (RNA only)   (DNA only)

Uracil monophosphate (UMP) is an intermediate in pyrimidine biosynthesis; molecules containing cytidine and thymine are derived from UMP. Again, the groups that differ from uracil are circled. (Note that cytidine, like adenine, also has an extra double bond in the ring.)

NUCLEOSIDES AND NUCLEOTIDES

Nucleo*S*ide = *S*ugar + Base (purine or pyrimidine)

Nucleo*T*ide = The *T*otal *P*ackage: sugar + base + *P*hosphate, *T*oo.

The carbons in the pentose are numbered beginning with the anomeric carbon, which is consistent with what you learned in Chapter 7. So as not to confuse the pentose carbons with those of the bases, the pentose carbon numbers carry a prime, such as 3′ and 5′.

The sugar in nucleotides can either be ribose (for RNA) or deoxyribose (for DNA). The base is connected to the anomeric carbon. The phosphate is connected to the 5′ carbon. In deoxyribose, the 2′-OH is replaced by an H.

Since the bases don't have any chiral carbons, the bases can be drawn "facing" the left or right. For example, these two structures of adenine are the same:

Point of attachment → to the sugar

Point of attachment ← to the sugar

However, once the bases are attached to the pentose, how they are drawn will represent either *syn* or *anti* conformations. Purines occur as both *syn* and *anti*, but pyrimidines are typically *anti* (the base's C=O interferes with the pentose).

Anti-guanosine-5′-monophosphate   *Syn*-guanosine-5′-monophosphate

NUCLEOSIDES (RIBOSE + BASE)

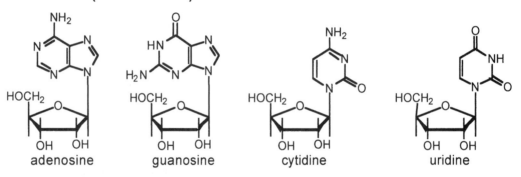

adenosine   guanosine   cytidine   uridine

DEOXYNUCLEOSIDES (DEOXYRIBOSE + BASE)

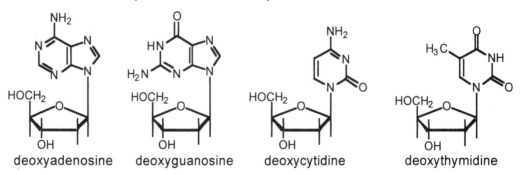

deoxyadenosine   deoxyguanosine   deoxycytidine   deoxythymidine

NUCLEOTIDES (RIBOSE + BASE + PHOSPHATE)

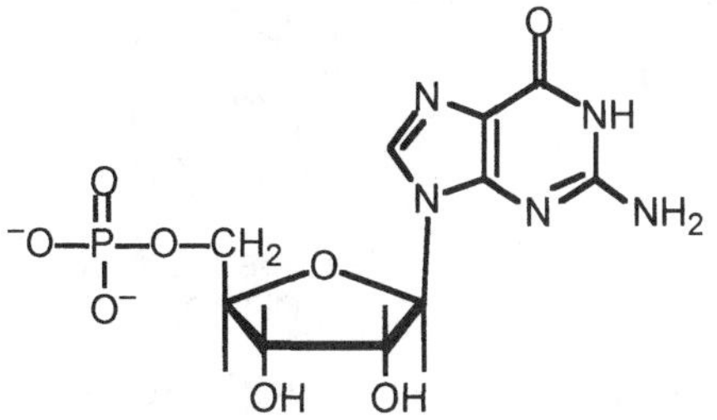
O
N
NH
N
N
NH2
O
-O—P—O—CH2
O-
O
OH OH

guanosine-5′-monophosphate
GMP

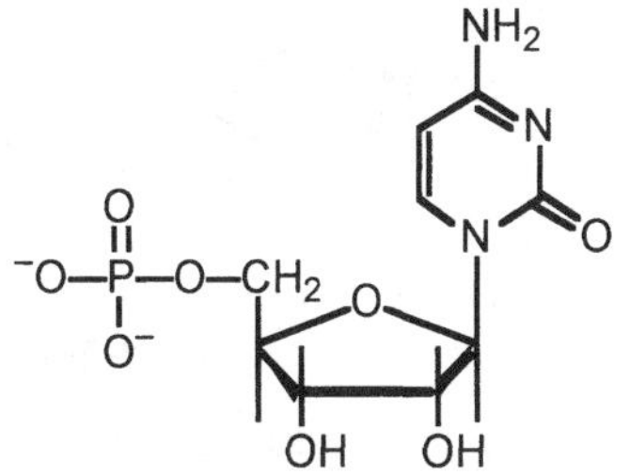
NH2
N
N
O
O
-O—P—O—CH2
O-
O
OH OH

cytidine-5′-monophosphate
CMP

For the deoxynucleotides, the OH at the 2′ carbon is replaced by an H, and the abbreviations are designated by a small "d", e.g., dGMP, dCMP.

As an exercise, draw the structures of the nucleotides dAMP, UMP, and dTMP. Check your structures with those in Figure 14.23 (pp. 487-8).

NUCLEOTIDE BIOSYNTHESIS: THE REAL CHALLENGE LIES IN MAKING THE BASE.

Nucleotides can be synthesized in *de novo* pathways (meaning "from scratch") or in "salvage" pathways. Many nucleic acids, such as RNA, are "turned over," meaning they are synthesized and degraded. Rather than having to synthesize the purine nucleotide *de novo*, the cell can recycle the bases. This makes sense when considering how much energy (ATP equivalents) is expended in *de novo* biosynthesis.

For *de novo* biosynthesis of...

...purines: the base is built onto ribose-5-phosphate.

...pyrimidines: the base is made first and *then* attached to ribose-5-phosphate.

Where does each atom in the purine or pyrimidine ring come from?

PURINES

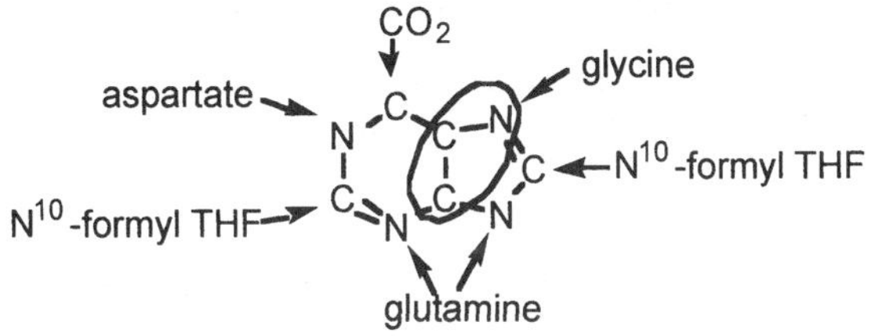
CO2
glycine
aspartate
N10-formyl THF
N10-formyl THF
glutamine

PYRIMIDINES

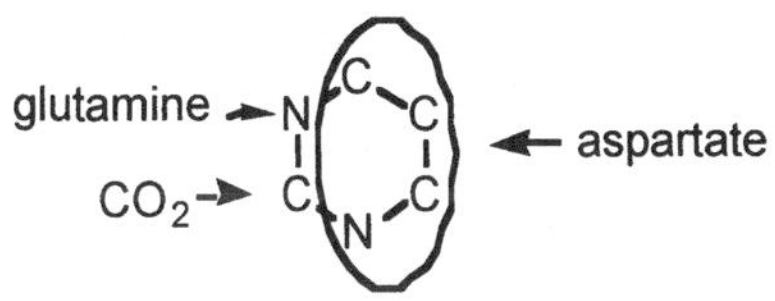
glutamine
CO2
aspartate

STUDY TIP: Write out your own condensed version of Figure 14.27, The Synthesis of Inosine-5′-Monophosphate (p.545), and Figure 14.30, Pyrimidine Nucleotide Synthesis. With different colored pens, color the atoms that come from each of the sources shown above. Check your final structure to make sure that it agrees with labels above. This exercise is indispensable in helping to simplify these complex pathways.

PURINES: RINGS ARE BUILT ONTO AN ACTIVATED RIBOSE (PRPP).

PURINE *DE NOVO* PATHWAYS:

1. Ribose-5-phosphate is activated by converting it to PRPP:

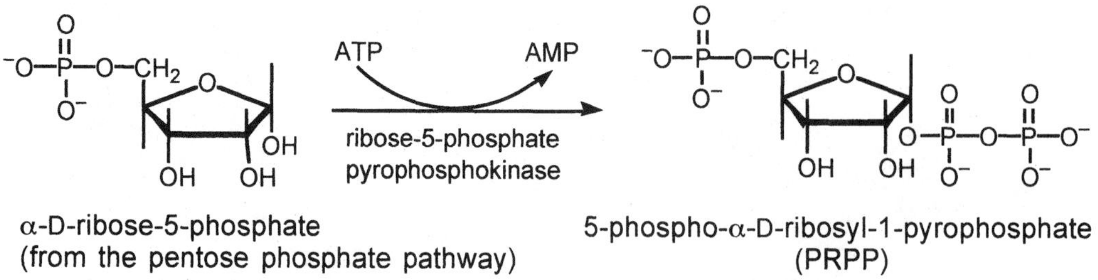

α-D-ribose-5-phosphate
(from the pentose phosphate pathway)

5-phospho-α-D-ribosyl-1-pyrophosphate
(PRPP)

Note that the anomeric carbon of ribose-5-phosphate and PRPP are in the α-configuration. This changes in the next step.

2. PRPP's pyrophosphate group is displaced by the amide nitrogen of glutamine to form 5-phospho-β-D-ribosylamine (and glutamate).

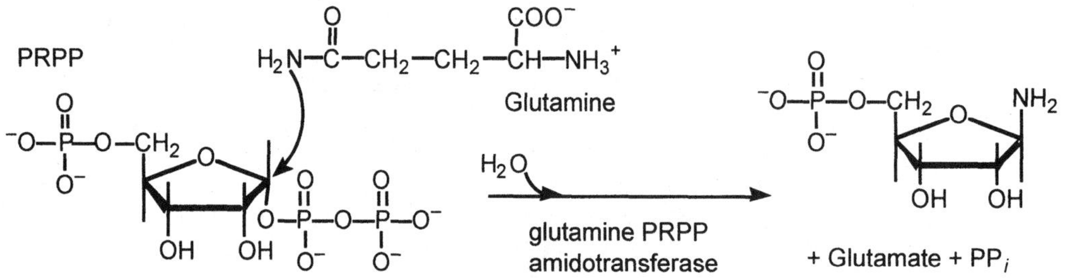

3. Nine further reactions add various groups and close the rings to form IMP. The sources of these groups are listed in the order that they react: Gly, N10-formyl-THF, Gln, (5-membered-ring closure), CO2, Asp, (loss of fumarate), N10-formyl-THF, (ring closure to form IMP). Phew!

4. Further reactions change IMP into either AMP or GMP.

PURINE SALVAGE PATHWAYS USE PRPP AND THE FREE BASE → NUCLEOTIDES

Hypoxanthine-guanine phosphoribosyltransferase (HGPRT) catalyzes nucleotide synthesis using PRPP and either hypoxanthine or guanine:

$$\text{Guanine} + \text{PRPP} \rightarrow \text{GMP} + \text{PP}_i$$

$$\text{Adenine} + \text{PRPP} \rightarrow \text{AMP} + \text{PP}_i \qquad \text{Adenine phosphoribosyltransferase}$$

PYRIMIDINES: THE BASE IS CREATED FIRST, THEN ATTACHED TO THE SUGAR

1. Carbamoyl phosphate is formed from glutamine and HCO_3-. This reaction uses one ATP for the phosphate and a second ATP to drive the reaction. It's catalyzed by the cytoplasmic enzyme carbamoyl phosphate synthetase II, a key regulatory enzyme in mammals. It's inhibited by UTP (the product), and stimulated by purine nucleotides (so that the numbers of pyrimidines and purines made are balanced).

2. Carbamoyl phosphate then condenses with aspartate (by aspartate transcarbamoylase).

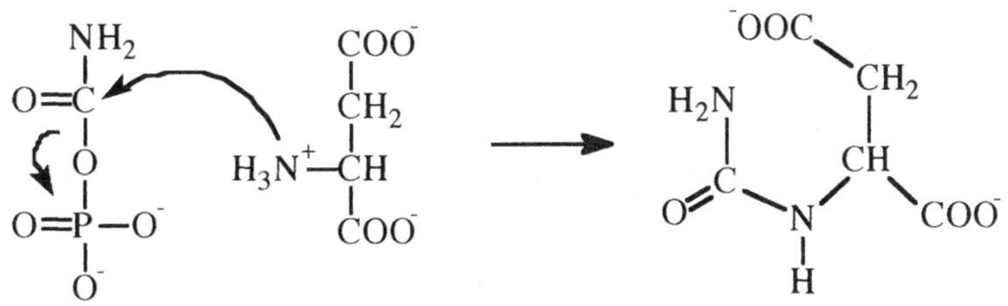

carbamoyl phosphate + aspartate $\rightarrow$ carbamoyl aspartate + P_i

3. The ring is closed by dihydroorotase to form dihydroorotate.

4. Dihydroorotate is then oxidized (hydrogens are removed to form a double bond) by the catalyst dihydroorotate dehydrogenase to form orotate.

5. The ring is attached to PRPP by orotate pyrophosphoribosyl transferase to form orotidine-5′-monophosphate (OMP).

6. A CO_2 is removed to form uridine monophosphate (UMP) (catalyzed by OMP decarboxylase).

UMP serves as a precursor for the other pyrimidine nucleotides.

UMP is phosphorylated twice by ATP to form UTP.

$$UMP + ATP \rightarrow UDP + ADP$$

$$UDP + ATP \rightarrow UTP + ADP$$

The amide nitrogen from glutamine replaces the carboxyl group on UTP to form cytidine triphosphate (CTP).

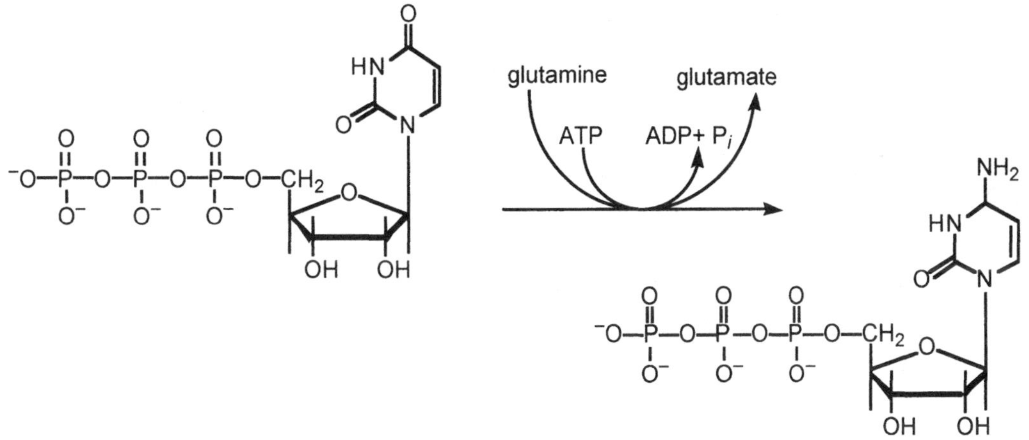

DEOXYRIBONUCLEOTIDES: REDUCTION OF RIBONUCLEOTIDE DIPHOSPHATES

Ribonucleotide reductase replaces the 2′-OH with an H, converting a ribonucleotide *di*phosphate into a deoxyribonucleotide diphosphate.

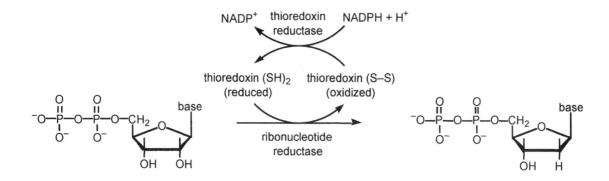

WHAT ABOUT THYMIDINE?

The only nucleotide that's missing is deoxythymidine monophosphate (dTMP).

Note that the only difference between uracil and thymine is a methyl group. The transformation of UMP to dTMP is: UMP → UDP → dUDP → dUMP → dTMP.

Ribonucleotide reductase catalyzes UDP → dUDP. Conversion to dUMP is a simple dephosphorylation. Adding the methyl group to dUMP requires N5,N10-methylene THF and the enzyme thymidylate synthase to obtain dTMP:

dUMP N^5,N^{10}-methylene THF   dihydrofolate   dTMP

thymidylate synthase

NADPH donates its electrons (and hydrogens) to the dihydrofolate produced to regenerate THF (tetrahydrofolate).

Since cancer cells grow rapidly, they require large amounts of deoxynucleotides for DNA synthesis. Many anti-tumor or anti-cancer agents block these steps.

For example, methotrexate and aminopterin block the reduction of DHF to THF, and fluorodeoxyuridine monophosphate (FdUMP) inhibits the methylation of dUMP to form dTMP.

HEME: A COMPLEX FE-CONTAINING PORPHYRIN RING MADE FROM GLYCINE & SUCCINYL-COA

How are the biosynthesis pathways for heme and chlorophyll similar? How are they different?

The pathways begin with different starting material. Heme begins with glycine and succinyl-CoA; chlorophyll synthesis begins with glutamate.

Their paths cross when they each form ALA (δ−aminolevulinate). The pathways are identical from ALA to protoporphyrin IX.

Inserting Fe^{2+} into protoporphyrin IX forms protoheme IX (heme).

Inserting Mg^{2+} forms Mg-protoporphyrin. After methylation and several light-dependent reactions, chlorophyll is produced.

CHAPTER 14: SOLUTIONS TO REVIEW QUESTIONS

14.1   a.   essential amino acid – an amino acid that cannot be synthesized by an animal and must be obtained from its diet

b.   nitrogen balance – a physiological condition in which the body's nitrogen intake is equal to the nitrogen loss

c.   *de novo* pathways – pathways in which molecules such as amino acids or nucleotides are synthesized from simple precursors

d.   biogenic amine – amino acids or amino acid derivatives that act as neurotransmitters

e.   vitamin B_{12} – a complex cobalt-containing molecule obtained from the diet that is required for the conversion of homocysteine to methionine and the isomerization of malonyl-CoA to form succinyl-CoA.

14.2   Transamination reactions represent a method of conserving valuable nitrogen reserves. The reaction is reversible and can be used to convert α-keto acids produced by metabolic reactions to α-amino acids that may be in short supply. Surplus amino acids present in larger amounts than required by current metabolic needs are used as the source of the amino groups.

14.4   Nitrogen-fixing organisms solve the problem of oxygen inactivation in several ways. These are: (1) anaerobic organisms live only in anaerobic soil and are not faced with the problem of oxygen inactivation, and (2) other organisms physically separate oxygen from the nitrogenase complex. For example many of the cyanobacteria produce specialized nitrogenase-containing cells called heterocysts. The thick cell walls of the heterocysts isolate the enzymes from atmospheric oxygen. In addition, legumes produce an oxygen binding protein called leghemoglobin, which traps oxygen before it can interact with the nitrogenase complex.

14.5   a.   An analogue is a compound that is similar in structure to a naturally occurring molecule.

b.   Pernicious anemia is a disease caused by a vitamin B_{12} deficiency; resulting symptoms include low red blood cell counts (anemia), weakness, and various neurological disturbances.

c.   An excitatory neurotransmitter is a signal molecule released from neurons that opens sodium channels, which promotes the depolarization of the membrane in another cell. The result of this action potential depends on the type of postsynaptic cell. If it's a muscle cell, then contraction occurs, and if it's another neuron, then neurotransmitter molecules are released into the synapse.

d.   An inhibitory neurotransmitter is a signal molecule released from neurons that opens chloride channels, which inhibits the formation of an action potential by making the membrane potential in the postsynaptic cell more negative.

e.   A retrograde neurotransmitter is a molecule that is released from a postsynaptic cell and diffuses back to and promotes an action in a presynaptic cell.

14.7   The Schiff base forms when pyridoxal phosphate reacts with an amino acid, which loses a proton to form a carbanion. The free electrons of the carbanion are in conjugation with the positively charged nitrogen of the pyridinium ion. As electrons flow to the positively charged nitrogen, the double bond system reorganizes itself to quench the charge on the nitrogen.

14.8   The concentrations of isoleucine (a) and valine (d) are typically higher in the blood leaving the liver than in the nutrient pool entering enterocytes because the liver preferentially exports these essential amino acids. The amino nitrogens of BCAA are used by other tissues for the synthesis of the nonessential amino acids. (Note that this question concerns concentrations, not total amounts. Since the liver draws amino acids from the blood to synthesize proteins, and the liver cannot replenish these EAA via synthesis, lower numbers of Ile, Leu, and Val molecules leave the liver than enter enterocytes. However, since the liver transports BCAA preferentially, their *concentration* in the blood flowing from the liver is higher.) The concentrations of serine (c) and alanine (e) are typically higher in the nutrient pool entering the enterocytes than in the blood leaving the liver because the liver uses alanine and serine to manufacture glucose for transport. Since glutamine (b) is a primary energy source for enterocytes, the concentration of glutamine is predicted to be higher in the nutrient pool entering the enterocytes. (*Note:* Recall that BCAA = branched chain amino acids and EAA are essential amino acids.)

14.10 Neurotransmitters are either excitatory or inhibitory. Excitatory neurotransmitters (e.g., glutamate and acetylcholine) promote the depolarization of a postsynaptic cell. Inhibitory neurotransmitters (e.g., glycine) inhibit action potentials in postsynaptic cells, i.e., they make the membrane potential more negative.

14.11 a. Glutamine is produced in the following reactions:

α-Ketoglutarate + NADH + NH_3 + H^+ → Glutamate

Glutamate + NH_3 + ATP → Glutamine + ADP + P_i

b. Methionine is produced in the following series of reactions:

L-Aspartate + ATP → β-Aspartylphosphate + P_i + ADP

β-Aspartylphosphate + NADPH + H^+ → Aspartate β-semialdehyde

Aspartate β-semialdehyde + NADPH + H^+ → Homoserine

Homoserine → Cysteine → Homocysteine → Methionine

c. Homoserine produced as in part b above reacts as follows:

Homoserine + ATP → Phosphohomoserine + ADP + H^+

Phosphohomoserine + H_2O → Threonine + P_i

d. Glycine is produced in the following series of reactions:

3-Phosphoglycerate + NAD^+ → 3-Phosphohydroxypyruvate + NADH + H^+

3-Phosphohydroxypyruvate + Glutamate → 3-Phosphoserine + α-Ketoglutarate

3-Phosphoserine + H_2O → Serine + P_i

Serine + THF → Glycine + N^5,N^{10}-methylene THF + H_2O

e. Serine + L-Homocysteine → Cystathione + H_2O

Cystathione + H_2O → Cysteine + α-Ketobutyrate + NH_4^+

14.13 a. LSD – lysergic acid diethylamide – a hallucinogenic drug that is believed to compete with serotonin for specific brain cell receptors

b. alkaloid – a class of naturally occurring molecules that have one or more nitrogen-containing rings; many of the alkaloids have medicinal and other physiological effects

c. *anti*-adenosine – a specific configuration of adenosine in which the purine ring system is located away from the ribose ring; in other words, the nitrogen atoms that are part of the purine's 6-membered ring are as far away from ribose as possible; adenosine is a nucleoside in which adenine is bonded via a β-*N*-glycosidic linkage to the C-1 of ribose

    d. PRPP – 5-phospho-α-D-ribosyl-1-pyrophosphate – the product of the first step of the *de novo* synthesis of purine nucleotides

    e. Lesch-Nyhan syndrome – a devastating disease caused by a deficiency of HGPRT (hypoxanthine-guaninephosphoribosyltransferase), which catalyzes nucleotide synthesis using PRPP and either hypoxanthine or guanine; symptoms include excessive uric acid production and certain neurological symptoms

14.14 The two most prominent one-carbon carriers are THF (tetrahydrofolate, the biologically active derivative of folic acid) and SAM (S-adenosylmethionine). THF plays important roles in the synthesis of several amino acids and the nucleotides. SAM contains an activated methyl thioether group, and is a major methyl donor in the synthesis of numerous biomolecules, such as phosphatidylcholine, epinephrine, and carnitine. Refer to Figure 14.19 (The Tetrahydrofolate and S-Adenosylmethionine Pathways) for an example of processes in which both THF and SAM participate.

14.16 The ten essential amino acids in humans are isoleucine, leucine, lysine, methionine, phenylalanine, threonine, tryptophan, and valine. In addition, histidine and arginine are essential for infants. These amino acids are essential because they cannot be synthesized in required amounts by humans and must be included in the diet.

14.17 a. Thymidine-5′-monophosphate is a nucleotide. (Note that the "deoxy" is omitted. It's assumed because thymidine contains exclusively deoxyribose, not ribose.)

    b. Adenosine is a nucleoside. (This conformation is "*syn*-adenosine".)

    c. Guanine is a purine.

    d. Cytosine is a pyrimidine.

    e. Adenosine-5′-triphosphate (ATP) is a nucleotide. (Again, this conformation is "*syn*-adenosine-5′-triphosphate.)

14.19 No. Perhaps surprisingly, the larger purine does not have similar steric interactions with the pentose. Even though the pyrimidine has only one ring, its carbonyl group interferes sterically with the pentose ring to hinder the formation of the *syn* conformation. In contrast, the shape and functionality of the purine rings are such that steric interactions with the pentose are minimal when in the *syn* conformation, in spite of the overall larger size of the purines.

14.20 Pyrimidine nucleosides occur predominantly in the *anti* conformation. Steric hindrance between the pentose sugar and the carbonyl oxygen at C-2 of the pyrimidine ring prevents free rotation around the N-glycosidic bond.

14.22 Five ATP are required to synthesize a purine by the *de novo* pathway. Only one ATP is required if a purine molecule is recovered by the salvage pathway thus saving a total of four ATP.

14.23 In the γ-glutamyl cycle glutathione is excreted from the cell. γ-Glutamyl-transpeptidase converts GSH to a γ-glutamylamino acid and Cys-Gly. The γ-glutamylamino acid is transported into the cell where it is converted to 5-oxoproline and the free amino acid. 5-oxoproline is eventually reconverted to GSH. The Cys-Gly is transported into the cell where it is hydrolyzed to cysteine and glycine. The location of the γ-glutamyltransferase on the plasma membrane facilitates the transport process because its function is linked to transport. The transported glutamylamino acid is immediately converted to 5-oxoproline and the free amino acid, therefore driving the reaction in favor of transport (uptake of amino acids).

14.25 Depressed synthesis or the total absence of glycine causes the accumulation of 5-phospho-β-D-ribosylamine, the other substrate of the enzyme phosphoribosylglycinamide synthase.

14.26 In ping-pong reactions, the first substrate must leave the active site before the second can enter. In the reaction of alanine with α-ketoglutarate to produce pyruvate and glutamate the following steps take place: (1) the alanine enters the active site and transfers the amino group to pyridoxal phosphate, (2) water enters the reaction site and hydrolyses the Schiff base to produce pyridoxamine phosphate and pyruvate, (3) pyruvate diffuses from the active site, (4) α-ketoglutarate, the second substrate, enters the reaction site and forms a Schiff base with the pyridoxamine phosphate, (5) water hydrolyzes the Schiff base to give pyridoxal phosphate and glutamate, and (6) glutamate diffuses out of the active site.

14.28 The biologically active form of folic acid, referred to as tetrahydrofolate or THF, is shown below. It is formed by the reduction of folic acid with NADPH, a reaction that is catalyzed by tetrahydrofolate reductase.

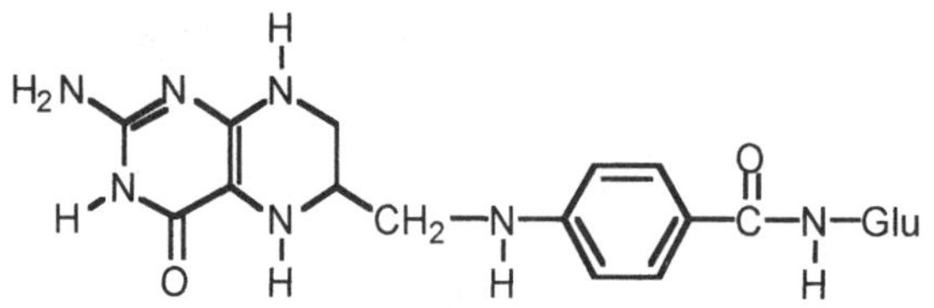

14.29 a. γ-Aminobutyric acid is an inhibitory neurotransmitter.

b. Tetrahydrobiopterin is an essential cofactor in the hydroxylation of certain amino acids.

c. Oxaloacetate is a citric acid cycle intermediate and the α-keto acid in transamination reactions involving aspartate.

d. Phosphoribosylpyrophosphate is the source of the ribose moiety in nucleotide synthesis pathways.

e. S-adenosylmethionine is a methyl transfer agent.

14.31 Nitrogen fixation is highly endergonic. One possibility of a naturally occurring method of fixing nitrogen is via lightning. The tremendous amount of energy in a lightning bolt splits N_2 molecules, which allows them to react with oxygen to form nitric oxide. Nitric oxide dissolves in, and/or reacts with water in the atmosphere to form nitrates that eventually "drop" to the earth as rain.

14.32 In the biosynthesis of uracil, the carbon atom from CO_2 becomes the carbonyl carbon flanked by nitrogen atoms in the pyrimidine ring. (Note that CO_2 is shown as HCO_3^- in Figure 14.30, Pyrimidine Nucleotide Synthesis.)

14.34 In the absence of melanin, skin is vulnerable to sunburn and other forms of light-induced skin damage. A lack of melanin is caused by a defective enzyme in the pathway that converts tyrosine, a precursor of L-DOPA, into melanin.

14.35 The source of the hydrogen gas produced by the nitrogen reductase system is the same as the source of electrons: NAD(P)H.

CHAPTER 14: SOLUTIONS TO THOUGHT QUESTIONS

14.37 Purine and pyrimidine bases are recycled in salvage pathways. Instead of being degraded to form precursors for catabolic pathways they are, instead, oxidized and excreted as nitrogenous waste products.

14.38 Running is a physical activity that requires the rapid metabolism of both fatty acids and glucose. The most rapidly utilized source of glucose is blood glucose. Although certain amino acids can be absorbed easily in the intestine and used as substrates in gluconeogenesis in the liver, this process is slower than the immediate absorption of glucose into the blood.

14.40 Protoporphyrin is photosensitive and accumulates in the skin. Exposure to light generates cytotoxic ROS. By reducing the exposure of the skin to light (covering and dim lights) the effect is eliminated.

14.41 The hydroxyl radical extracts a hydrogen atom from glutathione to form water in a reaction catalyzed by glutathione peroxidase.

$$GSH + \cdot OH \rightarrow GS \cdot + H_2O$$

Two molecules of oxidized glutathione then react to form GSSG. GSH is regenerated from GSSG by glutathione reductase using NADPH as the source of hydrogen atoms.

$$GSSG + NADPH + H^+ \rightarrow 2GSH + NADP^+$$

14.43 The transamination of lysine:

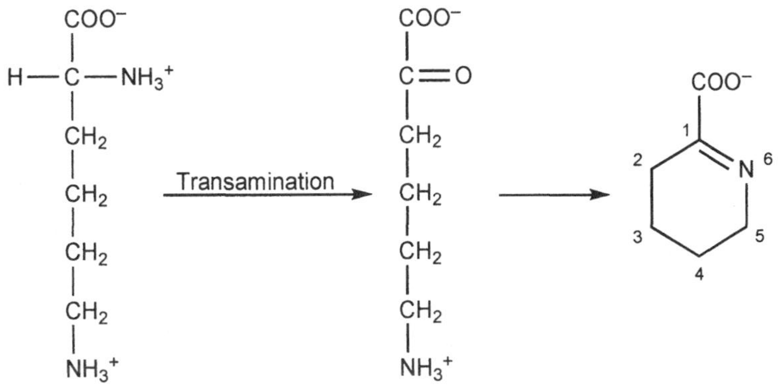

If lysine were transaminated, the ε-amino group of the new keto acid (derived from lysine) would cyclize to form an intramolecular Schiff base. Consequently, the α-keto acid required to produce lysine by transamination cannot exist in appreciable quantities. Therefore, lysine cannot be produced by this reaction.

14.44 Arginine is normally synthesized by the urea cycle. In small children, the urea cycle is not fully functional. Consequently, arginine must be obtained from external sources.

14.46 Breaking the bond perpendicular to the pyridinium ring generates a *p* orbital that can then interact with the p system of the pyridinium ring. The resulting delocalization of the charge stabilizes the carbanion.

14.47 The labeled carbon in proline is indicated by an arrow in this structure:

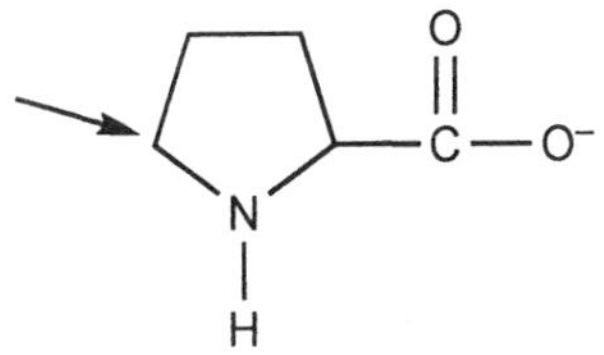

14.49 The carbon that originated at C-2 of pyruvate is indicated by an arrow.

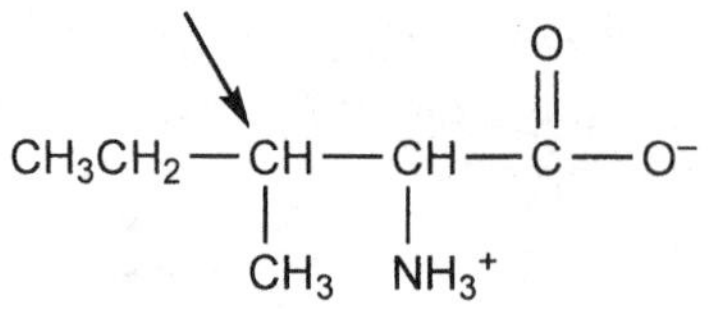

$$CH_3CH_2-\overset{\displaystyle CH_3}{\underset{|}{CH}}-\overset{\displaystyle NH_3^+}{\underset{|}{CH}}-\overset{\displaystyle O}{\overset{\|}{C}}-O^-$$

14.50 The conversion of cytosine to uracil is as follows:

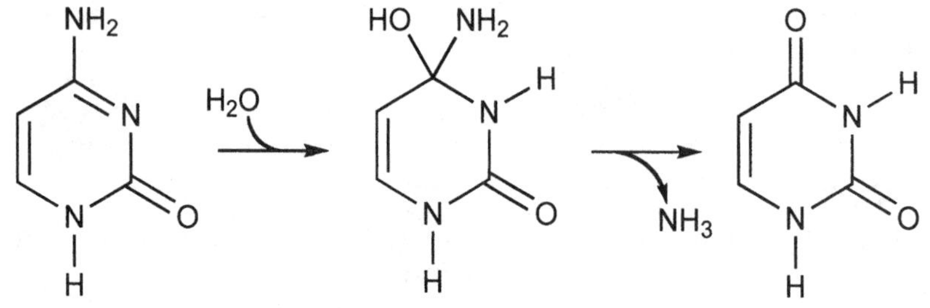

Cytosine →(H₂O) →(NH₃) Uracil

15 Nitrogen Metabolism II: Degradation

Brief Outline of Key Terms and Concepts

OVERVIEW

AMMONOTELIC; UREOTELIC; URICOTELIC

15.1 PROTEIN TURNOVER

Animals are constantly synthesizing and degrading nitrogen-containing molecules such as proteins and nucleic acids. **PROTEIN TURNOVER** is believed to provide cells with metabolic flexibility, protection from accumulations of abnormal proteins, and the timely destruction of proteins during developmental processes.

Protein motifs common in proteins that tend to have a shorter **HALF-LIFE** include the **PEST** sequence and in cyclins, the cyclin destruction box. Other markers include basic or bulky hydrophobic N-terminal residues or oxidized residues.

Most cellular proteins are degraded by the **UBIQUITIN PROTEASOMAL SYSTEM**. In **UBIQUINATION**, **UBIQUITIN** (a small, highly conserved protein) is covalently linked to worn out or damaged proteins, or short-lived regulatory proteins. Most proteins are then transferred to **PROTEASOMES** to be cleaved into peptide fragments, which are degraded by cytoplasmic **PROTEASES** to amino acids, which enter the **AMINO ACID POOL**.

15.2 AMINO ACID CATABOLISM

Amino acids are classified as **KETOGENIC** or **GLUCOGENIC** on the basis of whether their carbon skeletons are converted to fatty acids or to glucose. Several amino acids can be classified as both ketogenic and glucogenic because their carbon skeletons are precursors for both fat and carbohydrates.

DEAMINATION

In general, amino acid degradation begins with deamination. Most deamination is accomplished by **TRANSAMINATION** followed by **OXIDATIVE DEAMINATION** to produce NH_4^+. Although most deaminations are catalyzed by glutamate dehydrogenase, other enzymes also contribute to NH_4^+ formation.

UREA SYNTHESIS

Ammonia is prepared for excretion by the enzymes of the **UREA CYCLE**. Aspartate and CO_2 also contribute atoms to urea.

CONTROL OF THE UREA CYCLE

Short-term control is via substrate availability. Carbamoyl phosphate synthetase I is allosterically activated by *N*-acetylglutamate, the production of which is activated by Arg. Transcription of urea cycle enzymes are activated by glucagon and repressed by insulin.

CATABOLISM OF AMINO ACID CARBON SKELETONS

The α-amino acids are degraded to form: acetyl-CoA, acetoacetyl-CoA, pyruvate, α-ketoglutarate, succinyl-CoA, and oxaloacetate.

15.3 DEGRADATION OF SELECTED NEUROTRANSMITTERS

The degradation of neurotransmitters is critical to the proper functioning of information transfer in animals. The amine neurotransmitters such as acetylcholine, the catecholamines, and serotonin are among the best-researched examples.

15.4 NUCLEOTIDE DEGRADATION

NUCLEASES degrade the nucleic acids to **OLIGO-NUCLEOTIDES**. (**DEOXYRIBONUCLEASES** degrade DNA; **RIBONUCLEASES** degrade RNA.) **PHOSPHODIESTERASES** convert oligonucleotides to mononucleotides. **NUCLEOTIDASES** remove phosphoryl groups to convert nucleotides to nucleosides. **NUCLEOSIDASES** hydrolyze nucleosides to form free bases and ribose or deoxyribose. **NUCLEOSIDE PHOSPHORYLASES** convert ribonucleosides to free bases and ribose-1-phosphate.

PURINE CATABOLISM

Cellular purines are converted to uric acid. Many animals (other than primates) produce enzymes that can degrade uric acid further.

PYRIMIDINE CATABOLISM

Pyrimidine bases are degraded to either β-alanine (UMP, CMP, dCMP) or β-amino-isobutyrate (dTMP).

15.5 HEME BIOTRANSFORMATION

The heme group of hemoproteins is first converted to biliverdin and then to bilirubin. After undergoing conjugation reactions in the liver, bilirubin is excreted in bile. Bilirubin is a potent antioxidant.

BIOCHEMISTRY IN PERSPECTIVE: DISORDERS OF AMINO ACID CATABOLISM; HYPERAMMONEMIA; GOUT

OVERVIEW

HOW DO VARIOUS ANIMAL SPECIES DISPOSE OF NITROGENOUS WASTE?

The nitrogen in amino acids is removed by deamination reactions (transamination and oxidative deamination) and converted to ammonia, which must be detoxified and/or excreted as fast as it's made. The form in which nitrogen is excreted depends on the availability of water.

AMMONOTELIC: many aquatic animals that can excrete ammonia, which dissolves in the surrounding water and is quickly diluted.

UREOTELIC: terrestrial animals (such as mammals) that convert ammonia into urea because they must conserve body water. Urea can be excreted without a large loss of water.

URICOTELIC: animals such as birds, certain reptiles, and insects, that convert ammonia to uric acid because they have more stringent water conservation requirements. In many animals, uric acid is also the nitrogenous waste product of purine nucleotide catabolism.

15.1 PROTEIN TURNOVER

Protein turnover is the continuous degradation and re-synthesis of proteins.

WHAT ROLE DOES PROTEIN TURNOVER PLAY IN CELLULAR METABOLISM?

Metabolic flexibility is afforded by relatively quick changes in the concentrations of key regulatory enzymes, peptide hormones, and receptor molecules.

Protein turnover also protects cells from accumulation of abnormal or damaged proteins. Growth and development processes depend upon timely destruction of proteins as much as on protein synthesis.

UBIQUITIN PROTEASOMAL SYSTEM (UPS)

UPS is a rapid, efficient, and elaborate mechanism to degrade proteins.

HOW ARE PROTEINS TARGETED FOR DEGRADATION?

The mechanisms by which proteins are targeted for destruction by ubiquitination or other degradative processes are not fully understood. However, the following features of proteins appear to signal their destruction:

- **N-TERMINAL RESIDUES.** Very basic or bulky hydrophobic N-terminal residues tend to be very short-lived. N-terminal residues that are nonbulky and/or contain sulfur or –OH are more stable, with longer half-lives.

- **PEPTIDE MOTIFS. PEST SEQUENCES:** Pro-Glu-Ser-Thr. Proteins which have extended sequences containing proline, glutamate, serine, and threonine (PEST, the amino acid one-letter abbreviations) have fairly short half-lives (e.g., less than two hours). **CYCLIN DESTRUCTION BOX:** In cyclins, this set of homologous 9-residue sequences near the N-terminus marks these proteins for rapid ubiquination.

- **OXIDIZED RESIDUES.** Oxidized amino acid residues (i.e., residues which are altered by oxidases or attacked by ROS) promote protein degradation.

PROTEASOMES are massive proteolytic molecular machines that cleave proteins into peptide fragments with 7-8 amino acid residues. **CYTOPLASMIC PROTEASES** degrade these fragments into amino acids, which enter the **AMINO ACID POOL,** where they're available to be incorporated into new protein molecules.

UBIQUITIN is a small, highly conserved 76-residue protein.

UBIQUINATION, the attachment of ubiquitin to worn out, damaged, or short-lived regulatory proteins, occurs in several stages and involves three enzyme classes:
E1 (ubiquitin-activating enzyme)
E2 (ubiquitin-conjugating enzyme)
E3 (ubiquitin ligase; binds both E2 and protein)

15.2 AMINO ACID CATABOLISM CAN BE DIVIDED INTO THREE STAGES:

1. Deamination (removal of the α-amino group and transport to the liver)
2. Urea synthesis (to excrete nitrogen; occurs only in the liver)
3. Conversion of the carbon skeleton to one of six metabolic intermediates

DEAMINATION: TRANSAMINATION AND OXIDATIVE DEAMINATION

The first step in amino acid catabolism is almost always to remove the α-amino group by transamination, in which the α-amino group is transferred to α-ketoglutarate to form glutamate (Glu).

TRANSAMINATION:

EXCESS NH$_4^+$ IS CARRIED TO THE LIVER BY GLUTAMINE OR ALANINE:[1]

MUSCLE (ALANINE CYCLE):   ("α-kg" = α-ketoglutarate)

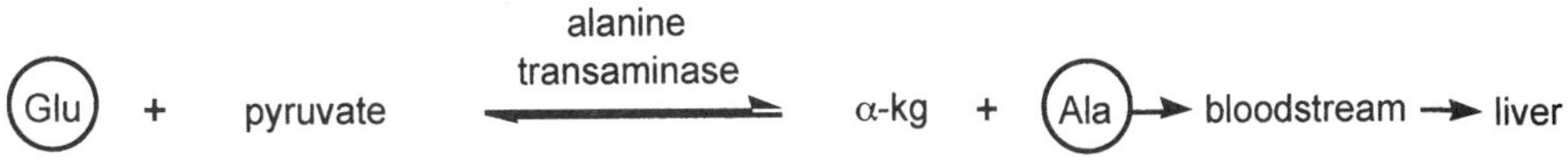

MOST OTHER TISSUES: OXIDATIVE DEAMINATION,[2] THEN GLUTAMINE SYNTHETASE

In most tissues (except the liver), two reactions (and two separate glutamates) are needed to transport NH$_4^+$ through the blood in the form of glutamine (Gln).

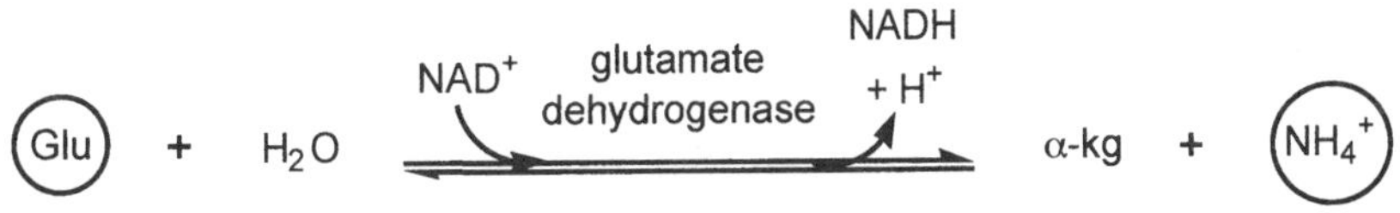

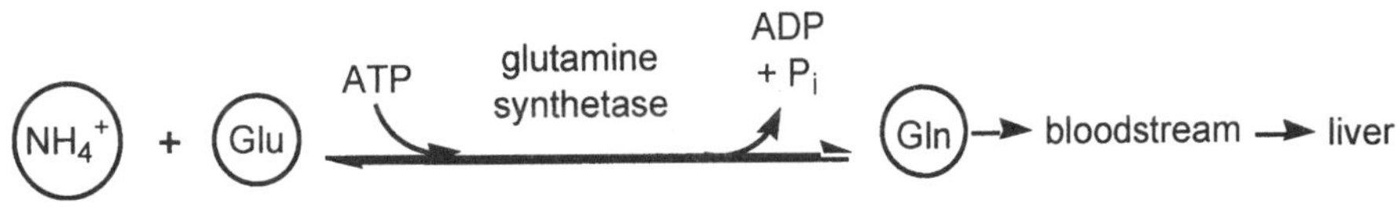

---

[1] Why doesn't glutamate carry nitrogen to the liver? One possible reason is that glutamate is also a neurotransmitter in the brain. If glutamate levels in the blood are elevated, this could cause the brain to "short circuit." In fact, this is an explanation for the headaches that some people experience as a result of consuming MSG (monosodium glutamate), a seasoning ingredient sometimes found in Asian cuisine.

[2] We saw the reverse of this reaction, reductive amination, in Chapter 14.

LIVER: FREES NH_4^+ FROM Gln (FROM OTHER TISSUES) AND Ala (FROM MUSCLE)

$$\text{(Ala)} + \alpha\text{-kg} \xrightarrow{\text{alanine transaminase}} \text{pyruvate} + \text{(Glu)}$$

Gln loses its NH_4^+ to regenerate Glu (in a reaction catalyzed by glutaminase). Glu is then oxidatively deaminated to release NH_4^+.

$$\text{(Gln)} + H_2O \xrightarrow{\text{glutaminase}} \text{(Glu)} + \text{(}NH_4^+\text{)}$$

$$\text{(Glu)} + H_2O \xrightarrow[\text{dehydrogenase}]{\text{glutamate}} \alpha\text{-kg} + \text{(}NH_4^+\text{)}$$

(with NAD^+ and $NADH + H^+$)

OTHER NH_4^+-GENERATING REACTIONS

1. Amino acid oxidases (in the liver and kidney)
2. Serine and threonine dehydratases (because serine and threonine can't be transaminated): Ser → pyruvate; Thr → α-ketobutyrate
3. Bacterial urease
4. Adenosine deaminase

THE UREA CYCLE OCCURS IN LIVER CELLS

THE OVERALL REACTION OF UREA SYNTHESIS:

$$CO_2 + NH_4^+ + Asp + 3ATP + 2H_2O \rightarrow$$
$$\text{urea} + \text{fumarate} + 2ADP + 2P_i + AMP + PP_i + 5H^+$$

THE UREA CYCLE: REACTION NOTES (SEE NEXT PAGE FOR DIAGRAM)

1. Carbamoyl phosphate synthetase I (Mitochondrial matrix)	$NH_4^+ + HCO_3^- + 2ATP \rightarrow$ Carbamoyl phosphate + 2ADP + P_i + $3H^+$
Similar to a reaction in pyrimidine biosynthesis Uses 2 ATP: one to activate HCO_3^-, and one to phosphorylate carbamate	
2. Ornithine transcarbamoylase (Mitochondrial matrix)	Carbamoyl phosphate + Ornithine → Citrulline + P_i
Carbamoyl phosphate has a high phosphate group transfer potential. Releasing its P_i drives this forward. Ornithine is an amino acid; in fact, it looks just like lysine but with one less –CH_2 group.	Citrulline → transport to cytoplasm
3. Arginosuccinate synthase (cytoplasm)	Citrulline + Asp → Arginosuccinate
Cleavage of PP_i (by pyrophosphatase) drives this forward. The α-amino of Asp adds the second N of urea. Uses 2 ATP equivalents, since 1ATP → AMP (not ADP)	
4. Arginosuccinate lyase (cytoplasm)	Arginosuccinate → Arg + Fumarate
Asp's N is left behind when fumarate is cleaved from argininosuccinate to form Arg. Fumarate's carbons came from Asp.	

5. Arginase (cytoplasm) hydrolyzes Arg
  Ornithine → transport back to mitochondrion
  Urea → transport (bloodstream) to kidneys → excretion
	Arg + H₂O → Ornithine + Urea

THE UREA CYCLE

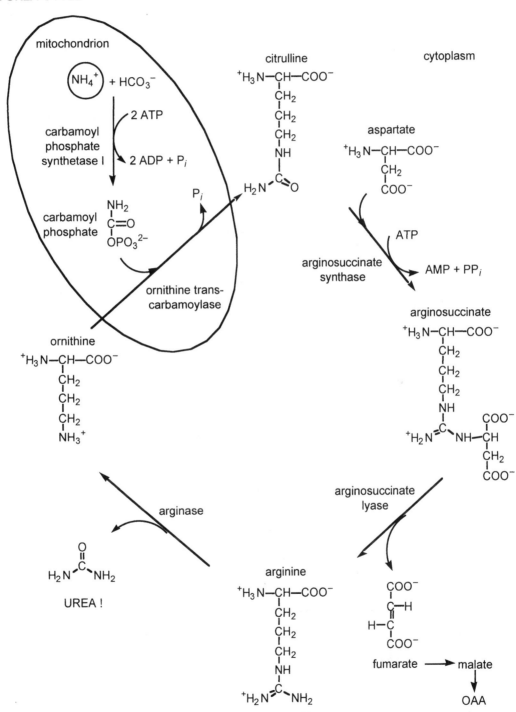

KREBS BICYCLE: THE UREA CYCLE – CITRIC ACID CYCLE CONNECTION

What happens to the fumarate? Where can those four carbons go?
Fumarate supplied by the urea cycle can enter the mitochondrion and the citric acid cycle: fumarate → malate → oxaloacetate (OAA). From OAA, three paths are possible, depending (of course) on the metabolic needs of the cell.

- Transamination   OAA + α-Amino acid → α-Keto acid + Asp → Urea cycle
- Gluconeogenesis: OAA → PEP →etc. → Glucose-6-phosphate → Glucose
- Citric acid cycle: OAA + Acetyl-CoA → Citrate → etc.→ Energy

CONTROL OF THE UREA CYCLE

Short term control is via substrate concentrations (availability). So, it makes sense that urea synthesis is stimulated when the diet is high in protein or during starvation, when muscle protein is degraded for energy. At these times, more NH_4^+ would need to be excreted.

Carbamoyl phosphate synthetase I is allosterically activated by *N*-acetylglutamate, an indicator of Glu concentration (Glu + acetyl-CoA → *N*-acetylglutamate). The production of *N*-acetylglutamate is activated by arginine.

The transcription of urea cycle enzymes is activated by glucagon and glucocorticosteroids, and repressed by insulin.

CATABOLISM OF AMINO ACID CARBON SKELETONS TO METABOLIC INTERMEDIATES

The carbon skeletons are eventually degraded to one of six intermediates. The degradation of the amino acid carbon skeletons are classified in terms of their converted end products. Acetyl-CoA and acetoacetyl-CoA are grouped together.

KETOGENIC AMINO ACIDS are degraded to form acetyl-CoA or acetoacetyl-CoA, which can be converted to either fatty acids or ketone bodies. Remember that no net synthesis of glucose can occur with acetyl-CoA or acetoacetyl-CoA.

GLUCOGENIC AMINO ACIDS are degraded to form intermediates that can be used in gluconeogenesis. The intermediates are pyruvate, α-ketoglutarate, succinyl CoA, fumarate, and oxaloacetate. These intermediates can then be catabolized by the citric acid cycle or converted to fatty acids, ketone bodies, or glucose.

Keep in mind the following reactions that can result in some overlap or confusion between groups:

pyruvate → acetyl-CoA

acetoacetate → acetoacetyl-CoA → acetyl-CoA

citric acid cycle reactions

PRODUCT OF AMINO ACID CATABOLISM	AMINO ACID	PATHWAY NOTES
Acetyl-CoA (KETOGENIC AMINO ACIDS)	Lys, Trp, Tyr, Phe, Leu	Acetoacetyl-CoA is an intermediate common to all five amino acids. acetoacetyl-CoA → acetyl-CoA Phe→Tyr→acetoacetate→acetoacetyl-CoA

PRODUCT OF AMINO ACID CATABOLISM	AMINO ACID	PATHWAY NOTES
Pyruvate	Ala, Cys, Thr, Gly, Ser	Thr→Gly→Ser→pyruvate (Acetyl-CoA is not an intermediate between these five amino acids and pyruvate.)
α-Ketoglutarate	Gln, Arg, Pro, His, Glu	Glu is an intermediate common to the other four amino acids. Glu → α-ketoglutarate
Succinyl CoA	Met, Ile, Val, Thr	Propionyl-CoA is an intermediate common to all four amino acids. In the TRANSSULFURATION PATHWAY, the sulfur atom in Met becomes the sulfur atom of cysteine in two reactions that involve SAM.
Oxaloacetate	Asp, Asn	
Fumarate	Tyr	Catabolism of Tyr forms both acetoacetate and fumarate.

15.3 DEGRADATION OF SELECTED NEUROTRANSMITTERS

WHAT ROLE DOES THE DESTRUCTION OF NEUROTRANSMITTERS PLAY IN THE FUNCTIONING OF NEURONS AND MUSCLE CELLS?

Neurotransmitter degradation in a timely and efficient fashion is required for maintaining precise information transfer in the nervous system. Degradation of neurotransmitters essentially results in the inactivation of the signal.

Acetylcholine: inactivated by acetylcholinesterase which hydrolyzes acetylcholine to acetate and choline

Catecholamines (epinephrine, norepinephrine, dopamine): inactivated by transport out of the synaptic cleft followed by monoamine oxidase (MAO) oxidation; also inactivated by methylations by catechol-O-methyltransferase (COMT); epinephrine is catabolized in nonneural tissue

Serotonin: after reuptake into nerve cells, MAO oxidation followed by aldehyde dehydrogenase oxidation

15.4 NUCLEOTIDE DEGRADATION

Digestion:

Nucleic acids + H_2O → Oligonucleotides	Nucleases (DNases, RNases)
Oligonucleotides + H_2O → Nucleotides	Phosphodiesterases
Nucleotide + H_2O → Nucleoside + P_i	Nucleotidases
Nucleoside + P_i → Base + Ribose-1-phosphate	Nucleosidases

Purines are degraded to uric acid.

Pyrimidines are degraded to NH_4^+, CO_2, and either β-alanine or β-aminoisobutyrate.

PURINE CATABOLISM

(HINT: Review nucleoside/nucleotide nomenclature.)
The structures are outlined on the following page.

Dephosphorylation of AMP, IMP, XMP, and GMP by 5'-nucleotidase forms the nucleosides adenosine, inosine, xanthosine and guanosine.

Adenosine is deaminated to form inosine, then the ribose is removed. (Note that adenine is not an intermediate.) AMP may also be deaminated first (to form IMP) then dephosphorylated (to form inosine). Inosine's ribose is removed to form the free base hypoxanthine, which is then oxidized to xanthine.

Guanosine loses its ribose to form the guanine, which is deaminated to form xanthine.

Finally, xanthine is oxidized to uric acid by xanthine oxidase.

Animals other than primates can degrade uric acid into other nitrogen products such as urea and NH_4^+.

Look at the structure of uric acid. Note that the purine rings remain intact.

PYRIMIDINE CATABOLISM

As an exercise, compare the outline of pyrimidine catabolism (following purine catabolism) with Figure 15.17 in your text (p. 584). Write in the groups that are added and removed (NH_4^+, P_i) and the names of the enzymes (or, copy this figure onto a separate sheet of paper). Drawing careful comparisons between the molecular structures will help you to understand and learn this pathway. Note that the pyrimidine ring can be degraded, whereas the purine ring could not.

15.5 HEME BIOTRANSFORMATION

The porphyrin heme is degraded to form the excretory product bilirubin in a biotransformation process that involves the enzymes heme oxygenase and biliverdin reductase and UDP-glucuronosyltransferase. After undergoing a conjugation reaction, bilirubin is excreted as a component of bile.

The heme oxygenase-1 gene is inducible by a number of cellular stressors (free heme, LDLOX, UV radiation, ROS). The increase in CO, biliverdin, and bilirubin in the vasculature provides protection against the cytotoxic effects of oxidative stress.

PURINE CATABOLISM

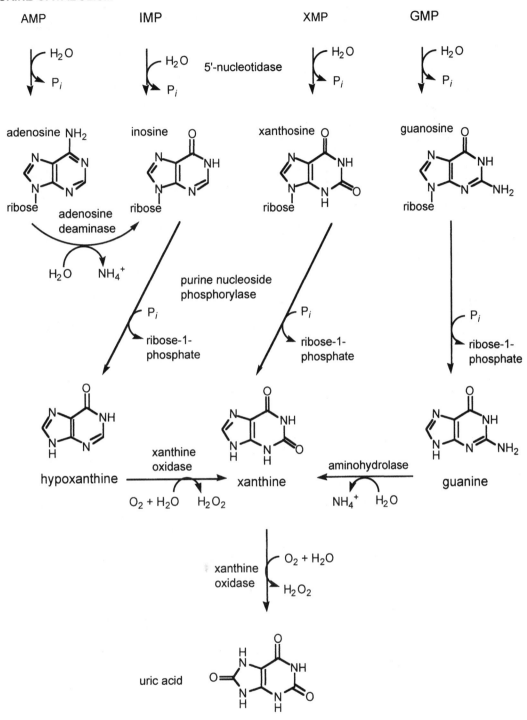

PYRIMIDINE CATABOLISM

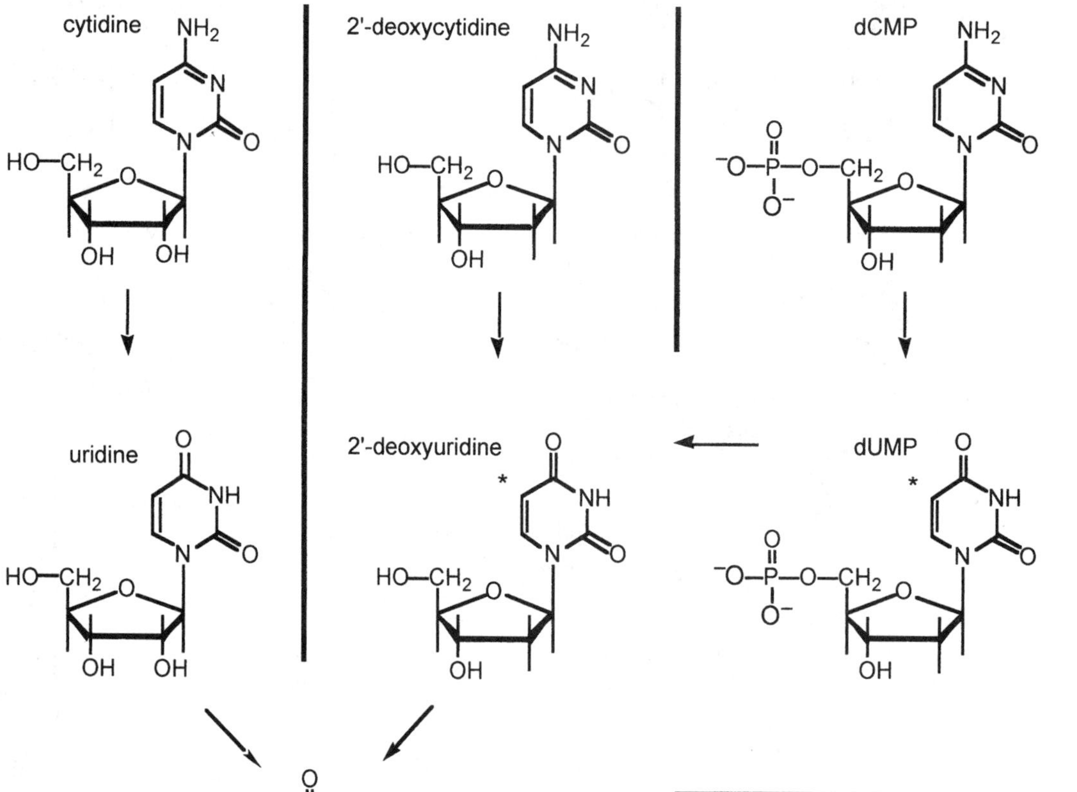

cytidine

2'-deoxycytidine

dCMP

uridine

2'-deoxyuridine

dUMP

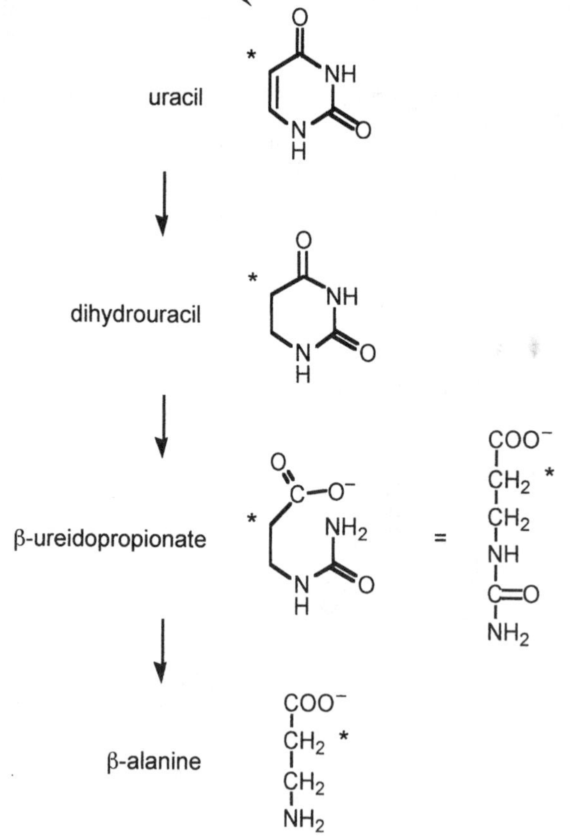

uracil *

dihydrouracil *

β-ureidopropionate *

β-alanine

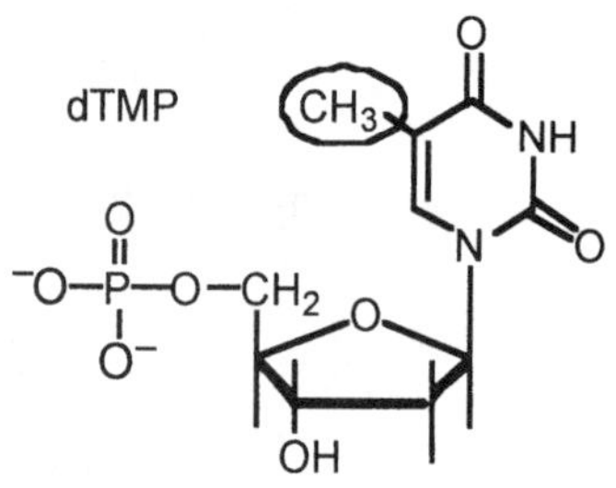

dTMP

Degradation of dTMP

Remember that the only difference between thymine and uracil is one little methyl group. So, the degradation of dTMP is the same as that of dUMP IF YOU ADD A METHYL GROUP at the carbon marked with " * ". The end product is β-alanine with a methyl group: β-aminoisobutyrate:

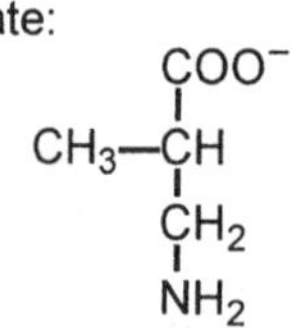

CHAPTER 15: SOLUTIONS TO REVIEW QUESTIONS

15.1 a. denitrification – the conversion of nitrate to atmospheric nitrogen

b. ammonotelic – animals that excrete ammonia directly, i.e., without conversion to less toxic molecules like urea or uric acid

c. protein turnover – a process in which the proteins of living organisms are constantly synthesized and degraded.

d. ubiquination – a eukaryotic mechanism whereby proteins destined to be degraded are first covalently linked to the small protein ubiquitin

e. humoral immune response – the production of antibodies by B lymphocytes

15.2 The major nitrogen-containing excretory molecules are ammonia, urea, uric acid, allantoin, and allantoate.

15.4 The structural features that apparently mark proteins for destruction are: (1) certain N-terminal amino acid residues (e.g., methionine or alanine), (2) Peptide motif sequences (e.g., amino acid sequences with proline, glutamic acid, serine and threonine), and (3) oxidized residues (amino acid residues whose side chains have been oxidized by oxidases or ROS).

15.5 a. glucogenic – a molecule whose carbon skeleton is a substrate in gluconeogenesis; this includes citric acid cycle substrates other than acetyl CoA, such as α-ketoglutarate, that can be converted to gluconeogenesis substrates

b. ubiquitin – a protein that is covalently attached by enzymes to other proteins destined to be degraded

c. mutation – any change in the nucleotide sequence of a gene

d. Kreb's bicycle – a biochemical pathway in which the aspartate required in the urea cycle is generated from oxaloacetate, an intermediate in the citric acid cycle

e. transsulfuration – a biochemical pathway that converts methionine to cysteine

15.7 Aspartate is converted to oxaloacetate via transamination, then catabolized by the citric acid cycle.

15.8 Purines are degraded to xanthine, which is oxidized to uric acid. Note that the structure of uric acid retains the original purine ring, which cannot be degraded by humans. A significant percentage of uric acid is excreted in the urine.

15.10 a. Ketogenic, b. Ketogenic, c. Glycogenic, d. Glycogenic, e. Glycogenic, f. Both.

15.11 Glutamate is converted to α-ketoglutarate by glutamate dehydrogenase or by transamination, and is then catabolized by the citric acid cycle.

15.13 In the muscle, pyruvate undergoes a transamination reaction and is converted to alanine. Alanine is then transferred to the liver, where it is reconverted to pyruvate by transferring its amino group to α-ketoglutarate to form glutamate. Glutamate is then oxidatively deaminated to form α-ketoglutarate and ammonia. The pyruvate and α-ketoglutarate are degraded by the citric acid cycle.

15.14 Alanine transfers its amino group to α-ketoglutarate to form pyruvate and glutamate. The glutamate is then oxidatively deaminated to form ammonia and α-ketoglutarate. The pyruvate and α-ketoglutarate are degraded by the citric acid cycle.

15.16 The NADH formed during the conversion of fumarate to aspartate results in the synthesis of approximately 2.5 ATP. Therefore, the net ATP requirement for urea synthesis (4 ATP – 2.5 ATP) is approximately 1.5 moles of ATP per mole of urea.

15.17 a. porphyrin – a family of compounds that contain a complex, highly conjugated system of four heterocyclic pyrrole rings that are connected together by methene carbons (=CH–) to form a larger ring structure; the four nitrogen atoms, located in the center, can form complexes with metal ions; the non-metal component of heme is an example of a porphyrin, as were the chlorophylls in Chapter 13

b. alkaptonuria – a genetic disease caused by a deficiency of homogentisate oxidase; symptoms include urine that contains large amounts of homogentisate that turns black when exposed to air, a gradual, uneven darkening of the skin, and development of arthritis in later life

c. L-DOPA – 3,4-dihydroxy-L-phenylalanine – a precursor of several neurotransmitters, including dopamine and norepinephrine)

d. cellular immunity – immune system processes mediated by T cells, a type of lymphocyte

e. biliverdin – a dark green pigment that is produced upon the oxidation of heme; biliverdin is excreted via conversion to bilirubin

15.19 The branched chain amino acids (Leu, Ile, Val) are metabolized primarily in muscle tissue, where they are principally used to synthesize nonessential amino acids.

15.20 In ubiquination, the major mechanism of protein degradation, ubiquitin (a small hsp) is covalently attached to lysine residues of a protein with structural features such as oxidized amino acid residues that mark it for destruction. Once the protein is ubiquinated it is degraded by proteases in ATP-requiring reactions.

15.22 a. uric acid – birds, reptiles and insects

b. urea – mammals

c. allantoate – bony fish

d. NH_4^+ – aquatic animals

e. allantoin – some mammals

15.23 a. α-methene carbon – an alkenyl carbon in heme that bridges two rings and is hydroxylated in the first reaction of heme catabolism

b. allantoate – an excretory molecule produced in bony fish by the hydration of allantoin by allantoicase

c. allantoin – an excretory molecule in some mammals that is produced by the oxidation of uric acid by urate oxidase

d. MAO – monoamine oxidase – an enzyme that catalyzes the oxidation of catecholamines, resulting in their inactivation; MAO is located in nerve endings

e. maple syrup urine disease – branched chain ketoaciduria – a serious disease in which a deficient branched-chain α-keto acid dehydrogenase complex causes the accumulation of leucine, isoleucine, and valine in the blood, which imparts the characteristic odor of maple syrup in the urine. Untreated, maple syrup urine disease causes death before 1 year of age.

15.25 Tyrosine is catabolized in five reactions to acetoacetate and fumarate. The resulting acetoacetate is then converted to acetoacetyl-CoA, and then to acetyl-CoA. The resulting acetyl-CoA and pyruvate are finally catabolized by the TCA cycle.

15.26 If one of the urea cycle enzymes is missing, ammonia (the nitrogen-containing substrate of the cycle) cannot be metabolized. If one of the enzymes is defective (i.e., it does not catalyze its reaction at an appropriate rate), ammonia is metabolized slowly. Under both circumstances the body's ammonia concentration is excessively high.

15.28 Nitrogen enters the urea cycle as NH_4^+ and aspartate. The two nitrogen atoms in urea most likely originated in the α-amino groups of amino acids. The pathways from α-amino acids to ammonium and aspartate are outlined below, with 1) and 2) providing most of the nitrogen in the urea cycle.

1) From muscle cell amino acids to NH_4^+ in the liver: In muscle cells, a transamination reaction with α-ketoglutarate produces glutamate (and an α-keto acid). Glutamate in muscle cells may transfer its amino group to pyruvate via another transamination to produce alanine, or it may be oxidatively deaminated to generate NH_4^+, which reacts with another glutamate to form glutamine. Alanine and glutamine transport the nitrogen atom from α-amino acids to the liver, where glutamate is regenerated: a) from alanine via transamination to α-ketoglutarate, or b) from glutamine via hydrolysis of the amide, which also forms a free NH_4^+. The glutamate produced from both alanine and glutamine is oxidatively deaminated to form NH_4^+.

2) From liver cell amino acids to aspartate: An α-amino acid and oxaloacetate react via transamination to form an α-keto acid and aspartate.

3) From serine and threonine to NH_4^+ in the liver: Since Ser and Thr cannot undergo transamination, they are deaminated by serine dehydratase and threonine dehydratase.

4) From urea present in the blood to NH_4^+ in the liver: Urea in the bloodstream diffuses into the intestinal lumen, where it is hydrolyzed by bacteria with urease to form ammonia, which diffuses back into the blood for transport to the liver.

5) From amino acids in the liver and kidney to NH_4^+ via the action of L-amino acid oxidases

6) From the C-6 amino group of adenine to NH_4^+ via the action of adenosine deaminase

15.29 Ammonia requires the most amount of water to be excreted safely, urea requires less, and uric acid requires the least. Because fish do not need to conserve water, and because toxic ammonia is soluble and quickly diluted by surrounding water, they excrete ammonia directly. Since mammals do need to conserve water, they convert ammonia to urea. Animals in greatest need of water conservation convert ammonia to uric acid.

15.31 Excess amino acids present after a high protein meal are eventually converted to glutamate in the liver (see the solution for Review Question 15.28). Excess glutamate is converted to N-acetylglutamate, which is an allosteric activator for carbamoyl phosphate synthetase I, the most critical enzyme of the urea cycle in that NH_4^+ is one of its substrates. Since all five enzymes are controlled by substrate concentrations, higher levels of aspartate would activate argininosuccinate synthase. (A continued high protein diet results in the activation of the synthesis of all five urea cycle enzymes.)

15.32 Foods high in purines include liver and sardines. Increased purine levels lead to increased uric acid synthesis. Increased uric acid levels increase the possibility of sodium urate crystal deposits in and around joints and within the kidney. These deposits cause the symptoms of gout.

15.34 Tryptophan is ketogenic because its catabolic pathway results in the formation of α-ketoadipate, which reacts further to form acetoacetyl-CoA, which is converted to acetyl-CoA. Tryptophan is glucogenic in that one of its catabolic reactions along this pathway also produces alanine, which is converted to pyruvate (a substrate for gluconeogenesis) via a transamination reaction.

15.35 A characteristic odor in the urine is the primary symptom of maple syrup urine disease. If untreated, symptoms include vomiting, convulsions, severe brain damage, mental retardation, and, often, death within an infant's first year. Maple syrup urine disease is caused by a deficient branched chain α-keto acid dehydrogenase complex, which converts α-keto acids to their acyl-CoA derivatives. Large quantities of the α-keto acids derived from branched-chain amino acids – Leu, Ile, and Val – accumulate in the blood.

15.37 Alkaptonuria is a genetic disease caused by a deficiency of homogentisate oxidase. In the absence of this enzyme, homogentisate accumulates and is excreted in the urine. Homogentisate turns black when exposed to air and is responsible for the black urine characteristic of this disease.

15.38 The amino acids involved in alkaptonuria are tyrosine and phenylalanine.

CHAPTER 15: SOLUTIONS TO THOUGHT QUESTIONS

15.40 The energy requirements of the urea cycle are closely linked to the energy-generating citric acid cycle. Recall that fumarate, a product of the urea cycle, is easily converted to oxaloacetate, the molecule that reacts with incoming acetyl CoA molecules. Both pathways together referred to as the Krebs bicycle occur within the mitochondrial matrix.

15.41 Once your diagram of the glycolysis and citric acid pathways is drawn, note that the linkage between the citric acid cycle and the purine nucleotide cycle is via fumarate, a citric acid cycle intermediate and aspartate, the product of oxaloacetate transamination.

15.43 Tetrahydrobiopterin is a cofactor in the oxidation of phenylalanine to form tyrosine. The sustained absence of this cofactor would result in a buildup of phenylalanine and the appearance of the symptoms of PKU.

15.44 5,6,7.8-Tetrahydrobiopterin (BH_4) is required in the synthesis of neurotransmitters such as norepinephrine and serotonin, which are produced in brain. BH_4 is required in the reaction catalyzed by tyrosine hydroxylase in which tyrosine is hydroxylated to form L-dopa (a precursor of several neurotransmitters including dopamine and norepinephrine), and the reaction catalyzed by tryptophan hydroxylase in which tryptophan is hydroxylated to form 5-hydroxytryptophan. L-Dopa and 5-hydroxytryptophan must be supplied to individuals lacking the capacity to synthesize BH_4 because the latter molecule does not cross the blood-brain barrier.

15.46
   a. Methylene ($-CH_2-$) groups

   b. Methyl groups

   c. Methyl groups

   d. Methyl groups

15.47 Because of the structural similarities to purine, caffeine is converted to a variety of derivatives by xanthine oxidase (e.g., 1-methyluric acid and 7-methylxanthine).

15.49 If parkin, an E3 (ubiquitin ligase) enzyme, is defective or inoperative the capacity of dopaminergic nerve cells to degrade proteins is compromised. The pace of protein turnover will slow and abnormal proteins will accumulate. Eventually such cells will die.

15.50 Atherosclerosis is caused by oxidative damage to the cells lining the arterial wall. HO1 gene expression is induced by free heme, oxidized LDL, and other risk factors for cardiovascular disease. HO1 produces bilirubin, a powerful antioxidant that acts to counter the effects of ROS.

15.52 The nitrogen at ring position 3 of uracil is released as ammonia, a substrate of the urea cycle when β-ureidopropionate, an intermediate formed during the degradation of uracil, is converted to β-alanine.

15.53 Primates lack urate oxidase, which converts uric acid to allantoin, a more soluble molecule. As a result uric acid crystals can build up and gout results.

15.55 The α-keto acids react with ammonia to produce amino acids thus sparing α-ketoglutarate, a citric acid cycle intermediate needed to generate energy.

16 Integration of Metabolism

Brief Outline of Key Terms and Concepts

16.1 OVERVIEW OF METABOLISM

Multicellular organisms require sophisticated regulatory mechanisms to ensure that all their cells, tissues, and organs cooperate, and use HORMONES to convey information between cells. ENDOCRINE HORMONES are secreted from cells that are distant from target cells.
TARGET CELLS; SIGNAL TRANSDUCTION; SECOND MESSENGERS; ENZYME CASCADE

To ensure proper control of metabolism, the synthesis and secretion of many mammalian hormones are regulated by a complex cascade mechanism ultimately controlled by the central nervous system.

Hormones: peptide or polypeptide, steroid, and amino acid derivatives

16.2 HORMONES AND INTERCELLULAR COMMUNICATION

PEPTIDE HORMONES
HYPOTHALAMUS; PITUITARY
Protection against overstimulation by hormones: FEEDBACK INHIBITION, DESENSITIZATION, DOWNREGULATION (e.g., INSULIN RESISTANCE)
Types of cell-surface receptors:
  G-PROTEIN-COUPLED RECEPTORS (GPCRs)
  RECEPTOR TYROSINE KINASES (RTKs)

GPCRs
  G PROTEINS and the action of the α subunit
  GUANINE NUCLEOTIDE EXCHANGE FACTOR (GEF)
  SECOND MESSENGERS: cAMP; cGMP; the Phosphatidylinositol Cycle, DAG, PIP$_3$, Ca^{2+}
RECEPTOR TYROSINE KINASES (RTKs) activate phosphorylation cascades when they undergo autophosphorylation triggered by the binding of signal molecules. RTKs do not directly involve the generation of a second messenger.

GROWTH FACTORS
EPIDERMAL GROWTH FACTOR (EGF)
PLATELET-DERIVED GROWTH FACTOR (PDGF)
SOMATOMEDINS
INSULIN-LIKE GROWTH FACTOR
CYTOKINES: INTERLEUKIN-2; INTERFERONS; TUMOR NECROSIS FACTORS (TNF)

STEROID AND THYROID HORMONE MECHANISMS
Nonpolar steroid and thyroid hormones diffuse through the lipid bilayer of cell membranes and bind to intracellular receptors. The hormone-receptor complex then binds to a DNA sequence, a HORMONE RESPONSE ELEMENT (HRE). The binding of a hormone-receptor complex to an HRE enhances or diminishes the expression of specific genes.

16.3 METABOLISM IN THE MAMMALIAN BODY: THE DIVISION OF LABOR
Each organ in the mammalian system contributes to the body's overall function, either by secreting specific hormones or by conducting a major metabolic pathway. Gastrointestinal Tract; Liver; Muscle; Adipose Tissue; Brain; Kidney

16.4 THE FEEDING-FASTING CYCLE
A variety of organs contribute via hormones and neurotransmitters to the acquisition and use of nutrients. The goal is to maintain balance between energy acquisition and energy expenditure. The hypothalamus contains the critical neural circuits that control appetite and satiety.

THE FEEDING PHASE
Food is consumed, digested, absorbed, and transported to the organs to be used or stored.
POSTPRANDIAL; CHYLOMICRON REMNANTS

THE FASTING PHASE
During fasting, several metabolic strategies maintain blood glucose levels.
POSTABSORPTION; POSTABSORPTIVE STATE

FEEDING BEHAVIOR
ARCUATE NUCLEUS (ARC) of the hypothalamus
NUCLEUS TRACTUS SOLITARIUS (NTS)

BIOCHEMISTRY IN PERSPECTIVE
- Diabetes Mellitus
- Obesity and the Metabolic Syndrome
- Available online: Hormone Methods

OVERVIEW

It's all about control – balancing anabolism and catabolism based upon the organism's energy needs, energy stores (e.g., ATP, glycogen, and triacylglycerols), and nutrient availability – and integrating these metabolic processes for optimal results.

Information transfer within an organism is critical for effective, efficient, and timely control, and is achieved by coordination of the nervous system with chemical signals (usually hormones). Descriptions of information transfer systems may vary somewhat depending upon the focus and depth of the discussion. Here are the major points that had been covered in previous chapters. The description in section 16.1 (below) is more detailed.

PROCESS that organisms use to receive and interpret information (**SIGNAL TRANSDUCTION**):
> **RECEPTION** – ligand binds to receptor
> **TRANSDUCTION** – ligand binding triggers a conformational change in the receptor that results in the conversion of the extracellular signal into an intracellular message
> **RESPONSE** – a cascade of events that involve covalent modifications; possible results: changes in enzyme activities and/or gene expression, cytoskeletal rearrangements, cell movement, or cell cycle progression (growth or division).

COMPONENTS of a simple system for information transfer:
> **PRIMARY SIGNAL** – the first messenger, usually a hormone; additional examples: neurotransmitter, growth factor, or cytokine
> **TARGET** – a specific **RECEPTOR**, usually bound to a membrane
> **TRANSDUCER SYSTEM** – converts the signal to a cellular response

16.1   OVERVIEW OF METABOLISM

STEADY STATE is when the rate of anabolic processes equals (approximately) the rate of catabolic processes, and as a result, the organism doesn't grow or change appreciably.

COMPONENTS AND PROCESS OF AN INTERCELLULAR COMMUNICATION SYSTEM VIA CHEMICAL SIGNALS:

> **PRIMARY SIGNAL** (the first messenger, usually a hormone)
> > **ENDOCRINE HORMONES**, secreted by specialized glandular cells, fall into three categories: peptide or polypeptide hormones, steroid hormones, and amino acid derivatives.
> > **GROWTH FACTORS** are hormone-like polypeptides and proteins. Unlike hormones, they're not synthesized by specialized glandular cells, and they specifically regulate growth, differentiation, and proliferation of various cells. **CYTOKINES** are proteins produced by blood-forming cells and immune system cells, and may stimulate or inhibit cell growth or proliferation.

> **TRANSPORT** to the **TARGET CELL**

> **RECEPTION:** A signal molecule binds to its specific **RECEPTOR**.

> **SIGNAL TRANSDUCTION:** Upon hormonal binding to a receptor, a conformational change transmits the signal across the membrane. As a result, second messengers may be released inside the cell. (Examples of second messengers include: cAMP, cGMP, DAG, IP_3, Ca^{2+}, and the inositol phospholipid system.) Most second messengers that bind to an enzyme cause a conformational transition that switches the enzyme from an inactive to an active form and often triggers an enzyme cascade that amplifies the signal.
> **ENZYME CASCADE:** Once activated by the second messenger, the enzyme modifies multiple copies of a number of *different* target enzymes. These newly-activated target enzymes may also modify multiple copies of *another* set of target proteins.

CELLULAR RESPONSE: Cellular functions are altered by changes in the activities of existing enzymes, rearrangements of the cytoskeleton, and/or altered gene expression.

16.2 HORMONES AND INTERCELLULAR COMMUNICATION

PEPTIDE HORMONES

SYNTHESIS AND SECRETION OF PEPTIDE HORMONES are regulated by the endocrine hormone cascade system that is ultimately controlled by the central nervous system.

First, a little anatomy:

HYPOTHALAMUS – an area in the brain that links the nervous and endocrine systems.

PITUITARY GLAND – attached to the hypothalamus by the pituitary stalk; consists of two lobes – the anterior lobe and the posterior lobe – that differ in both anatomy and function.

AXONS – long extensions of nerve cells (neurons)

ENDOCRINE HORMONE CASCADE SYSTEM:

The nervous system acquires and processes environmental information, and transmits this information to cells. The central nervous system sends sensory signals to stimulate the hypothalamus.

Then, one of two things happen:

→ The hypothalamus synthesizes and releases specific peptide-releasing hormones, which enter a specialized capillary bed that leads them to the anterior lobe of the pituitary.

Peptide-releasing hormones (TRH, GHRH, GnRH, CRH, or PRH)[1] target specific clusters of cells in the anterior lobe, and stimulate them to release (secrete) trophic ("turning" or "changing") hormones. These trophic hormones stimulate the synthesis and release of hormones from other endocrine glands.

Example:

The hypothalamus releases TRH (thyrotropin-releasing hormone), which stimulates the release of (you guessed it) thyrotropin from the anterior pituitary. Thyrotropin travels to the thyroid, where it stimulates the release of thyroid hormone (T_3, T_4). (For others, see Figure 16.4 of your text.)

→ Separate types of neurons that originate in the hypothalamus synthesize oxytocin or vasopressin, and package them into secretory granules that migrate down their axons to the posterior pituitary lobe. When an action potential reaches the nerve endings, exocytosis is stimulated, releasing these hormones into the bloodstream.

MECHANISMS THAT PROTECT AGAINST OVERSTIMULATION BY HORMONES

FEEDBACK INHIBITION – The hypothalamus and the anterior pituitary are controlled by the target cells they regulate.

DESENSITIZATION – Target cells decrease the number of cell surface receptors (or inactivate receptors) in response to changes in stimulation.

---

[1] Note that all of the peptide-Releasing Hormone acronyms end in "RH".

DOWN-REGULATION – the reduction in cell surface receptors (by endocytosis) in response to stimulation by specific hormones. The receptors may eventually be recycled back to the cell surface or degraded. An example is **INSULIN RESISTANCE**, in which a decrease in functional insulin receptors causes diabetes.

CELL-SURFACE RECEPTORS

GPCRs – G-PROTEIN-COUPLED RECEPTORS
RTKs – RECEPTOR TYROSINE KINASES (next page)

GPCRs – G-PROTEIN COUPLED RECEPTORS

GPCRs are composed of seven membrane-spanning helices, with an N-terminal segment as part of the ligand-binding site and a C-terminal segment that interacts with G-proteins on the cytoplasmic side of the membrane.

G PROTEINS (HETEROTRIMERIC GTP-BINDING PROTEINS)

Located on the cytoplasmic side of the cell membrane, G proteins transduce ligand binding to GPCRs into intracellular signals. G proteins contain 3 subunits (α, β, γ). While inactive, the $\beta\gamma$ complex binds to and inhibits the α subunit, which binds GDP.

GPCR ACTIVITY AND G PROTEIN ACTIVATION:

1. Ligand binding to the GCPR

   Examples of GPCR ligands: hormones (e.g., glucagon, TSH, the catecholamines, and the endocannabinoids), neurotransmitters (e.g., glutamate, dopamine, and GABA), neuropeptides (e.g., vasopressin and oxytocin), odorants and tastants, and light (rhodopsin).

2. Signal initiation

   Binding of the ligand causes a conformational change in the receptor, and the G protein binds to the occupied receptor. **GEF (GUANINE NUCLEOTIDE EXCHANGE FACTOR)** mediates the GDP/GTP exchange:

   $$\text{GDP-}\alpha\beta\gamma + \text{GTP} \xrightarrow{\text{GEF}} \text{GTP-}\alpha + \beta\gamma + \text{GDP}$$

   The activated GTP-α subunit dissociates from the $\beta\gamma$ complex and moves over the cytoplasmic surface of the membrane to activate an enzyme that synthesizes a specific second messenger, which initiates an enzyme cascade.

3. Primary signal termination: GTP hydrolysis: GTP-α → GDP-α

   GDP-α recombines with GPCR-$\beta\gamma$.

4. Secondary signal termination: removal of the second messenger.

EXAMPLES OF SECOND MESSENGERS: cAMP, cGMP, DAG, IP$_3$, Ca^{2+}, AND THE INOSITOL PHOSPHOLIPID SYSTEM.

cAMP $\text{ATP} \xrightarrow{\text{Adenylate cyclase}} \text{cAMP} + \text{PP}_i$

The G-protein **G$_S$** stimulates (and **G$_i$** inhibits) adenylate cyclase.

Upon ligand binding, G$_S$ binds to the occupied receptor.

GDP-α_S → GTP-α_S. The GTP-α_S subunit (the activated α subunit of G$_S$) detaches from $\beta\gamma$ and activates adenylate cyclase, which synthesizes cAMP.

The cAMP diffuses into the cytoplasm, where it binds to and activates PKA (cAMP-dependent protein kinase). PKA then phosphorylates key regulatory enzyme to change its activity.

Primary signal termination: $GTP\text{-}\alpha_S \rightarrow GDP\text{-}\alpha_S$

Secondary signal termination: $cAMP + H_2O \rightarrow AMP$, catalyzed by phosphodiesterase.

THE PHOSPHATIDYLINOSITOL CYCLE, DAG, and Ca^{2+} mediates the action of hormones and growth factors. (See Figure 16.8, page 602 in your text.)

DAG (DIACYLGLYCEROL) activates protein kinase C, which phosphorylates specific regulatory enzymes.

IP$_3$ (INOSITOL-1,4,5-TRIPHOSPHATE): Its receptor is a Ca^{2+} channel in the calcisome (SER). When activated, the channel opens and calcium ions flow through, and the action of Ca^{2+}-regulated proteins are affected.

THE PHOSPHATIDYLINOSITOL PATHWAY

The binding of certain hormones to their receptor activates the α-subunit of a G protein. The α-subunit then activates phospholipase C, which cleaves IP$_3$ from PIP$_2$, leaving DAG in the membrane.

$$PIP_2 \xrightarrow{\text{Phospholipase C}} IP_3 + DAG$$

DAG, acting with phosphatidylserine (PS) and Ca^{2+}, activates protein kinase C, which then phosphorylates key regulators. IP$_3$ binds to receptors on the SER, opening Ca^{2+} channels. Then Ca^{2+} moves into the cytoplasm and activates additional targets.

cGMP $$GTP \xrightarrow{\text{Guanylate cyclase}} cGMP + PP_i$$

TWO TYPES OF MEMBRANE-BOUND GUANYLATE CYCLASE:

ANF (atrial natriuretic factor) – lowers blood pressure – activates membrane-bound guanylate cyclase to produce cGMP, which activates the phosphorylating enzyme protein kinase G.

BACTERIAL ENTEROTOXIN – causes diarrhea – binds to and activates another type of membrane-bound guanylate cyclase.

RTKS – RECEPTOR TYROSINE KINASES

Like GPCRs: RTKs are a family of transmembrane receptors with an external domain that binds specific ligands. Upon ligand binding, a conformational change in the receptor protein transduces the signal to the cell's interior.

Unlike GPCRs: RTKs have a cytoplasmic catalytic domain with tyrosine kinase activity. Upon ligand binding, the tyrosine kinase domain autophosphorylates, initiating a phosphorylation cascade.

THE INSULIN RECEPTOR: A WELL-RESEARCHED EXAMPLE OF AN RTK

Structure: a transmembrane glycoprotein with 4 subunits connected by disulfide bridges. Two α-subunits extend out of the cell to bind insulin; two β-subunits extend through the membrane and contain a tyrosine kinase domain on the cytoplasmic side.

A SIMPLIFIED MODEL OF INSULIN SIGNALING (See Figure 16.10 in your text.)[2]
Insulin binds to its RTK, causing autophosphorylation and activation of its tyrosine kinase domain, which triggers a phosphorylation cascade that modulates various intracellular proteins.

This activated insulin receptor then phosphorylates several proteins, such as **insulin receptor substrate 1 (IRS1)**.

$$IRS1_{(inactive)} \xrightarrow{\text{Insulin Receptor (activated)}} IRS1_{(phosphorylated, active)}$$

This newly phosphorylated IRS1 then binds to and activates several proteins, including phosphatidylinositol-3-kinase, which catalyzes:

$$PIP_2 \xrightarrow{\text{Phosphatidylinositol-3-kinase}} PIP_3$$

PIP_3 activates PIP_3-dependent protein kinase, which initiates a phosphorylation cascade that leads to the activation of several kinases (e.g., PKB and PKC) which in turn trigger pathways that lead to changes in the expression of several genes.

In general, insulin induces processes that promote uptake and storage of nutrients. Some of the end results of insulin-receptor binding:
- Inhibits hormone-sensitive lipase in adipocytes;
- Induces transfer to the cell's surface of several types of protein that affect the uptake of nutrients (Examples: receptors for LDL and IGF-II; glucose transporter GLUT4, to the plasma membrane of adipocytes and muscle cells (and facilitated by activated PKB).

GROWTH FACTORS

EPIDERMAL GROWTH FACTOR (EGF) – a **MITOGEN** (stimulates cell division) for epidermal and gastrointestinal lining cells; triggers cell division when it binds to plasma membrane EGF receptors (transmembrane tyrosine kinases)

PLATELET-DERIVED GROWTH FACTOR (PDGF) – secreted by blood platelets during clotting; stimulates mitosis in fibroblasts and other cells during wound healing; promotes collagen synthesis in fibroblasts

SOMATOMEDINS – secreted by the liver into the bloodstream; mediate the growth-promoting actions of GH; include **INSULIN-LIKE GROWTH FACTORS I** and **II (IGF-I, IGF-II)**; bind to cell surface receptors that are also tyrosine kinases

CYTOKINES

INTERLEUKIN-2 (IL-2) – secreted by T cells after they've been activated by binding to a specific antigen-presenting cell; stimulates cell division so that numerous identical T cells are produced; regulates the immune system, promotes cell growth and differentiation.

INTERFERONS are growth inhibitors.
Type I protects cells from viral infection.
Type II inhibits cancer cell growth and also has several immunoregulatory effects.
TUMOR NECROSIS FACTORS (TNF), toxic to tumor cells, suppress cell division.

---

[2]

PIP_2	phosphatidylinositol 4,5-bisphosphate	PKB	protein kinase B
PIP_3	phosphatidylinositol 3,4,5-trisphosphate	IRS1	insulin receptor substrate 1
RTKs	receptor tyrosine kinases		

How do hormones, growth factors, and cytokines differ? (See p. 245 for the answer.)

STEROID AND THYROID HORMONES

Steroid and thyroid hormones switch genes on or off, changing the pattern of proteins that an affected cell makes.

STEROID HORMONES are lipid-soluble (but you knew that already). That means:

- Steroid hormones require transport proteins to travel through the bloodstream. Upon reaching the target cell, the hormone dissociates from its transport protein.

- A steroid hormone can diffuse across a cell membrane into a target cell and bind to a receptor protein in the cytoplasm (or in the nucleus). (Water-soluble hormones can't do that, so they bind to a receptor on the cell membrane surface.)

If in the cytoplasm, the hormone-receptor complex moves to the nucleus.

In the nucleus, it binds to specific sites on DNA. Specifically, the hormone-receptor complex binds to the base sequence of an **HRE** (HORMONE RESPONSE ELEMENTS = specific sites on DNA) via zinc finger domains, a cell's pattern and rate of gene transcription, ultimately affecting protein synthesis).

Since several HREs can bind to the same hormone-receptor complex, the expression of numerous genes can be altered simultaneously.

16.3 METABOLISM IN THE MAMMALIAN BODY: DIVISION OF LABOR

GASTROINTESTINAL TRACT

The small intestine aids in the digestion and absorption of nutrients such as carbohydrates, lipids, and proteins. The enterocytes of the small intestine absorb nutrients, then transport them into the blood and lymph. Glutamine supplies most of the enterocyte's energy needed for active transport and lipoprotein synthesis.

Protein hormones secreted by the GI tract include: ghrelin (Ghr), peptide YY (PYY), cholecystokinin (CCK), and glucagon-like peptide 1 (GLP-1). Ghr stimulates appetite; the rest of these hormones (and insulin) promote satiety.

LIVER: PLAYS A KEY ROLE IN CARBOHYDRATE, LIPID, AND AMINO ACID METABOLISM:

- monitors and regulates the chemical composition of blood and synthesizes several plasma proteins.
- distributes several types of nutrients to other parts of the body
- responsible for reducing fluctuations in nutrient availability caused by drastic dietary changes (such as intermittent feeding and fasting, or high-protein vs. high-carbohydrate diets)
- processes foreign molecules (a critically important protective role)

MUSCLE

The energy sources that are used to provide ATP for muscle contraction are: glucose from its own store of glycogen and from the bloodstream, fatty acids from adipose tissue, and ketone bodies from the liver. During fasting and prolonged starvation, some skeletal muscle protein is degraded to provide amino acids to the liver for gluconeogenesis.

The cardiac muscle in the heart must continuously contract to sustain blood flow throughout the body. To maintain its continuous operation, cardiac muscle relies mainly on fatty acids.

ADIPOSE TISSUE STORES ENERGY IN THE FORM OF TRIACYLGLYCEROLS...

...or degrades fat stores to generate energy-rich fatty acids and glycerol, depending upon whether nutrients are in excess or whether ATP synthesis is needed.

Adipokines are peptide hormones secreted by adipose tissue. Adipokines include leptin (promotes satiety) and adiponectin (activates AMPK, enhancing insulin activity).

BRAIN ULTIMATELY DIRECTS MOST METABOLIC PROCESSES IN THE BODY

Much of the body's hormonal activity is controlled either directly or indirectly by the hypothalamus and the pituitary gland. Under normal conditions the brain uses glucose as its sole fuel. Under conditions of prolonged starvation, the brain can adapt by using ketone bodies as an energy source.

KIDNEY

HAS SEVERAL IMPORTANT FUNCTIONS:

- filtration of blood plasma to excrete water-soluble waste products,
- reabsorption of electrolytes, sugars, and amino acids from the filtrate,
- regulation of blood pH, and
- regulation of the body's water content.

Energy needed for transport processes is provided largely by fatty acids and glucose. Under normal conditions, the small amounts of glucose which are formed by gluconeogenesis are used only within certain kidney cells. The rate of gluconeogenesis increases during starvation and acidosis.

16.4 THE FEEDING-FASTING CYCLE

Consuming food intermittently is possible because of elaborate mechanisms for storing and mobilizing energy-rich molecules derived from food. Regulation of opposing pathways ensure that they don't occur simultaneously.

THE FEEDING PHASE: FOOD IS CONSUMED, DIGESTED, AND ABSORBED...

Absorbed nutrients are transported to various organs where they are either used or stored. During the feeding phase, hormones such as gastrin, secretin, and cholecystokinin stimulate the secretion of various digestive enzymes or aids such as bicarbonate and bile. This phase is regulated by interactions between enzyme-producing cells of the digestive organs, the nervous system, and several hormones.

LIPIDS

Chylomicrons transport lipid molecules from the small intestine, through the lymph and the bloodstream to target tissues. CHYLOMICRON REMNANTS (chylomicrons after most triacylglycerols have been removed) are taken up by the liver where they're reused or degraded. Elevated fatty acids in the blood promotes lipogenesis in adipose tissue.

GLUCOSE

As glucose moves from the small intestine to the liver via the blood, β-cells in the pancreas are stimulated to release insulin. Insulin release triggers several processes that ensure nutrient storage (see table, below). Insulin also influences amino acid metabolism. For example, insulin promotes the transport of amino acids into cells and stimulates protein synthesis in most tissues.

THE FASTING PHASE

EARLY POSTABORPTIVE STATE

Glucagon is released as blood glucose and insulin levels return to normal. Glucagon prevents hypoglycemia by promoting glycogenolysis and gluconeogenesis in the liver.

PROLONGED FAST

Strategies to maintain blood glucose levels include:

* Norepinephrine stimulates lipolysis in adipose tissue, releasing fatty acids that provide an alternative energy source.
* Glucagon increases gluconeogenesis by using amino acids from muscle.

STARVATION

Metabolic changes occur; further fatty acids from adipose tissue are converted into ketone bodies in the liver.

FEEDING-FASTING CYCLE: EFFECTS OF INSULIN AND GLUCAGON		
	POSTPRANDIAL STATE AFTER A MEAL HIGH BLOOD NUTRIENT LEVELS **INSULIN** PROMOTES NUTRIENT STORAGE	**POSTABSORPTIVE STATE** AFTER A FAST LOW BLOOD NUTRIENT LEVELS **GLUCAGON** PREVENTS HYPOGLYCEMIA
BLOOD	Lowers blood glucose	Raises blood glucose
LIVER	Glucose → Glycogen Fatty acids, glycerol → Fats	Glycogen → Glucose Gluconeogenesis → Glucose
MUSCLE	Amino acids → Protein Glucose → Glycogen Glucose uptake	Protein → Amino acids (Ala, Gln)
ADIPOSE TISSUE	Fatty acids, glycerol → Fats Glucose uptake	Fats → Fatty acids, glycerol

FEEDING BEHAVIOR

Feeding behavior is the complex mechanism by which animals, including humans, seek out and consume food. Appetite and satiety in humans are regulated by hormonal and neural signals from peripheral organs (e.g., the GI tract and adipose tissue).

Neuronal pathways of the autonomic nervous system (such as the vagus nerve) continuously supply the brain with information related to the status of the body's internal organs. Also, the mammalian brain links appetite systems to taste, olfaction, and reward systems to create a powerful drive that ensures survival.

Neural circuits that control appetite are in the hypothalamus and in the brain stem:

ARC – *arcuate nucleus* – location in the hypothalamus of the primary neurons that control feeding behavior

NTS – *nucleus tractus solitarius*, location in the brain stem that integrates information with appetite-regulating signals from the GI tract.

REGULATION OF FEEDING BEHAVIOR: FACTORS THAT AFFECT APPETITE[3]		
	INHIBIT APPETITE *Stop, you're full!*	**STIMULATE APPETITE** *Eat! Eat!*
Release of Peptide Hormones	PYY (cells in the GI tract) CCK (cells in the GI tract) Insulin (pancreatic β-cells) Leptin (adipose tissue)	Ghr (cells in the stomach and small intestine) (Also: falling leptin levels caused by weight loss)
Appetite-Regulating Neurons in the **ARC** (in the hypothalamus)	Activated POMC neurons by leptin (insulin to a lesser extent) Inhibition of NPY/AgRP neurons by leptin and PYY	Activated NPY/AgRP neurons (by Ghr) Inhibition of POMC neurons
AMPK mediates when hormonal signals differ from nutrient signals. *(That looks really good, but I'm not hungry.)*	Inhibition of AMPK occurs when hormones such as leptin and insulin bind to their cell-surface receptors, resulting in the inhibition of NPY/AgRP neurons.	Activation of AMPK (by molecules like Ghr) triggers signal transduction events that cause neurons to fire and release NPY and AgRP neurotransmitter molecules that bind to cell-surface receptors in neurons in ARC (and other hypothalamic regions)
Sensory input	Nauseating sight, smell, taste	Appetizing sight, smell, taste

Signals from AgRP/NPY and POMC neurons are communicated via second-order neurons to other parts of the hypothalamus and then other brain centers. Signals from these centers are then sent to the NTS, which integrates this information with appetite-regulating signals from the GI tract and other organs.

---

[3] NPY (neuropeptide Y); AgRP (agouti-related peptide); PYY (peptide YY)
POMC (pro-opiomelanocortin) neurons produce α-MSH (α-melanocyte-stimulating hormone).

CHAPTER 16: SOLUTIONS TO REVIEW QUESTIONS

16.1   a. Ketoacidosis is a condition in which large amounts of ketone bodies occur in blood.

        b. In hyperlipoproteinemia, blood levels of lipoprotein are high.

        c. A target cell is a cell that responds to a specific chemical signal molecule.

        d. Growth factors are a series of polypeptides and proteins that regulate the growth, differentiation, and proliferation of various cells.

        e. Chylomicron remnants are chylomicrons from which triacylglycerol molecules have been removed.

16.2   a. kidney, b. liver, c. intestine, d. brain, e. adipose tissue, f. liver.

16.4   a. Corticotropin stimulates steroid synthesis in the adrenal cortex.

        b. Insulin promotes general anabolic effects, including glucose uptake by some cells and lipogenesis.

        c. Glucagon promotes glycogenolysis and lipolysis.

        d. Oxytocin stimulates uterine muscle contraction.

        e. LH stimulates the development of cells in the ovaries and testis and the synthesis of sex hormones.

16.5   NADPH, which is formed during the pentose phosphate pathway and reactions catalyzed by isocitrate dehydrogenase and malic enzyme, is used as a reducing agent in a wide variety of synthetic reactions (e.g. amino acids, fatty acids, sphingolipids and cholesterol). The degradation of some of these molecules (e.g., fatty acids and the carbon skeletons of the amino acids) results in the synthesis of NADH, a major source of cellular energy via the mitochondrial electron transport system.

16.7   The major recognized second messengers are (1) cAMP (generated from ATP by adenylate cyclase) stimulates changes in cellular activities by activating several protein kinases, (2) cGMP (generated from GTP by guanylate cyclase) activates protein kinase G, (3) diacylglycerol (DAG, a product of phospholipase C) activates protein kinase C, (4) inositol-1,4,5-triphosphate (IP3, a product of phospholipase C) that triggers the release of Ca^{2+} from the calcisome and (5) calcium ions which regulate the activities of numerous cellular activities when they bind to calcium-dependent regulatory proteins.

16.8   Phorbol esters, found in croton oil, activate protein kinase C, an action that stimulates cell growth and division. However, unlike DAG, the molecule that they mimic, phorbol esters continue to activate protein kinase C for a prolonged time. This circumstance provides the affected cell with an advantage over unstimulated cells. Phorbol esters may transform a cell previously exposed to a carcinogenic initiating event into a cancerous cell whose unrestrained proliferation creates a tumor.

16.10   In uncontrolled diabetes, large amounts of glucose are excreted in the urine. Excessive urine flow caused by the large amounts of water that are excreted along with the glucose dehydrates the body. Dehydration then usually triggers the thirst response.

16.11   a. vasopressin – maintains blood pressure and water balance

        b. FSH – follicle-stimulating hormone – in ovaries, promotes ovulation and estrogen synthesis; in testes, promotes sperm development

        c. leptin – induces satiety and is secreted into the bloodstream primarily by adipose tissue

d. ghrelin – stimulates appetite (food intake)

e. adiponectin – enhances glucose-stimulated insulin secretion and cellular responses to insulin

16.13 a. mitogen – a substance that stimulates cell division

b. phorbol ester – a potent tumor promotor found in croton oil (see Solution 16.8, above)

c. heat shock protein – a protein synthesized in response to stress (e.g., high temperature)

d. postprandial – the phase in the feeding-fasting cycle immediately after a meal; blood nutrient levels are relatively high

e. enteric – intestinal

16.14 Anabolic steroid hormones change the expression of a specific set of genes that code for proteins (e.g., enzymes) that would increase protein synthesis in skeletal muscle (among other metabolic changes).

16.16 The functions of the kidney include (1) the excretion of water-soluble waste products, (2) the reabsorption of electrolytes, amino acids, and sugars from the urinary filtrate, (3) regulation of pH, and (4) the regulation of the body's water content. In diabetes mellitus, the ability of the kidney to reabsorb glucose is overwhelmed and glucose spills over into the urine. The presence of glucose in the urine compromises the water recovery function of the kidney and dehydration results. This greatly affects the kidney's ability to maintain electrolyte and pH balance.

16.17 During the initial phase of a prolonged fast, blood glucose and insulin levels fall, and glucagon release is triggered. Glucagon acts to prevent hypoglycemia by promoting glycogenolysis and gluconeogenesis. The amino acids derived from muscle protein are a major source of the carbon skeleton substrates in gluconeogenesis.

16.19 The metabolism of glutamine and glutamate generates ammonia which leaves the kidney in the urine, taking with it a proton. This process, along with the active transport of protons down a sodium gradient and into the kidney tubules, helps to maintain the blood pH at 7.4.

16.20 a. hypothalamus – an area in the brain that integrates the nervous and endocrine systems; upon stimulation by neural signals, it synthesizes and releases specific peptide-releasing hormones

b. orexigenic – appetite-stimulating

c. metabolic syndrome – a cluster of clinical disorders that include obesity, hypertension, dyslipidemia, and insulin resistance

d. dyslipidemia – abnormal blood lipid levels; the most common condition is hyper-lipidemia (high blood levels of total cholesterol and TG and low HDL levels)

e. interleukin – an example of cytokines (proteins produced by blood-forming cells and immune system cells); stimulates cell division in T cells; secreted by T cells that had been activated by an antigen-presenting cell

16.22 a. progesterone – promotes implantation of fertilized eggs and maintenance of pregnancy

b. TRH – thyrotropin-releasing hormone - stimulates the secretion of TSH and prolactin

c. gastrin – stimulates the secretion of stomach acid and pancreatic enzymes

d. TSH – (thyrotropin; thyroid-stimulating hormone) – stimulates the synthesis of thyroid hormone

e. interleukin – regulates the immune system by stimulating the division of T-cells in response to an antigen, promotes cell growth and differentiation

16.23 The most common sites on proteins that are phosphorylated during signal transduction cascades are the hydroxyl groups of serine and tyrosine.

16.25 Obesity contributes to the onset of diabetes by promoting tissue insensitivity to insulin. Specifically, this process begins as enlarged adipocytes release free fatty acids into the bloodstream and are taken up by other cells, causing the disruption of signal transduction pathways. Consequences include: lipotoxicity, high blood insulin levels, excess glucose production (liver), inhibited insulin-mediated glucose uptake by muscle, and further release of fatty acids from adipocytes that have developed insulin resistance. Adipose tissue in the obese develops a low-level chronic inflammation, and as a result, adipocytes experience ER stress and oxidative stress, and release inflammatory cytokines that increases insulin resistance further. All of these factors contribute to the onset of diabetes.

16.26 In addition to carbohydrate metabolism, insulin also impacts lipid and protein metabolism by triggering several processes that are anabolic and/or ensure nutrient storage. These processes include: stimulating fat synthesis in the liver and fat storage in adipocytes, promoting amino acid uptake by cells (especially liver and muscle cells), stimulating protein synthesis in most tissues, decreasing lipolysis, and promoting satiety by inhibiting NPY/AgRP neurons and activating POMC neurons in the hypothalamus.

CHAPTER 16: SOLUTIONS TO THOUGHT QUESTIONS

16.28 The hypothalamus regulates the function of the pituitary largely by secreting small amounts of specific releasing factors into a specialized capillary bed which directly connects the two structures. If the pituitary were to be transplanted to another part of the body, the concentration of the hypothalamic releasing factors in the blood that reaches the pituitary cells would be too low to affect their function.

16.29 The second messenger is an effector molecule synthesized when a hormone (the first messenger) binds. It stimulates the cell to respond to the original signal. Second messengers also allow the signal to be amplified.

16.31 The recognition by the conscious centers in the brain that danger is imminent results in a discharge of epinephrine from the sympathetic nervous system and the adrenal medulla. The large amounts of glucose and fatty acids that flood into blood as a consequence of epinephrine's stimulation of glycogenolysis and lipolysis have several effects on the body. One effect, high blood glucose levels, provides the energy required for rapid decision-making processes in the brain. In addition, large quantities of glucose and fatty acids are required for strenuous physical activity if a decision is made to run away from the danger.

16.32 Increased mobilization of fatty acids provides an alternate energy source for muscle, thereby sparing glucose for the brain. In addition, glucagon stimulates gluconeogenesis, a pathway that utilizes amino acids derived from muscle.

16.34 To ensure proper control of metabolism, powerful hormones are synthesized in small quantities. Hormones also elicit responses in only specific target cells. They are metabolized quickly to ensure the precision of metabolic regulation.

16.35 The storage of preformed hormone molecules in secretory vesicles allows for a rapid response of the producing cells to metabolic signals. As soon as the appropriate signal is received the vesicles fuse with plasma membrane and (via exocytosis) release their contents into the bloodstream.

16.37 Glucagon is involved in maintaining steady state levels of glucose in the body. Liver is the main organ responsible for this process. Muscle tissue is not involved in maintenance of glucose levels, hence there are no receptors for glucagon.

16.38 While muscle does not synthesize fatty acids it dose oxidize them. The product of the carboxylation of acetyl-CoA is malonyl-CoA. When cellular ATP and NADH levels are high, malonyl-CoA has the effect of inhibiting the oxidation of fatty acids because it inhibits carnitine acyl transferase I, the enzyme involved in the transport of fatty acids into the mitochondria.

16.40 The fatty acids components of triacylglycerol molecules cannot be converted into glucose molecules in animals because these organisms lack the glyoxylate cycle enzymes isocitrate lyase and malate synthase. These enzymes convert acetyl-CoA to malate, the citric acid cycle intermediate. (Refer to Figure 9.17). Malate is then converted to oxaloacetate, a substrate for gluconeogenesis.

16.41 Effects of starvation include the breakdown of muscle protein and the use of amino acids to generate energy. Large amounts of amino nitrogen are excreted as urea.

16.43 In a signal cascade the initial signal can be present in low concentrations. The message can then be amplified as the cascade progresses. In addition, an intricate multistage cascade mechanism provides opportunities for the integration of numerous cellular processes.

16.44 Several processes may trigger the same response. For example, glucagon is released into the blood when blood glucose levels are low between meals, and epinephrine, the hormone released under threatening circumstances, triggers the quick release of glucose into blood. The overlap in function of these and other sets of hormones allows an effective response to what may be subtle differences in physiological conditions.

16.46 Exercise promotes insulin-independent glucose uptake by muscle cells, which facilitates blood glucose control.

16.47 Blood glucose is the immediate source of energy for a large number of tissues (e.g. brain and muscle). As blood glucose levels fall the liver releases glucose derived from glycogen (17.2 kJ/g) or gluconeogenesis. The body's long-term energy is stored in triacylglycerol molecules (38.9 kJ/g), which are stored in significantly higher amounts than those of glycogen. The fat reserves of an average person may last for an extended period of time. The amino acids from muscle protein, and triacylglycerol molecules are mobilized more slowly than glucose. In the average body, there is sufficient glucose available from muscle and liver glycogen for short term needs.

16.49 During prolonged starvation ketone body levels will eventually rise to levels that cause acidosis, a condition that causes kidney damage.

16.50 The main source of energy for the brain is glucose. High insulin concentrations cause rapid intake of glucose by body cells. As a result brain function falters and coma ensues.

17 Nucleic Acids

Brief Outline of Key Terms and Concepts

OVERVIEW

GENETICS; MOLECULAR BIOLOGY; CENTRAL DOGMA

REPLICATION; GENE; GENOME
TRANSCRIPTION; TRANSCRIPT; TRANSCRIPTOME
TRANSLATION; PROTEIN; PROTEOME

GENE EXPRESSION; METABOLOME

17.1 DNA

STRUCTURAL FEATURES OF DNA
Two antiparallel polynucleotide strands, held together by hydrogen bonding between complementary base pairs, wind around each other to form a relatively stable right-handed double helix, with nitrogenous bases facing towards the interior and a negatively-charged sugar-phosphate backbone.

DNA STRUCTURE: THE NATURE OF MUTATION
DNA is vulnerable to certain types of disruptive force that can result in mutations, permanent changes in its base sequence.
POINT MUTATIONS: TRANSITION MUTATION; TRANSVERSION MUTATION;
MUTAGENIC FACTORS
TAUTOMERIC SHIFTS; DEPURINATION; DEAMINATION
THYMINE DIMERS
XENOBIOTICS: BASE ANALOGUES, ALKYLATING AGENTS, NONALKYLATING AGENTS, INTERCALATING AGENTS

DNA STRUCTURE: FROM MENDEL'S GARDEN TO WATSON AND CRICK
The model of DNA structure proposed by James Watson and Francis Crick in 1953 was based on information derived from the efforts of many individuals.
PHAGE (BACTERIOPHAGE); CHARGAFF'S RULES

DNA STRUCTURE: VARIATIONS ON A THEME
B-DNA; A-DNA (LESS HYDRATED); Z-DNA (ZIGZAG; LEFT-HANDED HELIX; SLIMMER)
Higher order structures: CRUCIFORMS; INVERTED REPEATS (palindromes); PROTOCRUCIFORMS; SUPERCOILING

DNA SUPERCOILING
NEGATIVE SUPERCOILING; POSITIVE SUPERCOILING; NICK; TOPOISOMERASES; DNA GYRASE

CHROMOSOMES AND CHROMATIN
Prokaryotic chromosome: a supercoiled circular DNA molecule complexed to a protein core. Eukaryotic chromosome: a single linear DNA molecule complexed with histones; NUCLEOHISTONE; NUCLEOSOME; ORGANELLE DNA

GENOME STRUCTURE
Each organism's GENOME organizes and stores the information required to direct living processes. Genomes from different types of organisms differ in their sizes and levels of complexity.
TANDEM REPEATS

17.2 RNA
RNA is involved in various aspects of protein synthesis and in the regulation of gene expression. The most abundant types of RNA are TRANSFER RNA, RIBOSOMAL RNA, and MESSENGER RNA.
TRANSFER RNA (tRNA) molecules have specific amino acids attached to them by specific enzymes and transport these to the ribosome, for incorporation into newly synthesized protein, where they are properly aligned during protein synthesis.
RIBOSOMAL RNAs (rRNA) are components of ribosomes, the sites of catalytic activity.
MESSENGER RNA (mRNA) contains within its nucleotide sequence the coding instructions for synthesizing a specific polypeptide.

NONCODING RNA
Several classes of NONCODING RNAs have diverse roles in genome regulation and protection. Important NONCODING RNAs include: miRNAs (micro RNAs), siRNAs (small interfering RNAs), snoRNAs, and snRNAs.

17.3 VIRUSES

THE STRUCTURE OF VIRUSES
Viruses are composed of nucleic acid enclosed in a protective coat. The nucleic acid may be a single- or double-stranded DNA or RNA. In simple viruses the protective coat, called a CAPSID, is composed of protein. In more complex viruses the NUCLEOCAPSID, composed of nucleic acid and protein, is surrounded by a membranous envelope derived from host cell membrane.
RETROVIRUS; REVERSE TRANSCRIPTASE

OVERVIEW

DNA contains all of an organism's genetic information[1] encoded as a series of nitrogenous bases (purines and pyrimidines).

GENETICS is the scientific investigation of inheritance.

MOLECULAR BIOLOGY is the study of gene structure and genetic information processing.

THE CENTRAL DOGMA: REPLICATION, TRANSCRIPTION, AND TRANSLATION

The central dogma describes the flow of genetic information from DNA to RNA to proteins.

DNA → DNA	REPLICATION	GENE	GENOME
DNA → RNA	TRANSCRIPTION	TRANSCRIPT	TRANSCRIPTOME
RNA → PROTEINS	TRANSLATION	PROTEIN	PROTEOME

Regulation of these processes controls GENE EXPRESSION.   METABOLOME

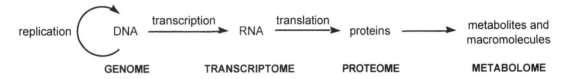

REVERSE TRANSCRIPTASE (in some viruses) is an important exception to the central dogma.

REVIEW: NUCLEOTIDES AND DINUCLEOTIDES

EXAMPLE OF A NUCLEOTIDE[2]

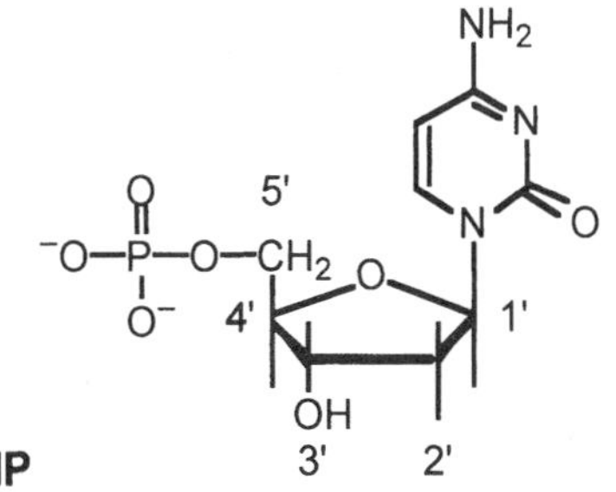

dCMP

Each 2'-deoxyribose has a nitrogenous base at its 1' carbon and a phosphoryl group at its 5' carbon. Nucleotides are connected by 3',5'-phosphodiester bonds.

Compare the following functional groups:

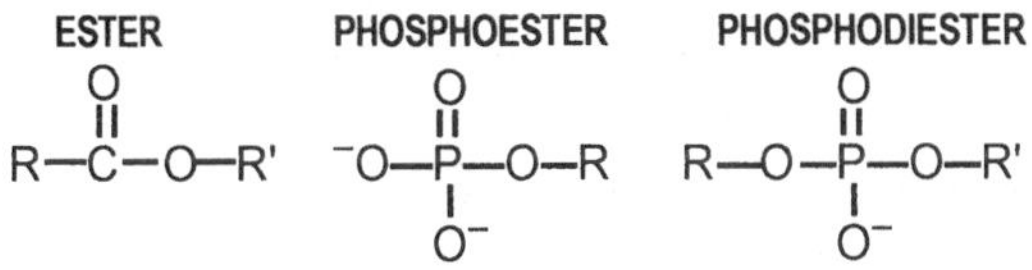

ESTER	PHOSPHOESTER	PHOSPHODIESTER

EXAMPLE OF A DINUCLEOTIDE

Note the phosphodiester bond between the 3'-OH on C and the 5'-phosphoryl group on A.

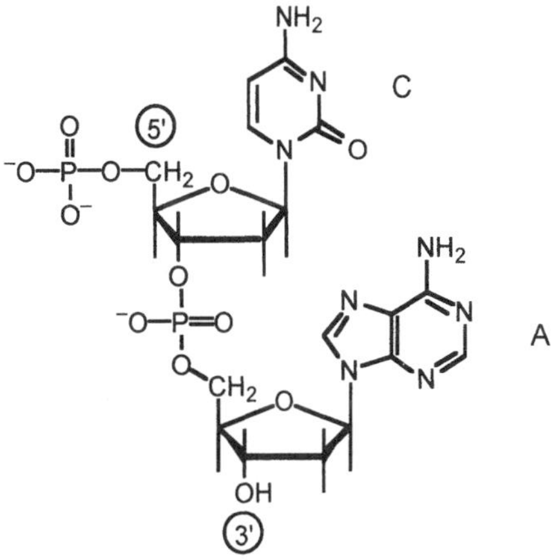

5'-CA-3'

---

[1] Some viruses use RNA rather than DNA for their genetic information.

[2] Remember the difference between a nucleoside and a nucleotide: the -Side is the Sugar + base, and the -Tide is the Total Package, with the Phosphate, Too. Review the structures of purine and pyrimidine bases, nucleosides, and nucleotides in Chapter 14.

When writing a DNA sequence, the order of nucleotides must be direction-specific.

DNA function is based upon the nucleotides being in a precise order, proceeding from the 5' end to the 3' end. (*5' end* and *3' end* refer to the open positions in the sugars at each end of the chain.)

The direction of one DNA strand can be noted as either 5'→3' or 3'→5'. If the direction is not specified, it's assumed to be 5'→3'.

Nucleic acid sequences are always 5'→3' unless noted otherwise.

For example, the strand on the right can be written as any of the following:

5'-CATG-3'   CATG   3'-GTAC-5'

5'-CATG-3'

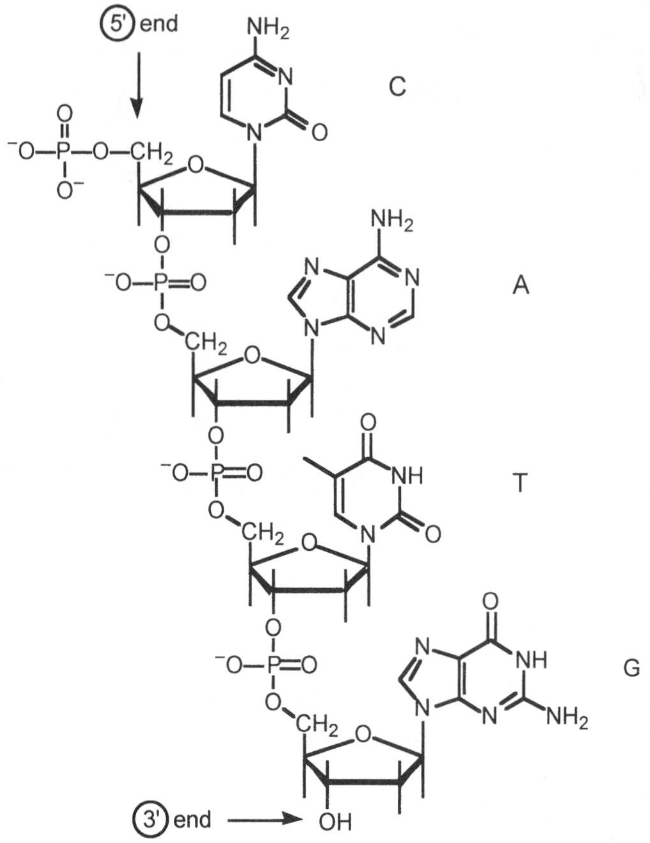

17.1 DNA: A RIGHT-HANDED, ANTIPARALLEL DOUBLE HELIX

Two **ANTIPARALLEL** polynucleotide strands, held together by hydrogen bonding between complementary base pairs, wind around each other to form a relatively stable right-handed **DOUBLE HELIX**.

Nitrogenous bases face towards the interior of the helix. The negatively-charged **SUGAR-PHOSPHATE BACKBONE** faces outwards.

Antiparallel chains run in opposite directions.   One strand is   5'→3'
                                                        and the other strand is   3'→5'.

BASE-PAIRING ALLOWS DNA TO TRANSFER INFORMATION.
Base-pairing brings two DNA strands together, and the chains are held together by hydrogen bonds between the bases.

BASE PAIR – bp – two bases that are on opposite strands of DNA and have hydrogen bonds between them

Possible DNA base pairs: A:::T (two hydrogen bonds, noted as AT)

C:::G (three hydrogen bonds, noted as GC)

Thymine (T)   Adenine (A)

AT base pair

Cytosine (C)   Guanine (G)

GC base pair

DNA STRANDS ARE COMPLEMENTARY

Because base-pairing is specific, the strands are complementary and can serve as templates for each other. So, if the sequence of one strand is known then the sequence of the other strand can be determined.

Complementary copies can be made from a single strand. One strand serve as the template for synthesis of the other strand.

5' AGCCTAGGTCG 3'
| | | | | | | | | | |
3' TCGGATCCAGC 5'

5' AGCCTAGGTCG 3'
| | |
AGC 5'

3' TCGGATCCAGC 5'
| | |
5' AGC

FORCES THAT STABILIZE THE DOUBLE HELIX

1. **HYDROPHOBIC INTERACTIONS:** The double helix structure allows the hydrophobic purine and pyrimidine rings to reside in the interior, away from water. Recall that hydrophobic interactions are driven by the significant increase in the entropy of the surrounding water, which increases the total entropy of the system.

2. **HYDROGEN BONDS:** Hydrogen bonding between bases, i.e., base pairing, is *very* specific. A only base-pairs with T, and C only base-pairs with G. A purine-purine would be too bulky, and a pyrimidine-pyrimidine base pair would create a gap. Also, the combinations AT and CG have the most effective hydrogen bonding. T cannot effectively hydrogen bond with G, and the same holds true for C and A. Try this for yourself. Draw the structures of T and G, and try to form hydrogen bonds between them.

3. **BASE STACKING:** The base pairs are parallel to each other and therefore allow van der Waal contacts between them. With just a few bases, the force is *very* weak. But when a DNA molecule has $\approx$10,000 bases, the cumulative effect is great.

4. **HYDRATION:** DNA binds a significant amount of water at its phosphoryl groups, 3'- and 5'-oxygens, and electronegative atoms in the nucleotide bases. B-DNA can bind about 18-19 H_2O molecules per nucleotide (measured under laboratory conditions).

5. **ELECTROSTATIC INTERACTIONS:** The pK_a of a phosphodiester is about 2, so at pH$\approx$7, DNA is a polyanion with many negative charges along its sugar-phosphate backbone. To relieve some of the repulsion between adjacent negative charges, the charged backbone is stabilized, or shielded, by cations such as Mg^{2+}, polyamines, and cationic proteins (proteins with basic amino acids such as arginine or lysine; e.g., histones) and H_2O.

DNA STRUCTURE: THE NATURE OF MUTATION

MUTATIONS are permanent changes in the base sequence of DNA molecules.

POINT MUTATIONS involve a single base pair. Types of point mutations:
TRANSITION MUTATION: a pyrimidine substitutes for another pyrimidine, or a purine substitutes for another purine.
TRANSVERSION MUTATION: a pyrimidine substitutes for a purine, or vice versa.

MUTAGENIC FACTORS: WHAT CAUSES MUTATIONS?

SPONTANEOUS CHEMICAL REACTIONS

TAUTOMERIC SHIFTS that occur during DNA replication can cause base mispairing and transition mutation. (Remember that tautomeric shifts are the spontaneous interconversion of amino and imino groups or of keto and enol groups.)

DEPURINATION REACTIONS: spontaneous hydrolysis that cleaves a purine from its sugar. One cause is the protonation of guanine's N-3 and N-7.

DEAMINATION REACTIONS: For example, C can change to U via deamination and a tautomeric shift. The resulting U would base-pair with A (rather than the G that was supposed to base-pair with C).

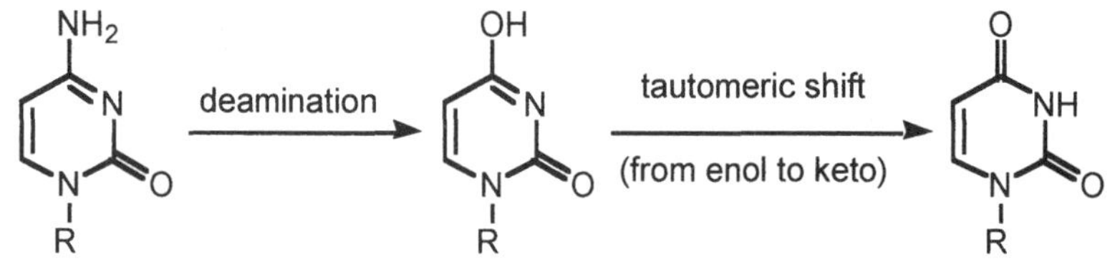

IONIZING RADIATION SUCH AS UV, X-RAYS, AND γ-RAYS

Free-radical mechanisms cause strand breaks, DNA-protein crosslinking, ring openings, base modifications (thymine glycol, 5-hydroxymethyl uracil, 8-hydroxyguanine), and the formation of pyrimidine dimers. **THYMINE DIMERS** are the most common, and are caused by UV energy absorption by double bonds. Thymine dimers distort the helix and stalls DNA synthesis.

XENOBIOTICS:

BASE ANALOGUES have structures that are similar to normal nucleotide bases, and can be inadvertently incorporated into DNA.

ALKYLATING AGENTS add carbon-containing alkyl groups that may result in a transversion mutation (methyl-guanine) or may promote tautomer formation. Examples: dimethylsulfate, dimethylnitrosamine, mitomycin C.

NONALKYLATING AGENTS modify DNA structure in other ways, and typically prevents base pairing. Examples: HNO_2 deaminates bases, and benzo[a]pyrene is metabolized to a highly reactive intermediate that forms adducts to bases.

INTERCALATING AGENTS can insert themselves between stacked base pairs, distorting the DNA double helix and causing deletion or insertion of base pairs (resulting in frame-shift mutations).

DNA STRUCTURE: FROM MENDEL'S GARDEN TO WATSON AND CRICK
HISTORICAL DEVELOPMENTS THAT LED TO THE DISCOVERY OF DNA STRUCTURE:

1865	The scientific revolution that eventually necessitated the determination of DNA structure began when Gregor Mendel discovered the basic rules of inheritance.
1869	Friedrich Miescher discovered "nuclein," later renamed nucleic acid.
1882-1897	The chemical composition of DNA was determined largely by Albrecht Kossel.
1928	Fred Griffith proposed the concept of transmission of genetic information between bacterial cells.
1944	Avery and McCarty demonstrated that the digestion of DNA by deoxyribonuclease inactivated the transforming agent, concluding that genetic information was carried by DNA.
1952	By using T2 bacteriophage, Hershey and Chase demonstrated the separate functions of viral nucleic acid and protein. These experiments reconfirmed DNA as the genetic material.

INFORMATION USED BY WATSON AND CRICK TO DETERMINE DNA STRUCTURE

1. The chemical structures and dimensions of the nucleotides had been elucidated.
2. Adenine:thymine and guanine:cytosine exist as 1:1 ratios in DNA. (**CHARGAFF'S RULES**)
3. X-ray diffraction studies performed by Rosalind Franklin indicated that DNA was a symmetrical molecule, in all likelihood a helix.
4. The diameter and pitch of the helix was estimated by Wilkins and Stokes.
5. Linus Pauling had recently shown that protein could exist as a helix.

DNA STRUCTURE: VARIATIONS ON A THEME

B-DNA
Right-handed helix; base pairs are at right angles to the backbone axis.

2.4	nm	diameter (This distance allows *only* purine-pyrimidine base pairing.)
10.3	bp	base pairs per turn of the helix (may vary slightly with pH and salt concentration)
3.3	nm	length per turn of the helix
0.33	nm	distance between adjacent base pairs

A-DNA
Right-handed helix; base pairs tilt 20° away from the horizontal. A-DNA occurs when DNA becomes partially dehydrated (about 13-14 H_2O/nucleotide, as opposed to B-DNA's 18-19 H_2O/nucleotide), as when extracted with solvents such as ethanol.

Z-DNA
Left-handed helix; base pairs are in a zigzag conformation. Base-pairs with alternating pyrimidine-purine bases are most likely to be in Z-DNA configuration.

B-DNA	A-DNA	Z-DNA	
2.4 nm	2.6 nm	1.8 nm	Helix diameter
10.3 bp	11 bp	12 bp	Base pairs per turn of the helix
3.3 nm	2.5 nm	4.5 nm	Helix rise (length per turn of the helix)

HIGHER-ORDER STRUCTURES:

CRUCIFORMS – crosslike structures; likely when a DNA sequence contains a palindrome – an inverted repeat. A protocruciform is a small bubble that eventually forms a cruciform.

DNA SUPERCOILING

SUPERCOILING packages large DNA into a compact form. One strand is nicked, the double helix is overwound or underwound, then the nicked strand is resealed.

NEGATIVE SUPERCOILING:

Underwound DNA twists to the right to relieve strain.

DNA winds around itself to form an interwound supercoil.

Negative supercoiling stores potential energy in the form of torque, which facilitates strand separation during replication and transcription.

POSITIVE SUPERCOILING:

Overwound DNA twists to the left to relieve strain.

DNA winds around a protein core to form a toroidal[3] supercoil.

Positive supercoiling interferes with replication and must be removed by topoisomerases such as DNA gyrase.

CHROMOSOMES AND CHROMATIN

PROKARYOTES:

CHROMOSOME: circular DNA that's extensively looped and coiled. A NUCLEOID is supercoiled DNA complexed with a protein core. Polyamines (e.g., spermidine and spermine) are polycations at cellular pH, and bind to DNA's sugar-phosphate backbone to overcome the charge repulsion between adjacent DNA coils.

EUKARYOTES:

CHROMOSOME: one linear DNA molecule complexed with histones; structural units are the nucleosomes. (Humans possess 23 pairs of chromosomes.)

HISTONES – a group of small basic proteins found in all eukaryotes; positively-charged side chains interact with DNA's negatively-charged sugar-phosphate backbone. Five classes of histones: H1, H2A, H2B, H3, H4. Nucleohistones[4] are histones that form complexes with DNA.

NUCLEOSOME – a complex of histones and DNA (about 145 bp):
- a positively supercoiled DNA segment (about 145 bp) that forms a toroidal coil around an octameric histone core.
- histone core: 2 H2A•H2B dimers and a $H3_2$•$H4_2$ tetramer; each histone contains a HISTONE FOLD (3 α-helices separated by 2 unstructured segments)
- H1, an additional histone that attaches to the coil in two places to hold the wrapping in place around the core.

A 60-bp segment of linker DNA connects adjacent nucleosomes.

Nucleosomes are coiled into **30-nm FIBERS**, which are further coiled to form **200- nm FILAMENTS**.

CHROMATIN – partially decondensed form of chromosomes; occurs when cell is not dividing (interphase). Chromatin compacts to form chromosomes. HETEROCHROMATIN; EUCHROMATIN

ORGANELLE DNA – in mitochondria and chloroplasts

---

[3] A toroid is the shape of a donut.

[4] "Nucleohistone" is an older term that was used to describe the DNA-protein complexes that resulted from laboratory separations of cellular components. The term "nucleosome" reflects recent research results regarding DNA-histone structure.

GENOME STRUCTURE

GENOME – a complete set of genetic information, encoded in the nucleotide base sequence of DNA. Genes are contained within genomes.

PROKARYOTIC VS. EUKARYOTIC GENOMES

FEATURE:	PROKARYOTIC GENOMES	EUKARYOTIC GENOMES
GENOME SIZE	Relatively small; fewer genes	Relatively large (but size doesn't imply complexity of an organism)
CODING CAPACITY; CODING CONTINUITY	Genes are compact and continuous Little to no noncoding DNA	Most DNA is noncoding. Most genes are discontinuous. Intergenic sequences ("junk DNA") do not code for gene products.
GENE EXPRESSION MECHANISMS	operon – set of functionally related genes, the transcription of which are regulated as a unit	Introns are interspersed between exons. introns – noncoding sequences exons – expressed sequences that encode for a gene product.
OTHER UNIQUE FEATURES	**PLASMIDS** - additional DNA, typically circular, that codes for biomolecules that provide a growth or survival advantage	Contains repetitive sequences: tandem repeats, interspersed genome-wide repeats

TYPES OF REPETITIVE SEQUENCES IN EUKARYOTIC GENOMES

TANDEM REPEATS (SATELLITE DNA)

- multiple copies arranged next to each other
- certain types play structural roles in centromeres[5] and telomeres[6]

TYPES OF RELATIVELY SMALL TANDEM REPEATS

MINISATELLITES – 10-100 bp repeat sequences; total = 10^2-10^5 bp

MICROSATELLITES – **SINGLE-SEQUENCE REPEATS, SSR** – 1-4 bp repeated 10-100 times; vary with individuals; used as markers in genetic disease diagnosis, forensics investigations; also called **SHORT TANDEM REPEATS, STR**

INTERSPERSED GENOME-WIDE REPEATS

- repetitive sequences scattered around the genome
- most result from transposition.
  In transposition, transposons (transposable DNA elements) excise themselves and insert at another site (or, involve RNA transposons or retrotransposons)

LINEs and SINEs: LONG and SHORT INTERSPERSED NUCLEAR ELEMENTS

LINEs are retrotransposons longer than 5000 bp. About one of every 1200 mutations are due to LINE insertions.

SINEs have less than 500 bp, and need a functional LINE sequence to undergo transposition. About 11% of the human genome are SINEs.

---

[5] Centromeres are structures that contain kinetochores and attach chromosomes to the mitotic spindle during mitosis and meiosis.

[6] Telomeres are structures at the end of chromosomes that buffer the loss of critical coding sequences after a round of DNA replication.

ALU ELEMENTS – the only SINE insertions linked to human disease;
– mediate chromosome rearrangements, insertions, deletions, and recombinations
– cause mutations that result in genetic diseases such as hemophilia B, Lesch-Nyhan disease, and Tay-Sachs disease

17.2 RNA: STRUCTURE, FUNCTION, AND SYNTHESIS (TRANSCRIPTION)

DIFFERENCES BETWEEN RNA AND DNA:

1. The sugar of RNA is ribose; so The 2′ –OH group makes RNA more reactive than DNA.
2. Uracil replaces thymine. So, the base pairs in RNA are: AU and GC.
3. RNA is a single strand (not a double helix), so it can coil back on itself and form unique and complex 3D structures. RNA does not follow Chargaff's rules.
4. RNA's 3D structure can form binding pockets, which gives some RNAs catalytic properties. Mg^{2+} is usually needed in RNA catalysis to stabilize transition states. Ribozymes are catalytic RNA molecules; most catalyze cleavage of itself or of other RNAs.

TYPES OF RNA:

Of the total RNA of the cell, ≈80% is rRNA, ≈15% is tRNA, ≈5% is mRNA.

1. **TRANSFER (tRNA)** - carries individual amino acids to ribosomes to be properly aligned and assembled into proteins.
2. **RIBOSOMAL (rRNA)** - an integral part of ribosomes.
3. **MESSENGER (mRNA)** - carries the genetic information for protein synthesis from the nucleus to the ribosomes. mRNA molecules are copies of DNA "genes," and contain the genetic information needed to synthesize specific polypeptides.
4. **NONCODING RNA (ncRNA)** – does not code for proteins or RNA.

TRANSFER RNA (tRNA): ADAPTER MOLECULES IN PROTEIN SYNTHESIS

tRNA are single-stranded polynucleotide chains about 75 nucleotides long that convert a nucleotide sequence into a specific amino acid.

AMINOACYL-tRNA SYNTHETASES link each amino acid to the **3′-TERMINUS** on its specific tRNA.

tRNA carries the amino acid to the ribosomes to be aligned, ordered, and assembled into proteins.

tRNA contains modified bases that help to stabilize and protect tRNA from degradation. These modifications can also occur in rRNA.

ribose	ribose	ribose	ribose	ribose
uridine	pseudouridine (ψ)	4-thiouridine	dihydrouridine	1-methyl guanosine
(shown for comparison)	Attachment to ribose is via a carbon atom.	S replaces a carbonyl O.		

THE SHAPE OF **tRNA** RESEMBLES A WARPED CLOVERLEAF WITH THE FOLLOWING FEATURES:

SPECIFIC – bonds to a specific amino acid

ANTICODON LOOP – contains a 3-bp sequence that is complementary to the DNA triplet code for a specific amino acid.

D LOOP – contains dihydrouridine

TψC LOOP – contains thymine, pseudouridine (ψ), and cytosine

VARIABLE LOOP – length varies between 4-5 to 20.

RIBOSOMAL RNA (rRNA) IS AN INTEGRAL COMPONENT OF RIBOSOMES

Ribosomes are cytoplasmic ribonucleoprotein structures that synthesize proteins and consist of 60-65% rRNA (the remainder is protein). Ribosomes consist of two subunits of unequal size, with several different kinds of rRNA and protein in each subunit.

MESSENGER RNA (mRNA) CARRIES DNA'S GENETIC INFO FOR PROTEIN SYNTHESIS.

mRNA specifies the order of amino acids in a protein.

CISTRON – a DNA sequence that contains coding info for a polypeptide plus signals needed for ribosome function

PROKARYOTIC VS. EUKARYOTIC mRNA:

POLYCISTRONIC (prokaryotes) vs. MONOCISTRONIC (eukaryotes): contain information for many (poly) or one (mono) polypeptide chain(s).

Prokaryotic mRNA isn't processed further, and can begin working right away. Eukaryotic mRNA is extensively modified, including:
- Capping the 5′ end with a 7-methylguanosine
- Splicing (removing introns)
- Attaching a poly (A) tail (polyadenylate) to the 3′ end

NONCODING RNA (ncRNA)

miRNA   micro RNA

siRNA   small interfering RNA

snoRNA   small nucleolar RNA

snRNA   small nuclear RNA – involved in splicing and other RNA processing; combine with proteins to form small nuclear ribonucleoprotein particles (snRNPs or snurps)

SPLICEOSOMES are molecular machines made of snRNPs and other proteins. Spliceosomes splice by excising introns from pre-mRNA and join the exons together.

17.3 VIRUSES

Viruses: obligate, intracellular parasites or mobile genetic elements?
Each virus contains a piece of nucleic acid within a protective coat. When a virus infects a host cell, its nucleic acid hijacks the cell's nucleic acid and protein-synthesizing machinery, making complete new viral particles that are released form the host cell (often rupturing the cell in the process). Or, the viral nucleic acid may insert itself into a host chromosome, resulting in cell transformation.

Viral infection disrupts cell function. By suppressing some cellular genes and activating others, the viral genome directs the host cell to produce new virus in a process that often results in cell death.

WHAT PROPERTIES OF VIRUSES MAKE THEM USEFUL RESEARCH TOOLS?

Because a virus subverts normal cell function to produce more viruses, a viral infection can provide unique insight into cellular metabolism. Several eukaryotic genetic mechanisms have been elucidated with the aid of viruses and/or viral enzymes. Viral research has also provided substantial information concerning genome structure and carcinogenesis. Viruses have also been invaluable in the development of recombinant DNA technology.

THE STRUCTURE OF VIRUSES

VIRIONS – complete viral particles = nucleic acid + capsid (+ membrane envelope in more complex viruses)
> **CAPSID** – a protein coat made of **CAPSOMERES** (interlocking proteins)
> **NUCLEOCAPSID** = nucleic acid + capsid

TYPES OF VIRAL GENOMES

Most common: double-stranded DNA (**dsDNA**), single-stranded RNA(**ssRNA**)
> Positive-sense RNA genome [(+)-ssRNA]
> Negative-sense RNA genome [(–)-ssRNA]: Viruses with (–)-ssRNA genomes must provide reverse transcriptase.

Also observed: single-stranded DNA (**ssDNA**), double-stranded RNA (**dsRNA**)

AFTER MANY YEARS OF RESEARCH AND EXPENDING BILLIONS OF DOLLARS, WHY IS AIDS STILL CONSIDERED AN INCURABLE DISEASE?

HIV is a retrovirus that causes AIDS. Retroviruses are a class of RNA viruses which possess a reverse transcriptase activity that converts their RNA genome to a DNA molecule. This DNA is then inserted into the host cell genome, causing a permanent infection.

Because the viral genome mutates frequently (i.e., its surface antigens become altered), a vaccine for the HIV virus has been difficult to develop. In the case of HIV, mutations occur because the reverse transcriptase doesn't have any proofreading capabilities. So, now and then it makes mistakes that are related to the surface antigens, which are continuously altered.

Similarly, that's why we need to get a flu shot every year. Flu viruses are very active, and mutate constantly because of continuous replication. Their enzymes make mistakes that are beneficial for the survival of the virus.

CHAPTER 17: SOLUTIONS TO REVIEW QUESTIONS

17.1
    a. genetics – the study of inheritance

    b. replication – the process by which DNA strands are copied

    c. transcription – the synthesis of RNA from a DNA template

    d. sugar-phosphate backbone – in polynucleotides, the 'backbone' is created by the 3',5'-phosphodiester bonds that join the 5'-hydroxyl group of deoxyribose or ribose to the 3'-hydroxyl group of the sugar unit of another nucleotide

    e. bacteriophage – a virus that attacks bacteria

17.2
    B-DNA, the right handed double helical structure discovered by Watson and Crick, in which there are 10.4 base pairs per helical turn (3.4 nm; diameter = 2.4 nm), occurs under humid conditions. A-DNA, a partially dehydrated molecule, possesses 11 base pairs per helical turn (2.5 nm; diameter 2.6 nm). The Z form of DNA is twisted into a left handed spiral with 12 base pairs per helical turn. Each helical turn occurs in 4.5 nm with a diameter of 1.8 nm.

17.4
    a. According to Chargaff's rules, there is a 1:1 ratio of adenine to thymine and guanine to cytosine, regardless of the source of DNA.

    b. hypochromic effect – a decrease in absorption intensity; used in the analysis of the nucleic acids, as it occurs upon basepairing and folding of the DNA

    c. transposon – a DNA segment that carries the genes necessary for transposition; sometimes the term is reserved for transposable elements that can also contain genes unrelated to transposition.

    d. Alu family – a group of short DNA sequences that constitute 11% of the human genome with over a million copies

    e. transcriptome – a complete set of RNA molecules that are produced within a cell under specified conditions.

17.5
    Biological processes that are facilitated by supercoiling include packaging of DNA into compact forms (i.e., chromosomes), DNA replication and transcription.

17.7
    Nucleosomes are the structural units of chromatin. Each nucleosome consists of a left-handed supercoiled DNA segment wound around eight histone molecules.

17.8
    a. satellite DNA – DNA sequences arranged next to each other that form a distinct band when genomic DNA is digested and centrifuged; the original term for tandem repeats

    b. splicing – the excision of introns during mRNA processing.

    c. chain-terminating method – a DNA sequencing method that uses 2',3'-dideoxynucleotide derivatives; without the 3'-OH, another nucleotide cannot be added, and chain growth stops; the result is polynucleotides of many different lengths that can be analyzed

    d. DNA denaturation – the separation of the DNA double helix into single strands.

    e. retrotransposon – a transposition mechanism that involves an RNA transcript; an RNA transposon

17.10
    The three major types of RNA are ribosomal RNA (a component of ribosomes), transfer RNA (each molecule transports a specific amino acid to the ribosome for assembly into proteins), and messenger RNA (each molecule specifies the sequence of amino acids in a polypeptide).

17.11
    The transition from B-DNA to Z-DNA can occur when the nucleotide base sequence is composed of alternating purine and pyrimidines (e.g., CGCGCG). Because alternate

nucleotides can assume different conformations (syn or anti), these DNA segments form a left-handed helix. The phosphate groups in the backbone of this DNA conformation zigzag hence the name Z-DNA.

17.13   a. telomere – structure at the ends of chromosomes that buffer the loss of critical coding sequences after a round of DNA replication.

b. minisatellite – tandemly-repeated DNA sequence of about 25 bp with total lengths between 10^2 and 10^5 bp.

c. DNA profile – consists of the unique pattern and number of repeats each in STR sequences used to identify individuals.

d. intron – a noncoding intervening sequence in a split or interrupted gene missing in the final RNA product

e. exon – a region in a split or interrupted gene that codes for RNA and ends up in the final product (e.g., mRNA).

17.14   According to Chargaff's rules, if a DNA sample contains 21% adenine then it also contains 21% thymine. If the A-T content is 42%, then the G-C content is 58%. Consequently the guanine and cytosine percentages are both 29% in the DNA sample.

17.16   a. transversion – a type of point mutation in which a pyrimidine is substituted for a purine or vice versa

b. epigenetic – modifications that affect DNA function indirectly without changing DNA structure, e.g., covalent modifications of histones that in turn modify the accessibility of DNA to transcription factors

c. intercalating agent – certain planar molecules that distort DNA by inserting themselves between the stacked base pairs of the double helix

d. LINE – long interspersed nuclear elements – retrotransposons with lengths greater than 5000 bp.

e. SINE – short interspersed nuclear elements – retrotransposons that are less than 500 bp.

17.17   The hierarchy from smallest to largest structure is: (d) nucleotide base pair, (c) nucleosome, (b) gene, and (a) chromosome.

17.19   a. transcriptome – the total set of RNA transcripts produced by a cell

b. noncoding RNA – RNA molecules that are not directly involved in protein synthesis

c. small nucleolar RNAs – (snoRNAs) – 70-1000 nt single stranded (ss) RNAs that facilitate chemical modifications of rRNA within the nucleolus

d. microsatellites – core DNA sequences of 2 to 4 bp that are tandemly repeated 10 to 20 times

e. small interfering RNA – a short noncoding RNA involved in RNA interference, a process used to inactivate or degrade RNA in order to: 1) inhibit or halt the expression of specific genes, and 2) to protect cells from viral RNA genomes

17.20   Polyamines are polycationic molecules that bind to the negatively charged phosphate groups along the DNA backbone. Positively charged polyamines help to stabilize the highly compressed structure of a chromosome by relieving the electrostatic repulsion between the negatively charged, tightly coiled DNA molecules. Examples of polyamines are spermine and spermidine.

17.22   RNA forms three-dimensional structures by coiling back on itself, allowing its single strand to form double-stranded regions with stabilizing features that are analogous to DNA, namely, hydrogen bonding between complementary base pairs (A-U, G-U, and

G-C) and base stacking. These double-stranded regions in RNA are primarily intrachain, but may occur between single strands from adjacent molecules. Additional stabilizing interactions may occur between double-stranded regions and free loops of RNA.

17.23 The stable inheritance of DNA methylation patterns from one cell generation to the next is made possible by a class of enzymes called the maintenance methyltransferases. They methylate the cytosines in CpG-rich regions in newly synthesized DNA strands. The enzymes methylate the cytosines on the new strand at sites opposite to the methylated cytosines on the parental strand.

17.25 When an aromatic hydrocarbon intercalates between two stacked base pairs, either adjacent base pairs are deleted or new base pairs are inserted. Intercalating agents may also cause chromosomes to break.

17.26 Intercalated hydrocarbons may cause frame-shift mutations during DNA replication.

17.28 Chargaff's rules do not apply to RNA because RNA is single-stranded.

17.29 Examples of modified tRNA bases include: pseudouridine (Ψ), 4-thiouridine, dihydrouridine (D), and 1-methylguanosine. The D loop, the TΨC loop (thymine, pseudouridine, and cytosine), and the variable loop incorporate these modified bases, and facilitate binding and alignment during protein synthesis.

CHAPTER 17: SOLUTIONS TO THOUGHT QUESTIONS

17.31 The principal structural difference between DNA and RNA is the 2'-OH group of ribose in RNA molecules. In DNA, which lacks the 2'-OH group in the deoxyribose sugar, hydrogen-bonded complementary strands can easily adopt the B-form double helix. In contrast, double-stranded regions of RNA molecules cannot adopt this conformation because of steric hindrance. Instead, they adopt the less compact A-helical form in which there are 11 bp per turn and the base pairs tilt 20° away from the horizontal.

17.32 The major and minor grooves of DNA arise because the glycosidic bonds in the two hydrogen bonded strands are not exactly opposite to each other.

17.34 RNA can coil back on itself to form complex three-dimensional structures.

17.35 Relaxed circular DNA with a nicked strand is less compact than supercoiled circular DNA and so has a lower effective density. As a result it will not migrate as far in the centrifuge tube as the supercoiled DNA.

17.37 Because nucleotide base sequences and amino acid sequences are such completely different "languages," a complex mechanism is required for the "translation" of one type of information into another. In the absence of any evidence to the contrary it does not appear likely that information expressed in proteins can be utilized to direct the synthesis of nucleic acids.

17.38 One of the principal reasons for the problems in the development of an AIDS vaccine is that the HIV genome mutates frequently. Consequently, the surface antigens of HIV also often change. Note that the antigens in a vaccine stimulate the immune system to produce antibodies that will bind specifically to antigens on the surface of disease-causing organisms.

17.40 The histones act to shield the DNA from the action of nucleases.

17.41 Each cell is constantly receiving information from its environment. Cells adapt to changing conditions as information in the form of nutrients, hormones, growth factors, and other types of molecules triggers changes in their molecular mechanisms that ultimately cause changes in gene expression. The transcriptome, the set of mRNA molecules produced under specified conditions, is a measure of the current status of gene expression.

17.43 At the crime scene, a forensic expert collects biological specimens such as blood, hair, and saliva. Once these specimens are delivered to the lab, they are analyzed and compared with the DNA of the victim. Any DNA not belonging to the victim is assumed to belong to a person, or persons, present during the time when the crime was committed. If a suspect is identified, his or her DNA profile (obtained from a swab of cheek cells or from a court-ordered blood sample) is compared with that obtained from crime scene specimens. If there is no obvious suspect, the crime scene specimens can be compared to the DNA profiles in the statewide database. This strategy has been remarkably successful in the identification of individuals later found guilty not only of recent murders, but also those from "cold cases" in which crime scene specimens had been preserved. The technology that makes this success possible includes PCR, RFLP, and STR-DNA analysis.

17.44 The protein components of the electron transport system are located in the inner membrane of mitochondria. If these components are damaged they need to be replaced immediately. If the genes coding for these molecules are in the mitochondria then they can be easily synthesized as needed. If the genes were in the cell nucleus a time-consuming signal mechanism would be required to initiate gene expression. The newly synthesized proteins would then have to diffuse back to the mitochondrion and be transported across the mitochondrial membrane.

17.46 The hydrogen bonds to water molecules formed by the atoms in a phosphodiester linkage are as follows:

The hydrogen bonds form between the hydrogen atoms in water and the unshared pairs of electrons on the oxygen atoms in the phosphodiester linkage.

17.47 When DNA molecules become alkylated, the chains break or misreading is facilitated, both of which can result in mutations. If such genetic changes cause the genes that control cell growth to be turned off, uncontrolled cell division will occur. Abnormal methylation of DNA sequences can result in epimutations.

17.49 DNA degrades with time. Ancient fossils would have little if any intact DNA with which to reconstruct the organisms. In addition, although intact DNA is vitally important for organismal function, it is only the operating system of an organism. Without access to a living example of such an organism, it would be impossible to reconstitute the physiological structure and functional properties that are unique to a species.

17.50 The enolization of the fluorouracil will promote the wrong hydrogen bonding pattern (i.e. it will hydrogen bond with guanine instead of adenine) and a transition mutation will result. As these mutations accumulate rapidly dividing cells (the majority of which are cancer cells) are fatally damaged.

18 Genetic Information

Brief Outline of Key Terms and Concepts

18.1 GENETIC INFORMATION: REPLICATION, REPAIR, AND RECOMBINATION

DNA REPLICATION is DNA synthesis.

SEMICONSERVATIVE REPLICATION MECHANISM – each strand of DNA serves as a template to synthesize a new strand. The **LEADING STRAND** is synthesized continuously; the **LAGGING STRAND** is synthesized in pieces, then covalently linked. Both syntheses are in the $5' \rightarrow 3'$ **DIRECTION**. **REPLICATION FACTORIES; REPLISOMES**

DNA SYNTHESIS IN PROKARYOTES
Replication requires enzymatic activities in: DNA unwinding (**TOPOISOMERASES, HELICASES, SSB; REPLICON, REPLICATION FORK, REPLICATION EYE**), **PRIMER** synthesis (**PRIMASE, PRIMOSOME**), ELONGATION (**DNA POLYMERASES, PROCESSIVITY, β-2-CLAMP, CLAMP LOADER**), supercoiling control and ligation (DNA **LIGASE**). Also: $3' \rightarrow 5'$ **EXONUCLEASE ACTIVITY**; $5' \rightarrow 3'$ **EXONUCLEASE ACTIVITY**

DNA SYNTHESIS IN EUKARYOTES
Significant differences between DNA replication in eukaryotes versus prokaryotes include: replication time and rate, replication origin numbers, **OKAZAKI FRAGMENT** size, and replication machinery structure. **PREINITIATION REPLICATION COMPLEX (preRC), ORIGIN OF REPLICATION COMPLEX (ORC), MCM COMPLEX (MINICHROMOSOME MAINTENANCE COMPLEX); REPLICATION LICENSING FACTORS (RLFs), REPLICATION PROTEIN A (RPA), REPLICATION FACTOR C (RFC); TELOMERASE, TELOMERE END-BINDING PROTEINS (TEBPs), TELOMERE REPEAT-BINDING PROTEINS (TRFs)**

DNA REPAIR
Each organism's survival depends on its capacity to repair DNA structural damage. Types of DNA repair mechanisms include **BASE EXCISION REPAIR, NUCLEOTIDE EXCISION REPAIR, PHOTOREACTIVATION (LIGHT-INDUCED) REPAIR**, and **RECOMBINATIONAL REPAIR**.

DNA RECOMBINATION
GENERAL RECOMBINATION – exchange of DNA sequences in homologous chromosomes **SITE-SPECIFIC RECOMBINATION**, the exchange of DNA sequences requires only short homologous sequences. DNA-protein interactions are responsible for the exchange of largely nonhomologous sequences. **DNA GLYCOSYLASE TRANSFORMATION, CONJUGATION, TRANSDUCTION**

TRANSPOSITION, the movement of genetic elements (**TRANSPOSONS**) from one place to another within a genome, can cause genetic changes such as insertions, deletions, and translocations. The movement of **RETROTRANSPOSONS**, found in large numbers in eukaryotic genomes, can cause disease or can provide opportunities for genetic diversity. **RETROELEMENT, RETROPOSON, TRANSPOSABLE ELEMENT**

18.2 TRANSCRIPTION (SYNTHESIS OF RNA)
During transcription, an RNA molecule is synthesized from a DNA template (**CODING STRAND**).

TRANSCRIPTION IN PROKARYOTES
In prokaryotes transcription involves a single RNA polymerase activity. Transcription is initiated when the RNA polymerase complex binds to a specific DNA sequence called a **PROMOTER**. **CONSENSUS SEQUENCE**

TRANSCRIPTION IN EUKARYOTES
Transcription is significantly more complex. In addition to chromatin-remodeling and RNA-processing reactions, gene transcription requires the binding of unique sets of **TRANSCRIPTION FACTORS** to **PROMOTER SEQUENCES**. Eukaryotic RNA transcripts undergo several processing reactions.

Regulation of transcription differs significantly between prokaryotes and eukaryotes. Transcription processes observed only in eukaryotes include RNA processing, such as **CAPPING, POLY(A) TAIL** synthesis, and **RNA SPLICING. SPLICEOSOME; ACCEPTOR SITE; SUPRASPLICEOSOME**

18.3 GENE EXPRESSION
CONSTITUTIVE GENES are routinely transcribed, whereas **INDUCIBLE GENES** are transcribed only under appropriate circumstances.

GENE EXPRESSION IN PROKARYOTES
In prokaryotes, inducible genes and their regulatory sequences are grouped into **OPERONS**.

RIBOSWITCHES are metabolite-sensing domains in mRNAs that regulate the transport or synthesis of certain types of molecules.

GENE EXPRESSION IN EUKARYOTES
Mechanisms that control gene expression include DNA methylation, histone covalent modification, and **CHROMATIN REMODELING**, RNA-processing reactions such as alternative splicing, and **RNA EDITING**, RNA transport, and translational controls.

OVERVIEW

Molecular machinery, consisting primarily of DNA-binding proteins, bend, twist, unwind and unzip DNA in replication and transcription.

DNA-BINDING PROTEINS tend to have:
- a twofold axis of symmetry and often form dimers.
- very specific DNA binding via noncovalent interactions between amino acids and the edges of nucleotide base pairs along the major groove.
- one of the following supersecondary structures: helix-turn-helix, helix-loop-helix, leucine zipper, or zinc finger.

18.1 GENETIC INFORMATION: REPLICATION, REPAIR, AND RECOMBINATION

DNA REPLICATION: DNA POLYMERASE MAKES COPIES OF DNA

GENERAL FEATURES OF DNA REPLICATION:

1. The two strands of DNA are separated by unwinding. **REPLICATION FORK**

2. **SEMICONSERVATIVE REPLICATION**: Each strand serves as a template for the synthesis of a complementary strand, so each new DNA molecule has one old strand and one new strand. (This was proven in the Meselson-Stahl experiment.) An existing DNA template is essential to provide the proper sequence of bases.

3. Deoxyribonucleotides are added to the 3′-OH of a pre-existing nucleic acid chain (either DNA or RNA) by forming new phosphodiester bonds.

4. Deoxyribonucleotide triphosphates (dNTPs) such as dATP, dGTP, dCTP, and dTTP, are the substrates for **DNA POLYMERASE** (with Mg^{2+} as a cofactor).

$$(DNA)_n + dNTP \rightarrow (DNA)_{n+1} + PP_i$$

5. **ELONGATION** of the new complementary strand proceeds *only* in the 5′→3′ direction. Because of this, and because elongation can only occur at the 3′-OH of a pre-existing chain, one of the new strands (the **LEADING STRAND**) can be synthesized continuously, but the other strand (the **LAGGING STRAND**) must be synthesized in short segments (**OKAZAKI FRAGMENTS**), with new RNA primer being synthesized as the replication fork opens.

LEADING STRAND
The 5′→3′ direction is *towards* the replication fork. dNTPs can react to add to the continuous, growing, complementary strand as the replication fork opens.

LAGGING STRAND
The 5′→ 3′ direction proceeds *away from* the fork. DNA can only be synthesized in Okazaki fragments as the replication fork opens.

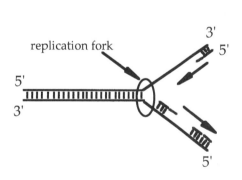

REPLICATION FACTORIES are specific nuclear or nucleoid compartments where DNA replication occurs.

Replication begins when there are enough copies of **DnaA** (a DNA binding protein) and when there's a high ATP/ADP ratio.

DNA SYNTHESIS IN PROKARYOTES (*E. COLI*)

1. **DNA UNWINDING UTILIZES THREE TYPES OF PROTEINS:**
   TOPOISOMERASES work ahead of the replication fork to relieve torque. (See supercoiling control, # 5, below.)
   HELICASES unwind the DNA at the fork (and require ATP).
   SSB (SINGLE-STRANDED BINDING PROTEIN) stabilizes and protects single-stranded DNA segments.
   Prokaryotic replication begins at oriC (initiation site on *E. coli*) and proceeds in two directions, forming a **REPLICATION EYE**. **REPLICON** - a DNA molecule or segment that contains an initiation site and regulatory sequences

2. **PRIMERS** – short RNA segments needed to initiate replication – are synthesized by **PRIMASE**. An RNA **PRIMER** provides a $3'$-OH group, to which DNA polymerase can add its first nucleotide. The leading strand only needs one primer per replication fork, but the lagging strand needs a primer for each Okazaki fragment.
   PRIMOSOME: a multienzyme complex containing primase (an RNA polymerase) and several auxiliary proteins.

3. **DNA SYNTHESIS IS CATALYZED BY DNA POLYMERASES**
   REPLISOME – the prokaryotic DNA replicating machine – DNA unwinding proteins, the primosome, and two copies of the pol III holoenzyme.
   DNA POLYMERASES are large multienzyme complexes that catalyze the formation of phosphodiester bonds between dNTPs in the $5' \rightarrow 3'$ direction.
   DNA POLYMERASE III (POL III) (in prokaryotes) has at least 10 subunits.
   CORE POLYMERASE - 3 subunits: α, ε, θ: α subunit forms the phosphodiester bonds ($5' \rightarrow 3'$ polymerase); ε subunit has $3' \rightarrow 5'$ exonuclease activity (see below); θ subunit function is unknown.

   SUBUNIT τ forms a dimer of 2 cores.

   β_2-CLAMP (β-PROTEIN or **SLIDING CLAMP PROTEIN**; 2 subunits) – forms a ring around the template DNA strand; promotes **PROCESSIVITY** by preventing the dissociation of the polymerase and the template DNA during replication.

   γ COMPLEX (5 subunits) – transfers the β_2-clamp to the core polymerase;.

   The RNA primers used in DNA replication need to be removed and replaced by DNA. This is where the $5' \rightarrow 3'$ exonuclease activity of **DNA POLYMERASE I (POL I)** comes into play. Pol I degrades RNA primers ahead of it as the enzyme synthesizes DNA. RNA primer sequences are replaced with DNA sequences.

   OTHER DNA POLYMERASES:
   DNA POLYMERASE I (pol I, Kornberg enzyme) – DNA repair enzyme; also removes RNA primer
   DNA POLYMERASE II (pol II) can reinitiate DNA synthesis beyond gaps caused by damaged segments.

   EXONUCLEASE REMOVES NUCLEOTIDE(S) FROM AN END OF A STRAND

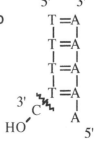

   $3' \rightarrow 5'$ EXONUCLEASE ACTIVITY functions as a "proofreader." If pol III adds an incorrect base to the newly synthesized strand, the $3' \rightarrow 5'$ exonuclease activity removes the mistake and lets the polymerase try again. All three DNA polymerases have $3' \rightarrow 5'$ exonuclease activity.

   $5' \rightarrow 3'$ EXONUCLEASE ACTIVITY (POL I) removes primers or repairs damaged DNA.

Replication ends when replication forks meet at the **TER REGION** (termination site).

4. **DNA LIGASE JOINS DNA FRAGMENTS** by catalyzing phosphodiester bond formation. Discontinuous DNA synthesis of the lagging strand requires DNA ligase to join the newly synthesized fragments.

5. **SUPERCOILING CONTROL: DNA TOPOISOMERASES** prevent the DNA strands from tangling, which would prevent further unwinding of the double helix.

   DNA topoisomerases change the supercoiling of DNA by breaking one or both strands, passing the DNA through the break and rejoining the strands. This functions to relieve torque.

   Controlled supercoiling can facilitate DNA unzipping.

   TYPE I TOPOISOMERASES: make transient single-strand breaks in DNA

   TYPE II TOPOISOMERASES: make transient double-strand breaks

   DNA GYRASE: a prokaryotic type II topoisomerase that helps to separate the replication products and to create the negative supercoils needed for genome packaging

DIFFERENCES BETWEEN PROKARYOTIC AND EUKARYOTIC DNA REPLICATION ...

... appear to be related to the size and complexity of eukaryotic genomes.

TIMING OF REPLICATION: Eukaryotic cells only replicate during a specific period of time during their cell cycle (the S phase). (The eukaryotic cell cycle includes times of rest - G0, G1, G2 - in addition to the S phase and the cell division phase.) In contrast, prokaryotes replicate throughout most of their cell cycle.

REPLICATION RATE: Because of the complex structure of chromatin, DNA replication is significantly slower in eukaryotes (about 50 nucleotides/second per replication fork) than in prokaryotes (about 500).

MULTIPLE REPLICONS are used by eukaryotes to compress the replication of their large genomes into short time periods. Instead of replisomes, eukaryotes have **REPLICATION FACTORIES** – immobilized sites that contain a large number of replication complexes. DNA is threaded through these complexes as it's synthesized.

OKAZAKI FRAGMENTS: Eukaryotic Okazaki fragments are significantly shorter (100-200 nucleotides) than prokaryotic fragments (1000-2000 nucleotides).

EUKARYOTIC REPLICATION ENZYMES

- Five types of eukaryotic DNA polymerase: α, β, δ, ϵ, γ

- **DNA POLYMERASE α:** initiates synthesis of both leading and lagging strands

- **DNA POLYMERASE δ:** two complexes: one synthesizes the leading strand, and one synthesizes the lagging strand; binds to PCNA, a sliding clamp protein that acts like β-protein in E. coli

- **REPLICATION PROTEIN A (RPA)** (acts like SSB): keeps DNA strands separated during DNA synthesis

- **FEN1 (MF1)** - removes RNA primer from each Okazaki fragment; activity is associated with the δ complex.

- **DNA LIGASE** joins Okazaki fragments

- **DNA POLYMERASE β:** involved in DNA repair

- **DNA POLYMERASE ϵ:** $3' \rightarrow 5'$ exonuclease activity; other functions unknown

- **DNA POLYMERASE γ:** catalyzes mitochondrial genome replication

- **EUKARYOTIC TYPE II TOPOISOMERASES** catalyze only the removal of superhelical tension (refer to the description of prokaryotic DNA gyrase, above).

DNA Repair Mechanisms[1]

DNA LIGASE can repair breaks in the phosphodiester linkages.

PHOTOACTIVATION REPAIR (also called **LIGHT-INDUCED REPAIR**) of pyrimidine dimers: DNA photolyase[2] has flavin and pterin chromophores that use energy from visible light to cleave pyrimidine dimers, restoring them to two separate pyrimidines and leaving the phosphodiester bonds intact.

EXCISION REPAIR: Incorrect bases are removed (excised) and replaced with the correct ones. Excision repair involves a series of enzymes.

A **REPAIR ENDONUCLEASE** (also, **EXCISION NUCLEASE** or **EXCINUCLEASE**) detects a distorted DNA segment, then cuts the damaged DNA and removes a single-stranded sequence about 12 nucleotides[3] in length. The enzyme pol I replaces the nucleotides in the gap left by the excised segment, and DNA ligase seals the break in the phosphodiesterase backbone.

RECOMBINATIONAL REPAIR: Post-replication repair of DNA. Damaged DNA interrupts replication; the replication complex detaches from the DNA and re-initiates after the damaged site. This results in a gap in the daughter strand. This gap is repaired by an exchange of the corresponding segment of the homologous DNA (this process is called recombination). After recombination, DNA polymerase and DNA ligase complete the repair process.

DNA RECOMBINATION PRODUCES NEW COMBINATIONS OF GENES (& FRAGMENTS): VARIATIONS THAT MAKE EVOLUTION POSSIBLE

DNA recombination is the rearrangement of DNA sequences by exchanging segments from different molecules. **GENERAL RECOMBINATION, SITE-SPECIFIC RECOMBINATION**

GENERAL RECOMBINATION (requires precise pairing of homologous DNA molecules)

1. Pairing of two homologous DNA molecules

2. Nicking: Two of the DNA strands (one in each molecule) are cleaved.

3. Crossover: The two strand segments cross over, forming a Holliday intermediate

4. Sealing nicks: DNA ligase seals the cut ends

5. Branch migration caused by base pairing exchange leads to the transfer of a segment of DNA from one homologue to the other.

6. A second series of DNA strand cuts occurs.

7. DNA polymerase fills any gaps, and DNA ligase seals the cut strands.

FORMS OF INTERMICROBIAL DNA TRANSFER (IN BACTERIA) THAT INVOLVES GENERAL RECOMBINATION:

1. **TRANSFORMATION** - naked DNA fragments enter through an opening in the cell wall and are introduced into the bacterial genome

---

[1] (At this point, it might be helpful to review the types of DNA mutations that were discussed in Chapter 17. Remember what causes pyrimidine dimers and which bonds needs to be cleaved to separate them? ☺)

[2] DNA photolyase is not present in humans. ☹

[3] 27-29 in eukaryotes

2. **TRANSDUCTION** - bacteriophage inadvertently carry bacterial DNA to a recipient cell; after recombination, the cell uses the transduced DNA

3. **CONJUGATION** - an unconventional sexual mating: A donor cell synthesizes a sex pilus (via a specialized plasmid) that attaches to the surface of the recipient cell. The pilus transfers a fragment of the donor's DNA, which can undergo recombination or exist in plasmid form.

SITE-SPECIFIC RECOMBINATION

Site-specific recombination requires only short segments of homologous DNA [ATTACHMENT (ATT) SITES or INSERTIONAL (IS) ELEMENTS] of DNA homology, and depends more on protein-DNA interactions than on sequence homology. *Example*: Integration of bacteriophage λ into the *E. coli* chromosome (See Figure 18.25, p. 695 of your text.)

TRANSPOSITION – a variation of site-specific recombination – transposable elements (certain DNA sequences) are moved from one chromosome or chromosomal region to another; differs from site-specific recombination in that a specific protein-DNA interaction occurs on only one of the 2 recombining sequences. The recombination of the 2nd DNA sequence is nonspecific.

TRANSPOSONS - transposable elements (JUMPING GENES) that can jump between bacterial chromosomes, plasmids, and viral genomes

IS ELEMENTS - INSERTION ELEMENTS – bacterial transposons that consist only of a gene that codes for a transposase (transposition enzyme) flanked by inverted repeats (short palindromes)

COMPOSITE TRANSPOSONS – bacterial transposons that contain additional genes, several of which may code for antibiotic resistance (!)

1. REPLICATIVE TRANSPOSITION – transposon inserts a replicated copy, stays in its original site

2. NONREPLICATIVE TRANSPOSITION – transposon is cut out of its original (donor) site and inserted into the target site; the donor site must be repaired

EUKARYOTIC TRANSPOSONS: many contain LTR (LONG TERMINAL REPEATS or DELTA REPEATS); many mechanisms involve an RNA intermediate and resemble the replicative phase of a retrovirus

18.2 TRANSCRIPTION: RNA SYNTHESIS

CATALYZED BY **RNA** POLYMERASE: $NTP + (NMP)n \rightarrow (NMP)n{+}1 + PP_i$.

NONTEMPLATE STRAND = plus (+) strand; also called the coding strand because it has the same base sequence as the RNA transcription product (except U substitutes for T)

TEMPLATE STRAND = minus (–) strand

The direction of the gene = the direction of the coding strand. So, polymerization proceeds from the 5'-end to the 3'-end of both the coding strand and the gene.

TRANSCRIPTION IN PROKARYOTES

RNA POLYMERASE IN **E. COLI**: core enzyme catalyzes RNA synthesis; sigma factor binds transiently to the core enzyme, and allows it to bind both the correct template strand and the proper site to initiate transcription
Pribnow box - region 10 nucleotides before the transcription initiation site

STAGES OF TRANSCRIPTION IN *E. COLI*:

1. **INITIATION**: RNA polymerase binds to a promoter (a specific DNA sequence)

   A short DNA segment near the Pribnow box unwinds.

   The first nucleoside triphosphate binds to the RNA polymerase complex, beginning transcription; then attacks the second NTP to form the first phosphodiester bond.

   After the transcribed sequence is about 10 nucleotides long, the conformation of RNA polymerase complex changes.

   The σ factor detaches, RNA polymerase affinity for the promoter site decreases and the   initiation phase ends.

2. **ELONGATION PHASE:** Core RNA polymerase converts to an active transcription complex, and binds several accessory proteins.

   DNA unwinds ahead of the transcription bubble. (Topoisomerases resolve positive and negative supercoiling ahead of and behind the bubble.)

   Elongation continues until a termination sequence is reached.

3. **TERMINATION**: Termination sequences contain palindromes, and their RNA transcripts form a stable hairpin turn.

PRODUCTS OF TRANSCRIPTION:
   mRNA is used immediately for protein synthesis.
   tRNA and rRNA require post-transcriptional processing.

TRANSCRIPTION IN EUKARYOTES: SIGNIFICANTLY MORE COMPLEX THAN IN PROKARYOTES

UNIQUE FEATURES OF EUKARYOTIC TRANSCRIPTION:
1. RNA polymerase activity: requires three nuclear RNA polymerases:
       RNA polymerase I: transcribes large rRNAs
       RNA polymerase II: transcribes precursors of mRNA and most snRNAs
       RNA polymerase III: transcribes precursors of tRNAs and 5S rRNA
   Eukaryotic RNA polymerases can't initiate transcription. Various transcription factors must be bound at the promoter before transcription can begin.

2. **PROMOTERS**: larger, more complicated, and more variable than prokaryotic promoters. Many promoters for RNA polymerase II contains consensus sequences (**TATA BOX**) about 25-30 bp upstream from the initiation site.
   CAAT BOX, GC BOX are examples of sequences upstream that bind transcription factors and affect the frequency of transcription initiation.
   ENHANCERS are regulatory sequences that may be thousands of bp away from the gene, but affect activity of promoters.

3. **POSTTRANSCRIPTIONAL PROCESSING:** Eukaryotic mRNAs are extensively processed (unlike prokaryotic mRNAs, which typically have little to no post-transcriptional processing).

   WHY MODIFY MRNA?   – to help stabilize mRNA
                             – to help mRNA transport out of the nucleus.

279

EUKARYOTIC MRNA MODIFICATIONS:

CAPPING the 5′-phosphate end with a 7-methylguanosine to protect the 5′-end from exonucleases and to promote mRNA translation. (The first two nucleotides of the transcript are methylated at the 2′–OH.)

ATTACHING A POLY A TAIL (a polyadenylate with 100-250 As) to the 3′-end protects from the action of 3′,5′-exonucleases and promotes mRNA export to the cytoplasm

SPLICING - removing INTRONS (DNA sequences that intervene and occur within a particular gene); Each intron is excised as a LARIAT (a configuration that resembles a loop); then the exons are joined.

> EXONS - coding sequences (DNA sequences that designate the amino acid sequence)
>
> Not all of the DNA sequences in eukaryotes code for amino acids. Some DNA sequences serve regulatory or structural roles. So, the removal of introns is necessary to synthesize an mRNA that will be translated into a continuous (and correct) amino acid sequence
>
> SPLICEOSOME – where splicing happens; multi-component structure with several snRNAs, several proteins
>
> RIBOZYME – catalytic RNA that exhibits self-splicing; found in several organisms.

18.3 GENE EXPRESSION

...how cells produce the genes they need, when they need them.

CONSTITUTIVE or HOUSEKEEPING GENES are transcribed routinely because their products are needed for cell function.

INDUCIBLE GENES are expressed (turned on) only under certain circumstances.

OPERONS control inducible genes. Operons are groups of structural and regulatory genes that are linked.

Most gene expression mechanisms involve DNA-protein binding.

DEREPRESSION: Inhibit a repressor to activate a gene. (It's like taking your foot off the brake in a stopped car that's pointed downhill. It might not be the same as hitting the gas, but the car still moves forward!)

GENE EXPRESSION IN PROKARYOTES

REGULATION OF LACTOSE[4] METABOLISM IN *E. COLI* CELLS BY THE LAC OPERON SUMMARY: When lactose is present, β-galactosidase converts a few molecules to allolactose, which turns the lac operon on. The lac operon codes for lactose metabolism enzymes until the lactose supply is consumed. Then, the repressor protein can once again bind to the operator site, and the lac operon turns off.

lac operon	structural genes Z, Y, and A that code for lactose enzymes; plus a control element - a promoter site that contains the CAP site and overlaps the operator site
CAP site	where the CAP protein binds

---

[4] Lactose is the disaccharide galactose β(1,4) glucose (but you knew that!)

operator site	DNA sequence that binds a repressor protein and helps to regulate adjacent genes
repressor gene i	codes for the lac repressor protein when lactose is absent; directly adjacent to lac operon
lac repressor	protein that binds to the operator and prevents binding of RNA polymerase to the promoter (that is, it turns the lac operon off)
allolactose	β-1,6-isomer of lactose, and an inducer. When allolactose binds to the lac repressor, its conformation changes and it dissociates from the operator. When the operator is free of the repressor, the lac operon is turned on, and transcription of the structural genes can begin. In the absence of inducer, the lac operon remains off.

Glucose is a preferred carbon and energy source for *E. coli*. An organism that has both glucose and lactose will use glucose first. Only after the glucose is gone will the lac operon enzymes be synthesized. This is how it happens:

When glucose is depleted (and the cell needs energy), cAMP levels in the cell rise.[5] cAMP binds to CAP (catabolite gene activator protein), which binds to the lac promoter. CAP-promoter binding increases the affinity of RNA polymerase for the lac promoter, thus promoting transcription and activating lactose metabolism.

GENE EXPRESSION IN EUKARYOTES: HOW ARE EUKARYOTIC GENES REGULATED?

GENOMIC CONTROL
MOST COMMON REGULATORY CHANGES:
DNA METHYLATION: *Example*: methylation of cytosines in certain 5′-CG-3′ sequences turns off genes

HISTONE ACETYLATION: acetylation of lysine residues in H3 and H4 reduces their affinity for DNA; histone acetylation promotes genes expression

TRANSCRIPTIONAL CONTROL IS HEAVILY INFLUENCED BY:
CHROMATIN STRUCTURE: heterochromatin (too condensed to do transcription) vs. euchromatin (less condensed; varying levels of transcription activity)

GENE REGULATORY PROTEINS: bind to DNA to activate or repress genes
Mechanisms of gene regulatory proteins:
- competitive DNA binding of transcription factor proteins
- masking the activation surface
- direct interaction with (binding to) transcription factors

GENE REARRANGEMENTS regulate certain genes, and may be involved in cell differentiation. *Example*: rearrangement of antibody genes in B lymphocytes.

GENE AMPLIFICATION by repeated rounds of replication within the amplified region (It's like setting your CD player to automatically repeat.) This occurs when the need for specific gene products is unusually high. For example, during the early developmental stages of fertilized eggs, the huge demand for protein synthesis requires amplification of rRNA genes.

BIOCHEMISTRY IN PERSPECTIVE: CARCINOGENESIS: TERMS: PROTOONCOGENE ; GTPASE-ACTIVATING PROTEIN, GUANINE NUCLEOTIDE EXCHANGE FACTOR; MITOGEN; ONCOGENE; TUMOR PROMOTER

---

[5] cAMP again?!! Food for thought: one way to tie a number of chapters together is to list out the regulatory effects of cAMP in various types of cells and pathways.

ALTERNATIVE RNA PROCESSING TO CONTROL GENE EXPRESSION:
ALTERNATIVE SPLICING - joining of different combinations of exons ultimately results in different proteins. Example: tissue-specific forms of α–tropomyosin, a structural protein produced in various tissues.

ALTERNATIVE SITES TO ATTACH POLY A TAILS

LENGTH OF POLY A TAILS: Longer tails give more stability to mRNA, resulting in increased opportunity for translation.

RNA EDITING: certain bases are chemically modified, deleted, or added. *Example*: Deaminating cytosine to produce uracil changes a CAA codon for glutamine into a UAA codon, which is a stop signal, producing a shorter version of the protein.

mRNA TRANSPORT CONTROL THROUGH NUCLEAR PORE COMPLEXES
NUCLEAR EXPORT SIGNALS: capping, association with or presence of specific proteins, CBP (cap-binding protein)

TRANSLATIONAL CONTROL: ALTERING PROTEIN SYNTHESIS
…allows eukaryotic cells to respond to various stimuli (e.g., heat shock, viral infections, and cell cycle phase changes)

COVALENT MODIFICATION of translation factors (nonribosomal proteins that aid translation) alters translation rate or enhances translation of specific mRNAs.

SIGNAL TRANSDUCTION: RESPONDING TO ENVIRONMENTAL SIGNALS BY ALTERING GENE EXPRESSION; INVOLVES SIGNAL MOLECULES
Gene expression changes are initiated by ligand binding to a receptor (cell surface or intracellular).
Complicating features of intracellular signal transduction mechanisms include:
1. Each type of signal may activate one or more pathways.
2. Signal transduction pathways may converge or diverge.

Checkpoints in cell cycle phases prevent the cell from entering the next phase until conditions are optimal and specific signals are received. The mechanism of progression: alternating synthesis and degradation of cyclins, a group of regulatory proteins that bind to and activate cyclin-dependent protein kinases (Cdks). Cdks phosphorylate a variety of proteins that signal the cell past a checkpoint to the next phase of mitosis.

REGULATION OF CELL DIVISION:
POSITIVE CONTROL: Growth factors bind to specialized cell surface receptors, initiating a cascade of reactions that induces two classes of genes:
> **EARLY RESPONSE GENES** are rapidly activated. *Example*: protooncogenes - normal genes that, if mutated, can promote carcinogenesis
> **DELAYED RESPONSE GENES** are induced by activities of transcription factors and other proteins that were produced or activated during the early response phase. Examples of delayed response gene products: Cdks, cyclins, and other components needed for cell division.

NEGATIVE CONTROL: tumor suppressor genes (*Examples*: Rb gene and p53 gene); Apoptosis (programmed cell death) occurs if too much DNA damage has occurred and/or if DNA repair mechanisms are incomplete.

BIOCHEMISTRY IN THE LAB: GENOMICS KEY TERMS: GENOMICS; FUNCTIONAL GENOMICS, RECOMBINANT DNA TECHNOLOGY, COSMID, ELECTROPORATION, TRANSFECTION, TRANSGENIC ANIMAL VECTOR, BACTERIAL ARTIFICIAL CHROMOSOME, YEAST ARTIFICIAL CHROMOSOME, COLONY HYBRIDIZATION TECHNIQUE, MARKER GENE, POLYMERASE CHAIN REACTION CDNA LIBARY, SHOTGUN CLONING, CHROMOSOMAL JUMPING, CONTIG, DNA MICROARRAY, ANNOTATION, BIOINFORMATICS

CHAPTER 18: SOLUTIONS TO REVIEW QUESTIONS

18.1   a. recombination – the rearrangement of DNA sequences involving the exchange of segments from different molecules

          b. replisome – the protein complex that replicates DNA molecules

          c. replicon – a unit of the genome that contains an origin for initiating replication

          d. retroelement – retrotransposons or retroposons – a DNA sequence that is inserted into a target DNA sequence via a transposition mechanism that involves an RNA transcript

          e. riboswitch – a metabolite-sensing domain and gene expression regulator located in the 5′-untranslated region of mRNAs; acts as a feedback inhibitor to prevent the production of a protein needed to synthesize or import a specific molecule that is already present; conformational changes resulting from binding a small metabolite results in transcription termination, translation prevention, or a self-cleavage reaction

18.2   Negative supercoiling facilitates the unzipping of DNA during the initiation phase of replication.

18.4   Briefly, prokaryotic DNA replication consists of DNA unwinding, RNA primer formation, DNA synthesis catalyzed by DNA polymerase and the joining of Okazaki fragments by DNA ligase. Prokaryotic DNA replication differs from the eukaryotic process in that prokaryotic replication is faster, and in prokaryotes the Okazaki fragments are longer.

18.5   a. spliceosome – a large complex composed of proteins and snRNA in which exons are spliced together during RNA processing

          b. telomerase – a ribonucleoprotein with reverse transcriptase activity, used to extend the 3′ strand of the telomere

          c. transposition – the movement of a piece of DNA from one site in a genome to another

          d. tumor promoter – a molecule that provides cells a growth advantage over nearby cells

          e. inverted repeat – a sequence that is a reversed complement of another sequence that occurs 'downstream'; inverted repeats without intervening sequences constitute a palindrome

18.7

18.8   a. replication licensing factor – RLF – proteins that bind to the ORC and complete the structure of the preRC

        b. primosome – a multienzyme complex involved in the synthesis of the RNA primers at various points along the DNA template strand during *E. coli* DNA replication

        c. processivity – the prevention of frequent dissociation of a polymerase from the DNA template

        d. exonuclease – an enzyme that removes nucleotides from the end of the polynucleotide strand

        e. consensus sequence – the average of several similar sequences; for example, the consensus sequence of the –10 box of *E. coli* promoter is TATAAT

18.10 Viruses can cause mutations that affect the expression of protooncogenes by inserting their genomes into host cell regulatory sequences, thereby inactivating them.

18.11 "Jumping genes" is the popular name for transposons. First discovered by Barbara McClintock, transposons (transposable elements) are DNA sequences that can move around the genome.

18.13 Genetic recombination promotes species diversity. General recombination, a process in which segments of homologous DNA molecules are exchanged, is most commonly observed during meiosis. In site-specific recombination, protein-DNA interactions promote the recombination of nonhomologous DNA. Transposition is an example of site-specific recombination.

18.14 Most mutations are silent. Of those that do affect the functioning of an organism, most are deleterious because of the complex nature of living processes. Change in the properties of any of the thousands of different gene products is potentially disruptive. Only on rare occasions does a mutation improve the viability of an individual organism.

18.16 In replicative transposition, a replicated copy of a transposable element is inserted into a new chromosome location in a process that involves the formation of an intermediate called a cointegrate. In nonreplicative transposition, sequence replication does not occur, that is, the transposable element is spliced out of its donor site and inserted into the target site. The donor site must be repaired.

18.17 In DNA, if cytosine is converted to uracil, which forms a base pair with adenine, an AT base pair is substituted for a GC base pair. Such a change in RNA is not as important because RNA molecules are short lived and disposable. In contrast, because DNA molecules are the cell's permanent repository of genetic information, any change in base sequence may affect an organism's viability.

18.19 a. RNA editing – the alteration of the base sequence in a newly synthesized mRNA molecule; bases may be chemically modified, deleted, or added

b. electroporation – a method of introducing a cloning vector into a host cell that involves treatment with an electrical current

c. annotation – the functional identification of the gene in a genome

d. acceptor site – the 3′ site that bonds to (accepts) the 5′ splice site during the final step of RNA splicing

e. polymerase chain reaction – a laboratory technique used to synthesize large quantities of specific nucleotide sequences from small amounts of DNA using a heat-stable DNA polymerase

18.20 In both cases DNA copies are produced. However, in DNA replication usually only one copy of each DNA molecule is synthesized and several proofreading mechanisms ensure accurate copying. PCR technology is designed to produce multiple copies of a DNA molecule and proofreading is limited to the DNA polymerase that is employed.

18.22 a. DNA ligase – an enzyme that links DNA segments; responsible for linkage of Okazaki fragment during DNA synthesis

b. DNA polymerase III – the major DNA synthesizing enzyme in prokaryotes

c. SSB proteins – DNA binding proteins in prokaryotes that keep DNA strands separated during DNA synthesis

d. primase – an RNA polymerase that synthesizes short RNA segments, called primers, that are required in DNA synthesis

e. helicase – an ATP-requiring enzyme that catalyzes the unwinding of duplex DNA during DNA synthesis

f. RPA – replication protein A; the primary homologue of SSB in eukaryotes

g. Okazaki fragments – short DNA strands that are synthesized during discontinuous replication of the lagging DNA strand as the leading strand is continuously replicated

18.23 a. Transcription factors are proteins that regulate or initiate RNA synthesis by binding to specific DNA sequences called response elements.

b. RNA polymerase is one of a group of enzymes that transcribe a DNA sequence into an RNA product.

c. A promoter is a DNA sequence immediately before a gene that is recognized by RNA polymerase and signals the start point and direction of transcription.

d. A sigma factor is a bacterial protein that facilitates the binding of the core enzyme of RNA polymerase to the initiation site during transcription.

e. An enhancer is a eukaryotic DNA sequence that can increase the expression of a gene.

    f. TATA box is a consensus sequence in eukaryotic DNA that occurs within promoters for RNA polymerase II.

18.25 In relatively simple genomes, such as those in bacteria, operons provide a convenient mechanism for regulating genes. Proteins required in the same metabolic pathway or functional process are synthesized together because their genes are controlled by the same promoter.

18.26 The amplification of a single DNA molecule during 5 cycles yields 2^5 or 32 molecules.

18.28 Mechanisms to regulate gene expression differently in different types of cells include: genomic control (including gene rearrangements and selective gene amplification), transcription initiation control, RNA processing, RNA editing, RNA transport control, translational control, and signal transduction-triggered gene expression. See Solution 18.29 for examples of these mechanisms.

18.29 Examples of genomic control includes gene rearrangements of antibody genes in B lymphocytes (see Figure 18.48) and the selective gene amplification of rRNA genes in oocytes. An example of RNA processing is the alternative splicing of the vertebrate tropomyosin mRNA gene, in which different combinations of its 13-15 exons result in protein isoforms that serve the needs of different cells types, namely, striated muscle, smooth muscle, fibroblast, and brain cells. Examples of RNA editing are found in intestinal cells and some brain neurons. Intestinal cell apolipoprotein B-100 mRNA undergoes a C→U conversion, resulting in apolipoprotein B-48, a truncated version produced for chylomicron particles.

18.31 The purpose of gene amplification is to rapidly produce multiple copies of specific gene products that are required in greater quantities during certain stages in a cell's development.

18.32 Gene amplification occurs via repeated rounds of replication within the amplified region.

18.34 RNA molecules are more reactive than DNA because of the presence of the 2′-OH group of ribose. In addition the complexity of its three-dimensional structures that result from single-stranded RNA coiling back on itself provides more opportunities for more diverse intermolecular interactions and binding than does DNA.

18.35 a. DnaA box – highly conserved 9-bp sites on DNA that bind to DnaA proteins to begin replication

    b. MCM complex – a protein complex considered to be the major DNA helicase in eukaryotes and a component of the preRC

    c. contig – one of a set of overlapping DNA sequences used to identify the base sequence of a region of DNA

    d. shotgun method – shotgun cloning – a cloning technique in which genomic libraries are created by the random digestion of a genome

    e. RFC – replication factor C – a clamp loader protein that controls the attachment of DNA polymerase δ to each DNA strand

18.37 The function of telomere end-binding proteins is to bind to GT-rich telomere sequences as part of the process that sequesters and stabilizes telomeres.

18.38 The function of telomere repeat-binding proteins is to secure the 3′ overhang as part of the process that sequesters and stabilizes telomeres.

CHAPTER 18: SOLUTIONS TO THOUGHT QUESTIONS

18.40 In the Meselson-Stahl experiment **all** of the nitrogen was ^{15}N. To accomplish the same effect with a carbon isotope all of the carbon in the medium would have to be isotopically pure in both phases of the experiment. This would be prohibitively expensive.

18.41 DNA replication time is calculated as follows:

$$\frac{150,000,000 \text{ base pairs}}{50 \text{ bases/s}} = 3 \times 10^6 \text{ s} = 34.5 \text{ days}$$

Consequently, approximately one month is required for this DNA replication. Eukaryotic DNA synthesis is significantly faster than expected because each chromosome contains multiple replication units (replicons).

18.43 Mustard gas cross links the strands in DNA with permanent covalent bonds.

18.44 The increase in pathogenicity of the streptococcus organism may result from the incorporation of a segment of the virus genome into the streptococcus genome. The genome of the current virulent streptococcus and the DNA of preserved specimens can be compared with the use of a probe for the viral toxin gene. The presence or absence of the viral sequence in modern streptococcus can be determined.

18.46 Recall that phorbol esters mimic the action of DAG, the normal cell metabolite that activates protein kinase C (PKC). PKC initiates a phosphorylation cascade that results in the activation of numerous molecules involved in cell growth and division, including jun and fos, which then combine to form AP-1. AP-1 is a transcription factor whose presence promotes cell division. Its formation causes an affected cell to have a growth advantage over nearby cells. Because phorbol esters are tumor promoters any exposure to them increases the risk that initiated cells may progress toward a cancerous state.

18.47 Antibiotic resistance arises because the overuse of antibiotics acts as a selection pressure, i.e., they provide a growth advantage for disease-causing organisms that possess resistant genes. So-called superbugs are organisms that are resistant to several types of antibiotics because they possess plasmids containing several resistant genes. If the circumstances that cause antibiotic resistance continue, antibiotics may eventually become ineffective against most infectious diseases.

18.49 Errors that occur during DNA replication have the potential to become permanent if repair processes fail. Errors made during transcription affect only a few RNA molecules and are temporary.

18.50 Gene amplification, the selective duplication of certain genes, can occur via a reverse transcriptase–mediated event. The creation of one or more cDNAs from an mRNA is followed by insertion of these sequences into the genome.

18.52 The DNA polymerase enzyme complex is a sophisticated machine that is composed of a large number of fragile components. By keeping the complex stationary, cells protect it from mechanical damage that would occur if it moved through the crowded nucleoplasm.

18.53 The riboswitch must have a binding site for the metabolite. This binding usually triggers a conformational change that blocks translation. So the riboswitch must also have a sequence that changes shape.

19 Protein Synthesis

Brief Outline of Key Terms and Concepts

PROTEIN SYNTHESIS
　TRANSLATION of nucleotide base sequences into the amino acid sequence of polypeptides.
　POSTTRANSLATIONAL MODIFICATION
　TARGETING

19.1 THE GENETIC CODE

THE GENETIC CODE
– a mechanism by which ribosomes translate nucleotide base sequences into the primary sequence of polypeptides.
– consists of 64 codons: 61 codons that specify the amino acids and 3 STOP CODONS.
CODON; READING FRAME; OPEN READING FRAME

CODON-ANTICODON INTERACTIONS
During TRANSLATION, the genetic code is translated through base-pairing interactions between **mRNA** CODONS and tRNA ANTICODONS. The WOBBLE HYPOTHESIS explains why cells usually have fewer tRNAs than expected.

AMINOACYL-TRNA SYNTHETASE REACTION
– links a tRNA to its amino acid, using 2 ATP equivalents. ACTIVATION; MIXED ANHYDRIDE; tRNA LINKAGE; COGNATE TRNA; PROOFREADING SITE

19.2 PROTEIN SYNTHESIS

TRANSLATION consists of three phases: INITIATION, ELONGATION, and TERMINATION. Each phase requires several types of PROTEIN FACTOR.
　INITIATION: INITIATOR TRNA; POLYSOME
　ELONGATION: TRANSPEPTIDATION, TRANSLOCATION
　TERMINATION: RELEASING FACTOR
POSTTRANSLATIONAL MODIFICATIONS
TARGETING

PROKARYOTIC PROTEIN SYNTHESIS is a rapid process involving several protein factors.
Most prokaryotic gene expression appears to be regulated by transcription initiation.
INITIATION
　INITIATION COMPLEX ; SHINE-DALGARNO SEQUENCE
ELONGATION
　GUANINE NUCLEOTIDE EXCHANGE FACTOR (GEF); EF-TU PEPTIDYL TRANSFERASE CENTER, PRETRANSLOCATION SITE; POSTTRANSLOCATION STATE
TERMINATION
　RELEASING FACTORS, RIBOSOME RECYCLING FACTOR
POSTTRANSLATIONAL MODIFICATIONS prepare the polypeptide for its function, assist in folding, or target it to a specific destination, and include:
PROTEOLYTIC PROCESSING, CONJUGATION, METHYLATION,

PHOSPHORYLATION, and insertion of cofactors.
　SIGNAL PEPTIDES
TRANSLATIONAL CONTROL MECHANISMS
Prokaryotes and eukaryotes differ in their usage of translational control mechanisms. Prokaryotes:
• vary the SHINE-DALGARNO sequences; and
• use NEGATIVE TRANSLATIONAL CONTROL, i.e., the translation of a POLYCISTRONIC mRNA is repressed by one of its products.
In contrast, eukaryotes exhibit a wide range of eukaryotic translational control mechanisms, from
• global controls in which the translation rate of a large number of mRNAs is altered, to
• specific controls that alter the translation of a specific mRNA or small group of mRNAs.

EUKARYOTIC PROTEIN SYNTHESIS

COMPARED TO PROKARYOTIC PROTEIN SYNTHESIS, EUKARYOTIC PROTEIN SYNTHESIS IS SLOWER, MORE COMPLEX
• is slower, due in part to **mRNA** SECONDARY STRUCTURE, **mRNA** SCANNING;
• uses more protein factors (translation factors) at each step;
• has a more complex initiation mechanism;
　43S PREINITIATION COMPLEX ; CAP-BINDING COMPLEX (CBC); GUANINE NUCLEOTIDE ACTIVATION FACTOR; POLY(A) BINDING PROTEIN (PABP)
　includes a cap-binding protein;
• forms CIRCULAR mRNA POLYSOMES;
• KINETIC PROOFREADING
• has posttranslational processing and targeting mechanisms that are much more complicated.

POSTTRANSLATIONAL MODIFICATIONS (EUKARYOTES)
　PROTEOLYTIC CLEAVAGE (PROPROTEIN; PREPROPROTEIN); DISULFIDE BOND FORMATION, DISULFIDE EXCHANGE; PROTEIN SPLICING (INTEIN, EXTEIN); GLYCOSYLATION; LIPOPHILIC MODIFICATIONS; HYDROXYLATION; METHYLATION; CARBOXYLATION; PHOSPHORYLATION
TARGETING (EUKARYOTES)
　TRANSCRIPT LOCALIZATION;
　SIGNAL HYPOTHESIS , SIGNAL RECOGNITION PARTICLE; DOCKING PROTEIN
　COTRANSLATIONAL TRANSFER, TRANSLOCON, POSTTRANSLATIONAL TRANSLOCATION
TRANSLATION CONTROL MECHANISMS (EUKARYOTES)
　mRNA EXPORT; mRNA STABILITY; NEGATIVE TRANSLATIONAL CONTROL; INITIATION FACTOR PHOSPHORYLATION; TRANSLATIONAL FRAMESHIFTING
CONTEXT-DEPENDENT CODING REASSIGNMENT
PROTEOMICS: Goals, Tools

OVERVIEW

Translation of the nucleotide-base-sequence code results in a sequence of amino acids that form a polypeptide. Protein synthesis also includes posttranslational modification and targeting.

19.1 THE GENETIC CODE

GENETIC CODE   a coding dictionary that specifies a meaning for each base sequence

CODON   a three-base-sequence (triplet) in mRNA that codes for each amino acid. Example: "CCC" codes for proline.

Of the 64 triplets that are possible, 61 code for amino acids and 3 are **STOP SIGNALS** that terminate the growing polypeptide chain. (Stop codons: UAA, UGA, UAG)

PROPERTIES OF THE GENETIC CODE:

1. **SPECIFIC:** Each codon signals a specific amino acid. *Exception*: AUG codes for both Met and a start signal (initiating codon). Also, many of the codons that code for a specific amino acid have similar sequences. *Example*: UC? codes for Ser, whether the third base is U, C, A, or G.

2. **DEGENERATE:** many codons have the same meaning; that is, most amino acids are specified by more than one codon. (Only Met and Trp have one codon.)

3. **NONOVERLAPPING AND WITHOUT PUNCTUATION:** The code reads from the initiating codon (AUG) straight through to a stop codon, without repeating or skipping any bases.

READING FRAME: set of side-by-side codons in mRNA

OPEN READING FRAME: series of codons without a stop codon

Why are reading frames important? Since the codons are sequential and nonoverlapping, the actual amino acid sequence depends on the starting point for the first triplet. Each different starting point defines a unique potential protein. Changing the starting point changes the reading frame, which changes the final product – the amino acid sequence. Check out the effects of changing the reading frame for the following sequence:

$$5'-\text{A G G C A G A A C U A A C C A G G U C U A} -3'$$

| Frame 1: | AGG CAG AAC UAA CCA GGU CUA |
| | Arg Gln Asn Stop |

| Frame 2: | A GGC AGA ACU AAC CAG GUC UA |
| | Gly Arg Thr Ile Gln Val |

| Frame 3: | AG GCA GAA CUA ACC AGG UCU A |
| | Ala Glu Leu Thr Arg Ser |

4. **ALMOST UNIVERSAL:** Most codons have the same meaning in different species. (Exceptions: mitochondrial DNA, protozoa, yeast have some deviations from the general genetic code.)

CODON-ANTICODON INTERACTIONS

The genetic code is translated through base-pairing interactions between **mRNA CODONS** and **tRNA ANTICODONS**. The **WOBBLE HYPOTHESIS** explains why cells usually have fewer tRNAs than expected.

ANTICODON – a three-base sequence on tRNA that base-pairs with its complementary codon on mRNA.

Codon-anticodon base-pairings are **ANTIPARALLEL.**

tRNA has an anticodon on one side and a specific amino acid on the other.

> **EXAMPLE:** What is the anti-codon for the mRNA codon AGG?
>
> mRNA codon $5'$- A G G -$3'$
> base-pairs with
> tRNA anti-codon $3'$- U C C -$5'$
>
> Remember that base sequences are always written in the $5' \rightarrow 3'$ direction, so the final answer is: the tRNA anti-codon is $5'$-CCU-$3'$.

Careful (exam alert!):

Again, codon-anticodon base-pairings are antiparallel, but the base sequences are always written in the $5' \rightarrow 3'$ direction.*(...and oh yes, you can be sure that the incorrect $3' \rightarrow 5'$ order will be one of the options in a multiple-choice question!)*

THE WOBBLE HYPOTHESIS

One tRNA may recognize more than one codon, i.e., a tRNA can have multiple codon-anticodon interactions.

In a codon-anticodon interaction,

- the first two base pairings confer most of the specificity needed for protein synthesis.
- interactions between the third codon and anticodon nucleotides are not as strict, and non-traditional base pairs can occur.

That third anticodon nucleotide (in the $5'$-position) corresponding to the third codon nucleotide in the mRNA sequence (in the $3'$-position) is the "wobble" position.

The wobble hypothesis explains why cells usually have fewer tRNAs than expected. As few as 32 tRNAs (31 + a tRNA for initiating protein synthesis) are needed to translate all 61 codons.

AMINOACYL-TRNA SYNTHETASE REACTION:

amino acid + ATP + tRNA $\rightarrow$ aminoacyl-tRNA + AMP + PP_i

Aminoacyl-tRNA synthetases catalyze attachment of amino acids to tRNAs in two steps: **AMINO ACID ACTIVATION** and **tRNA LINKAGE**. This reaction is irreversible due to the hydrolysis of PP_i.

At least one aminoacyl-tRNA synthetase exists for each amino acid.

Many synthetases have a proofreading site to correct mistakes. Example: If isoleucyl-tRNA^{Ile} synthetase makes Val–tRNA, its proofreading site will fit the Val end (but not Ile) and hydrolyze the incorrect bond.

This reaction consumes *two ATP equivalents*, even though only one ATP reacts. This is because AMP (not ADP) is produced. AMP needs *two* phosphoryl groups to turn it back into an ATP. [Recall that AMP + ATP $\rightarrow$ 2 ADP]

LINKING AN AMINO ACID TO THE 3'-TERMINUS OF THE CORRECT tRNA:

1. ACTIVATION OF THE AMINO ACID: Amino acid + ATP $\rightarrow$ Aminoacyl-AMP + PP$_i$

2. tRNA LINKAGE: Aminoacyl-AMP + tRNA $\rightarrow$ Aminoacyl-tRNA + AMP

Linkage always happens at the 3' position on the ribose. The aminoacyl group can move between the 2' and 3' positions, but only the 3'-aminoacyl esters are used in protein synthesis.

19.2 PROTEIN SYNTHESIS: TRANSLATION

POLYSOME – an mRNA with several ribosomes bound to it. One mRNA can be read at the same time by several ribosomes.

P-SITE, A-SITE – two ribosomal sites for codon-anticodon interactions
   (P = peptidyl) (A = acyl)

Polypeptide synthesis
   – proceeds from the N-terminal to the C-terminal because the code is read in the 5'→3' direction.
   – requires GTP (as an energy source) plus protein factors.

PHASES OF TRANSLATION: INITIATION, ELONGATION, AND TERMINATION

INITIATION:
1. The **SMALL RIBOSOMAL SUBUNIT** binds an mRNA.
2. **INITIATOR tRNA** base-pairs with the initiation codon AUG on the mRNA.
3. The **LARGE RIBOSOMAL SUBUNIT** combines with the small subunit; the initiator tRNA is bound to the P-site.

ELONGATION:

1. A second aminoacyl-tRNA base-pairs to the A-site.
2. Peptidyl transferase catalyzes peptide bond formation (transpeptidation) – α-amino group of the A-site amino acid attacks the C=O of the P-site amino acid. Now both amino acids are on the A-site tRNA.
3. The P-site tRNA leaves.
4. **TRANSLOCATION** – the ribosome moves along the mRNA. The tRNA with the growing peptide chain moves to the P-site, and the next codon enters the A-site.

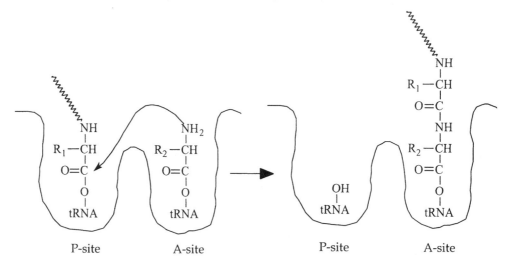

TERMINATION:

1. Stop codon enters the A-site.
2. **PROTEIN RELEASING FACTOR** binds to the A-site.
3. **PEPTIDYL TRANSFERASE** hydrolyzes the polypeptide–tRNA ester bond, releasing the polypeptide chain.
4. The ribosome releases mRNA and dissociates into its subunits.

POSTTRANSLATIONAL MODIFICATIONS – PURPOSES:

- – to prepare a polypeptide for its specific function
- – **TARGETING** – to direct a polypeptide to a specific location

PROKARYOTIC PROTEIN SYNTHESIS

PROKARYOTIC RIBOSOMES are 70S in size and composed of:
50S (large) subunit (that contains catalytic site for peptide bond formation)
30S (small) subunit (that serves as a guide for regulatory translation factors)

IF = INITIATION FACTOR   EF = ELONGATION FACTOR   RF = RELEASE FACTOR

The **SHINE-DALGARNO SEQUENCE** on the mRNA is located upstream from the initiation codon AUG. It binds to a complementary sequence on the small ribosomal subunit, and distinguishes between AUG as a start codon vs. AUG as a methionine codon. Each gene has its own start codon and its own Shine-Dalgarno sequence.

The initiating tRNA is *N*-formylmethionine-tRNA or fmet-tRNA^{fmet}. The formyl group also helps to distinguish the start amino acid from an internal methionine.

N-formylmethionine is synthesized on the charged tRNA.

Met → met-tRNA^{fmet} → *N*-formylmethionine tRNA (fmet-tRNA^{fmet})

INITIATION

1. IF-1 and IF-3 bind to the 30S subunit
    IF-1 binds to the A-site and blocks it.
    IF-3 prevents the 30S subunit from binding to the 50S subunit too soon.

2. Shine-Dalgarno sequence base-pairs to the 30S subunit, guiding the mRNA which also binds to the 30S subunit.

3. IF-2 (with a bound GTP) binds to the 30S subunit, promotes binding of the initiating tRNA, fmet-tRNA^{fmet}.

4. fmet-tRNA^{fmet} binds to the initiating codon (AUG) on mRNA.

5. GTP hydrolyzes to GDP + PP$_i$, and the large and small subunits join.

ELONGATION

1. The aminoacyl-tRNA needs to enter the empty A-site of the ribosome complex. An aminoacyl-tRNA binds *first* to EF-Tu-GTP, *then* to the A-site; GTP hydrolysis releases EF-Tu-GDP (and a P$_i$) from the ribosome.

    EF-Ts regenerates EF-Tu-GTP:

    EF-Ts + EF-Tu-GDP → EF-Ts/EF-Tu-GDP → EF-Ts/EF-Tu + GDP

    EF-Ts/EF-Tu + GTP → EF-Ts/EF-Tu-GTP → EF-Ts + EF-Tu-GTP

2. **PEPTIDYL TRANSFERASE** catalyzes peptide bond formation. Now both amino acids are on the A-site tRNA, and the tRNA on the P-site leaves.

3. **TRANSLOCATION**: The polypeptide-tRNA moves from the A-site to the P-site, and the ribosome moves so that the next codon on mRNA is in the A-site. This stage requires EF-G and GTP hydrolysis.

The elongation process repeats until the ribosome reaches a stop codon.

TERMINATION REQUIRES RELEASING FACTORS RF-1, RF-2, AND RF-3

There are no tRNAs for stop codons (UAA, UGA, UAG). Instead, **RELEASING FACTORS** (RFs) are used. RF-1 recognizes UAA and UAG, and RF-2 recognizes UAA and UGA. (RF-3 may promote RF-1 and RF-2 binding.) A GTP is hydrolyzed, and peptidyl transferase hydrolyzes the polypeptide-tRNA bond, releasing the polypeptide. The mRNA and tRNA dissociates from the ribosome, which separates into its subunits.

POSTTRANSLATIONAL MODIFICATIONS

1. **PROTEOLYTIC PROCESSING:** Cleavage reactions include the removal of formylmethionine and signal peptide sequences. **SIGNAL PEPTIDES** (leader peptides) are short peptides, typically near the N-terminus, that determine a polypeptide's destination.
2. **CONJUGATION** typically forms lipoproteins. Some instances of glycosylation (covalently linking carbohydrates to proteins) are also known.
3. **METHYLATION:** Protein methyltransferases use SAM to add methyl groups (reversibly) to amino acid residues of proteins. (Protein methylation plays a role in signal transduction.)
4. **PHOSPHORYLATION/DEPHOSPHORYLATION** are catalyzed by protein kinases and phosphatases; used in chemotaxis and the regulation of nitrogen metabolism.

TRANSLATIONAL CONTROL MECHANISMS IN PROKARYOTES

- Control of the rate of *transcription* initiation is the predominant method of translational control.
- Differences in Shine-Dalgarno sequences may cause different translation rates.
- Negative translational regulation (e.g., some ribosomal proteins inhibit the translation of their own or related operons).

EUKARYOTIC PROTEIN SYNTHESIS

EUKARYOTIC RIBOSOMES are 80S in size, with a 40S small subunit and a 60S large subunit. The number and size of rRNAs and ribosomal proteins also differ.

RATE OF SYNTHESIS = ~50 amino acids per minute = *much* slower than prokaryotes (prokaryotic translation rate = ~ 1,200 amino acids per minute

eIF = eukaryotic Initiating Factor *(at least 12)*
eEF = eukaryotic Elongation Factor
eRF = eukaryotic Releasing Factor

Eukaryotic initiation is more complex than prokaryotic initiation, in part because of:

1. **mRNA SECONDARY STRUCTURE:** Eukaryotic mRNA has a cap and a poly(A) tail, and its introns have been removed. Also, mRNA is made inside the nucleus, but the ribosome is outside the nucleus; mRNA might interact with cellular proteins before it gets there. (**RIBONUCLEOPROTEIN PARTICLE** = mRNA + protein)
2. **mRNA SCANNING:** Ribosomes bind to the capped 5′-end and move towards the 3′-end, searching for a translation start site (because there are no Shine-Dalgarno sequences in eukaryotic mRNA).

Eukaryotic elongation and termination are similar to those in prokaryotes.

FEATURES OF EUKARYOTIC TRANSLATION

INITIATION (refer to Figure 19.12, p. 756 of your text.)

1. Assembly of the **43S preinitiation complex**:

   small (40S) subunit | eIF-3 | eIF-1A | eIF-2-GTP | met-tRNA$_i$

   eIF-3 helps to prevent the 40S subunit from binding to the large (60S) subunit during this phase of initiation.

   eIF-2 is a GTP-binding protein[1] that mediates met-tRNA$_i$ binding to the 40S subunit.

   met-tRNA$_i$ is an initiating species of methionyl-tRNA^{met}.

---

[1] eIF-2B, a GEF, regenerates eIF-2-GTP from eIF-2-GDP when it releases its GDP.

2. At the same time, the **CAP-BINDING COMPLEX (CBC)** binds to the mRNA 5'-cap structure.

	CBC		
eIF-4A (helicase)	eIF-4E (translation initiation factor)	eIF-4G (scaffold protein)	mRNA 5'-cap

3. The helicase eIF-4B removes any secondary structures in the 5'-UTR that could interfere with initiation.

4. The 43S preinitiation complex binds to the mRNA (at the 5'-end).

5. eIF-G and a **poly(A)-binding protein (PABP)** interact and bring the 3'-poly(A) tail of the mRNA close to the 5'-capped end, causing mRNA to become circular.

6. The initiation complex begins to scan for the start codon, 5'-AUG-3'.

7. When the start codon is recognized, the initiation complex binds the 60S subunit (now dissociated from eIF-6), which triggers the release of the initiation factors eIF-2, eIF-3, eIF-4B, and eIF-1A. An eIF-5 subunit acts as a **GUANINE NUCLEOTIDE ACTIVATING PROTEIN (GNAP)** to hydrolyze the GTP bound to eIF-2; this hydrolysis is involved in the formation of the complete ribosome (Figure 19.13 in your text), the **80S COMPLEX**.

ELONGATION is very similar to prokaryotic elongation.

eEF-1α is analogous to EF-Tu. eEF-1β and eEF-1γ work together to regenerate eEF-1α-GTP from eEF-1α-GDP. (This is similar to the action of EF-Ts). In translocation, eEF-2 is analogous to EF-G.

KINETIC PROOFREADING: if incorrect pairing occurs between the A-site and the eEF-1α-GTP-aminoacyl-tRNA, the complex leaves the A-site.

TERMINATION is also very similar to prokaryotic termination.

GTP binds to eRF-3, which then forms a complex with eRF-1. This complex binds in the A-site when a stop codon (UAG, UGA, or UAA) enters.

GTP hydrolysis promotes dissociation of the eRFs from the ribosome; and peptidyl transferase hydrolyzes the polypeptide-tRNA bond, releasing the polypeptide. The mRNA and tRNA dissociates from the ribosome, which separates into its subunits.

EUKARYOTIC POSTTRANSLATIONAL MODIFICATIONS INCREASE THE STRUCTURAL AND FUNCTIONAL DIVERSITY OF PROTEINS.

- **PROTEOLYTIC CLEAVAGE:** hydrolysis of specific peptide bonds by proteases

  Purpose: a common regulatory mechanism

  Examples: removing N-terminal Met and signal peptides; conversion of inactive **PROPROTEINS** to their active forms (proinsulin → insulin; zymogens or proenzymes → enzymes) (**PREPROPROTEIN** = proprotein + signal peptide)

- **GLYCOSYLATION** adds sugar groups (secreted proteins have complex oligosaccharides; ER membrane proteins have high mannose species).

- **HYDROXYLATION** of the amino acids proline and lysine by prolyl-4-hydroxylase, prolyl-3-hydroxylase, and lysyl hydroxylase in the **RER**; substrate requirements are highly specific. For example, hydroxylated proline and lysine are required for the structural integrity of collagen and elastin.

- **PHOSPHORYLATION** for metabolic control and signal transduction

- **LIPOPHILIC MODIFICATIONS** covalently attach lipid moieties to proteins to enhance membrane binding capacity and/or certain protein-protein interactions. Most common are acylation (attaches a fatty acid) and prenylation.

- **METHYLATION** by methyltransferases may alter the cellular roles of certain proteins. Methylation of altered Asp residues promotes either the repair or the degradation of damaged proteins. Other examples of amino acid methylation are Lys, His, Arg.

- **CARBOXYLATION** of glutamyl residues forms γ-carboxyglutamyl residues, which increases a protein's sensitivity to Ca^{2+}-dependent modulation. This carboxylation requires vitamin K and NADPH.

- **DISULFIDE BOND (S–S) FORMATION:** in secreted proteins and certain membrane proteins, because reducing agents (like GSH) in the cytoplasm will reduce –S–S– bonds to –SH + HS–.

  DISULFIDE EXCHANGE – disulfide bonds rapidly migrate from one position to another until the most stable structure is achieved. (S–S bonds are not necessarily formed sequentially.)

- **PROTEIN SPLICING:** An **INTEIN** (internal section of a protein) is cut out of a protein, and the two flanking sections – the **EXTEINS** – are spliced together. Protein splicing is self-catalyzed, i.e., it requires no other enzymes, cofactors, or energy sources.

TARGETING DIRECTS THE PROTEIN TO ITS PROPER DESTINATION

TARGETING MECHANISMS

TRANSCRIPT LOCALIZATION: A specific mRNA binds to receptors in certain cytoplasmic locations. Translation of this localized mRNA results in a cytoplasmic protein gradient, that is, an asymmetrical distribution of this protein in the cytoplasm.

SIGNAL PEPTIDES are sorting signals that target polypeptides to their proper location (for secretion, or for use in the plasma membrane or any of the membranous organelles). Signal peptides help to insert the polypeptide that contains it into an appropriate membrane.

THE SIGNAL HYPOTHESIS

– was proposed to explain how polypeptides translocate across RER membrane.

SIGNIFICANCE OF THE SIGNAL HYPOTHESIS: It helps to explain the ability of proteins to be specifically targeted to their proper location. The fate of a targeted polypeptide depends on the location of the **SIGNAL PEPTIDE** and other signal sequences.

SIGNAL RECOGNITION PARTICLE (SRP) binds to a ribosome and interrupts translation. The SRP then mediates binding of the ribosome to RER via a **DOCKING PROTEIN (SRP RECEPTOR PROTEIN)**. Translation restarts, the growing polypeptide inserts into the membrane, and SRP is released.

TRANSLOCON – an integral membrane protein complex believed to mediate polypeptide translocation after SRP is released.

COTRANSLATIONAL TRANSFER – simultaneous translocation of a polypeptide during ongoing protein synthesis

POSTTRANSLATIONAL TRANSLOCATION – previously-synthesized polypeptides are pulled across the RER membrane by an ATP-binding peripheral translocon-associated protein (hsp70).

After a protein is in the RER, it typically undergoes initial posttranslational modifications, and is transferred to the Golgi complex via transport vesicles that bud off from the ER and fuse with the *cis* face of the Golgi membrane. Further modifications are made inside the Golgi complex. Transport vesicles exit from the *trans* face of the Golgi and move to target locations.

TRANSLATION CONTROL MECHANISMS IN EUKARYOTES

mRNA EXPORT

The spatial separation of transcription and translation that is afforded by the nuclear membrane appears to provide eukaryotes with the opportunity to control translation. Export through the nuclear pore complex is known to be a carefully controlled, energy-driven process whose minimum requirements include a 5'-cap and a 3'-poly(A) tail.

mRNA STABILITY = HOW WELL mRNA CAN AVOID DEGRADATION BY NUCLEASES

In general, the translation rate of any mRNA species is related to its abundance, which in turn is dependent on both its rates of synthesis and degradation. The length of the poly(A) tail is significant and affects its stability.

NEGATIVE TRANSLATIONAL CONTROL

The translation of some mRNAs is known to be controlled by the binding of repressor proteins to the 5'-ends of the mRNA. This effectively blocks ribosome binding and scanning.

INITIATION FACTOR PHOSPHORYLATION

The phosphorylation of eIF-2 in response to certain stimuli (e.g., heat shock, viral infections, and growth factor deprivation) has been observed to decrease protein synthesis. However, the translation of certain mRNA increases (e.g., hsp synthesis in response to heat shock).

TRANSLATIONAL FRAMESHIFTING

This process, often observed in retroviruses-infected cells, allows the synthesis of more than one polypeptide from a single mRNA.

BIOCHEMISTRY IN PERSPECTIVE:
EF-Tu: A MOTOR PROTEIN
CONTEXT-DEPENDENT CODING REASSIGNMENT
SELENOCYSTEINE; SECIS ELEMENT; PYRROLYSINE

BIOCHEMISTRY IN THE LAB: PROTEOMICS
GOALS OF PROTEOMICS:
- to study the global changes in the expression of cellular proteins over time.
- to determine the identity and the functions of all the proteins produced by organisms.

TOOLS OF PROTEOMICS include **TWO-DIMENSIONAL GEL ELECTROPHORESIS** and **MASS SPECTROMETRY.**

CHAPTER 19: SOLUTIONS TO REVIEW QUESTIONS

19.1   The genetic code is degenerate (several codons have the same meaning), specific (each codon specifies only one amino acid), and universal (with a few exceptions each codon always specifies the same amino acid). In addition, the genetic code is nonoverlapping and without punctuation (i.e., mRNA is read as a continuous coding sequence).

19.2   The observations upon which the wobble hypothesis is based are: (1) the first two base pairings in a codon-anticodon interaction confer most of the specificity required during translation, and (2) the interactions between the third codon and anticodon nucleotides are less stringent. Because of the "wobble rules," only a minimum of 31 tRNAs are required for the translation of all 61 codons.

19.4   a.   protein targeting – a series of mechanisms that directs newly synthesized polypeptides to their correct cellular locations.

b.   scanning – a mechanism that eukaryotic ribosomes use to locate a translation start site on an mRNA.

c.   codon – an mRNA triplet base sequence that specifies the incorporation of a specific amino acid into a growing polypeptide chain during translation or acts as a start or stop signal.

d.   reading frame – a set of contiguous triplet codons.

e.   disulfide exchange – a mechanism that facilitates the formation of disulfide bridges in newly synthesized proteins.

19.5   The major differences between prokaryotic and eukaryotic translation are speed (the prokaryotic process is significantly faster), location (the eukaryotic process is not directly coupled to transcription as prokaryotic translation is), complexity (because of their complex life styles, eukaryotes possess complex mechanisms for regulatory protein synthesis, e.g., eukaryotic translation involves a significantly larger number of protein factors than prokaryotic translation), and posttranslational modifications (eukaryotic reactions appear to be considerably more complex and varied than those observed in prokaryotes).

19.7   During the elongation phase of protein synthesis, the second aminoacyl-RNA becomes bound to the ribosome in the A site. Peptide bond formation is then catalyzed by peptidyl transferase. Subsequently, the ribosome is moved along the mRNA by a mechanism referred to as translocation.

19.8   The major mechanisms used by eukaryotes to control translation are mRNA export (transcription and translation are spatially separated; the export of processed mRNAs can be selectively blocked), mRNA stability (mRNAs have several destabilizing sequences that affect the molecule's longevity, i.e., its susceptibility to nucleases), negative translational control (the translation of some RNAs is specifically blocked by the binding of repression proteins near their 5'-ends), initiation factor phosphorylation (the phosphorylation of eIF-2 increases translation rate of certain mRNAs), and translocational frameshifting (certain RNAs have structural information that if activated results in a +1 or –1 change in reading frame; this allows for more than one polypeptide from a single mRNA).

19.10   A preproprotein is the inactive precursor of a protein with a removable signal peptide. A proprotein is an inactive precursor protein. A protein is a fully functional product of translation.

19.11   Kinetic proofreading is a mechanism that ensures that the correct codon-anticodon pairing occurs in the A site of ribosomes. In eukaryotes eEF-1a mediates the binding of aminoacyl-tRNAs to the A site. When the correct pairing occurs eEF-1a hydrolyses its bound GTP and subsequently exits the ribosome. If correct pairing does not occur the eEF-1a-GTP-aminoacyl complex leaves the A site, thereby preventing the incorporation of incorrect amino acids.

19.13 a. wobble – the first (5′) anticodon base position that can form nontraditional base pairs with the third base of the codon, allowing some tRNAs to pair with several codons; this wobble hypothesis explains why cells often have fewer tRNAs than expected

b. anhydride – the product of a condensation reaction between two carboxyl groups or two phosphate groups in which a molecule of water is eliminated

c. aminoacyl-tRNA synthetase – an enzyme that catalyzes the attachment of an amino acid to its cognate tRNA

d. cognate tRNA – the tRNA that associates with a specific amino acid

e. polysome – an mRNA with several ribosomes bound to it

19.14 Cotranslational transfer is a process in which nascent polypeptides are inserted through an intracellular membrane during ongoing protein synthesis. An integral membrane protein complex referred to as the translocon, mediates the transfer of polypeptides (each of which contain some hydrophilic residues) across the hydrophobic core of the membrane.

19.16 Synthesis of a secretory glycoprotein begins on a ribosome. An appropriate signal peptide mediates the translocation of the polypeptide into the ER lumen. The core N-linked oligosaccharides are then covalently linked to appropriate asparagine residues in the polypeptide in a reaction catalyzed by glucosyl transferase. Subsequently, the molecule is transferred in transport vesicles to the Golgi complex where additional glycosylation reactions occur. Eventually, the glycoprotein is incorporated into secretory vesicles which migrate to the plasma membrane. Secretion of the glycoprotein then occurs via exocytosis.

19.17 One of the most significant problems associated with predicting the three-dimensional structure of a polypeptide based solely on its primary structure is that the calculations based on the forces that drive the folding process (e.g., bond rotations, free energy considerations, and the behavior of the amino acids in aqueous environments) are extraordinarily complex.

19.19 In protein splicing an intervening peptide sequence is excised from a nascent polypeptide. A peptide bond is formed between the flanking amino terminal and carboxy terminal amino acid.

19.20 tRNA molecules are adaptor molecules because of their use to bind to specific amino acids and then to position those molecules in the ribosome according to the code on sequence of an mRNA. In other words, they bridge the gap between the base code of the nucleic acids and the amino acid sequence of polypeptides.

19.22 The proteins involved in the initiation of prokaryotic protein synthesis are: IF-1 (binds to the A site of the 30S subunit, blocking it during initiation), IF-2 (binds to the 30S subunit and promotes the binding of the initiating tRNA to the initiation codon of mRNA), and IF-3 (prevents the 30S subunit from binding prematurely to the 5OS subunit).

19.23 a. PABP – poly(A) binding protein – a scaffold protein that interacts with eIF-G to bring the 3′-poly(A) tail of the mRNA close to the 5′ capped end, forming a circular mRNA molecule that increases the overall efficiency of translation; PABP is a component of CBC

b. CBC – cap binding complex – eIF-4F – a complex that binds to the mRNA 5′-cap structure prior to the mRNA binding to the 43S preinitiation complex; the CBC consists of eIF-4A (a helicase), eIF-4E (a translation initiation factor), and eIF-G (a scaffold protein)

c. diphthamide – a unique modification of a histidine residue; occurs at a specific location in eEF-2 (see In-Chapter Question 19.4)

d. SECIS element – <u>se</u>leno<u>c</u>ysteine <u>i</u>nsertion <u>s</u>equence – a sequence element required in order to code for selenocysteine and located in the 3′-UTR of the mRNA for a selenocysteine-containing polypeptide

e. signal hypothesis – a mechanism that explains how secretory proteins are synthesized on ribosomes bound to the RER; a sequence of amino acid residues on the nascent polypeptide chain mediates the insertion of the molecule into the RER membrane

19.25 The large subunit contains the catalytic site for peptide bond formation. The small subunit serves as a guide for the translation factors required to regulate protein synthesis. Together the two subunits come together and form a molecular machine that polymerizes amino acids in a sequence specified by the base sequence in the mRNA molecule.

19.26 Answer d is the process of translation.

19.28 To ensure that proteins end up in a location appropriate to their function in a timely and predictable way, it is necessary to have a targeting mechanism. The signaling process begins with specific signal sequences, which determine where translation will be completed. Specific localization sequences and/or posttranslational modification of the product protein then ensures delivery of the protein to its target location.

19.29 The aminoacyl-tRNA synthetases correctly attach each amino acid to its cognate tRNA and proofread the product. This process increases the accuracy of protein synthesis.

19.31 The phases of protein synthesis during which each process occurs are as follows: a. Initiation, b. Elongation, c. Elongation, and d. Termination.

19.32 The total amount of nucleotide bond energy that is required to synthesize Lys-Ala-Ser-Val is equivalent to the hydrolysis of either 16 or 17 phosphoryl groups from ATP or GTP, determined as follows:

1 GTP to form the complete ribosome and initiate translation

8 ATP to create 4 aminoacyl-tRNAs; 2 ATP equivalents per amino acid

6 GTP to create 3 peptide bonds; 1 GTP for elongation and 1 GTP for translocation for each peptide bond formed

1 or 2 GTP to terminate translation; 1 GTP in eukaryotes, 2 GTP in prokaryotes

19.34 (Refer to Solution 19.33 in the appendix of your text.)

There are many correct answers to this question. The three-letter sequences listed below each amino acid in the first table below are the possible mRNA sequences that code for that specific amino acid. An asterisk "*" is used to designate any of the four possible nucleotides. For example, GC* is either GCU, GCC, GCA, or GCG. Choose one 3-letter sequence from each column to build an mRNA sequence that will code for this peptide.

Possible choices for the 5′→3′ mRNA codon base sequences for this peptide (* = A, G, U, or C):

5'-	Ala	Ser	Phe	Tyr	Ser	Lys	Lys	Leu	Ala	Asp	Val	Ile	-3'
mRNA: 5'-	GC*	UC*	UUU	UAU	UC*	AAA	AAA	UUA	GC*	GAU	GU*	AUU	-3'
	AGU	UUC	UAC	AGU	AAG	AAG	UUG		GAC		AUC	
	AGC			AGC			CU*				AUA	

One possible sequence is:

5'-	Ala	Ser	Phe	Tyr	Ser	Lys	Lys	Leu	Ala	Asp	Val	Ile	-3'
mRNA: 5'-	GCU	UCU	UUU	UAU	UCU	AAA	AAA	UUA	GCU	GAU	GUU	AUU	-3'

19.35   The deletion of a single DNA base will alter the amino acid sequence in the polypeptide produced from mRNA, and may truncate the polypeptide, depending upon the base in the Ser codon that was deleted. As noted in Solutions 19.33 and 19.34, many possible base sequences will code for this polypeptide. Using the solution from Solution 19.33:

5'-	Ala	Ser	Phe	Tyr	Ser	Lys	Lys	Leu	Ala	Asp	Val	Ile	-3'
DNA: 3'-	CGA	AGA	AAA	ATA	AGA	TTT	TTT	AAT	CGA	CTA	CAA	TAA	-5'

Removing the first base, A, from the second Ser residue results in the following altered DNA and mRNA base sequences and a new polypeptide that corresponds to this sequence.

DNA: 3'-	CGA	AGA	AAA	ATA	GAT	TTT	TTA	ATC	GAC	TAC	AAT	AA	-5'
mRNA: 5'-	GCU	UCU	UUU	UAU	CUA	AAA	AAU	UAG	CUG	AUG	UUA	UU	-3'
Amino Acid	Ala	Ser	Phe	Tyr	Leu	Lys	Asn	Stop				

CHAPTER 19: SOLUTIONS TO THOUGHT QUESTIONS

19.37   Despite considerable species differences in the amino acid and nucleotide sequences of ribosomal proteins and RNA, respectively, the overall three-dimensional structures of these molecules are remarkably similar. This similarity is presumably due to high selection pressure. In other words, ribosomal function is such an important factor in species viability that evolution has conserved their tertiary structure.

19.38   Because the accuracy of protein synthesis depends directly on codon-anticodon interactions, the specificity with which t-RNAs are linked to amino acids is critically important. The process in which the amino acid-tRNA synthetases catalyze the covalent binding of each of the t-RNAs with its correct amino acid has, therefore, been referred to as the second genetic code.

19.40   The three phases of protein synthesis are: (1) initiation (the small ribosomal subunit binds to an mRNA, an initiation tRNA and the large subunit), (2) elongation (polypeptide synthesis occurs as amino acids attached to tRNAs aligned in the P and A sites undergo transpeptidation), and (3) termination (polypeptide synthesis ends when a stop codon on the mRNA enters the A site). Translation factors perform a variety of roles. Some have catalytic functions (e.g., EF-TU and eEF-2 are GTP binding proteins which catalyze GTP hydrolysis), while others (e.g., IF-3 and eIF-4F) stabilize translation structures.

19.41 One possible codon sequence for the peptide sequence is GGUAGUUGUAGAGCU. The number of possible codons for the amino acids in this peptide sequence is as follows: glycine (4), serine (6), cysteine (2), arginine (6), and alanine (4). The total number of possible codon sequences for this peptide sequence is therefore 1152.

19.43 The nucleotide GTP is the source of the energy required to drive various steps in the translation mechanism.

19.44 Posttranslational modification reactions prepare polypeptides to serve their specific functions and direct them to specific cellular or extracellular locations. Examples of these modifications include proteolytic processing (e.g., removal of signal proteins), glycosylation, methylation, phosphorylation, hydroxylation, lipophilic modifications (e.g., N-myristoylation and prenylation), and disulfide bond formation.

19.45 Each Shine-Dalgarno sequence in a prokaryotic mRNA occurs near a start codon (AUG). The Shine-Dalgarno sequence provides a mechanism for promoting the correct alignment of the start codon on the ribosome (as opposed to a methionine codon) because it binds to a nearby complementary sequence in the 16S rRNA component of the 30S ribosome. Eukaryotic ribosomes identify the initiating AUG codon by binding to the capped 5′ end of the mRNA and scanning the molecules for a translation start site.

19.46 While you can go directly and predictably from a nucleotide sequence to one and only one amino acid sequence, the reverse is not true because of the degeneracy of the genetic code.

19.47 Sets of amino acids which may require proofreading include phenylalanine/tyrosine, serine/threonine, aspartate/glutamate, asparagine/glutamine, isoleucine/leucine, and glycine/alanine.

19.49 The principal factors that ensure accuracy in protein synthesis are codon-anticodon base pairing and the mechanism by which amino acids are linked to their cognate tRNAs. The level of accuracy of protein synthesis, while quite high, is still less than that achieved during replication or transcription.

19.50 A two subunit ribosome is essential to ensure that all of the required elements are in place before the translational process begins. This is a physical ordering process much like an assembly line; the parts must be in place before the enzymatic activities are set in motion.